Selected Titles in This Series

Volume

6 **Solomon Feferman, Editor**
The Collected Works of Julia Robinson
1996

5 **Freeman Dyson**
Selected Papers of Freeman Dyson with Commentary
1996

4 **Krystyna Kuperberg, Editor**
Collected Works of Witold Hurewicz
1995

3.2 **Richard E. Block, Nathan Jacobson, J. Marshall Osborn, David J. Saltman, and Daniel Zelinsky, Editors**
A. Adrian Albert Collected Mathematical Papers: Nonassociative Algebras and Miscellany, Part 2
1993

3.1 **Richard E. Block, Nathan Jacobson, J. Marshall Osborn, David J. Saltman, and Daniel Zelinsky, Editors**
A. Adrian Albert Collected Mathematical Papers: Associative Algebras and Riemann Matrices, Part 1
1993

2 **Robert C. Gunning, Editor**
Collected Papers of Salomon Bochner, Parts 1–4
1992

1 **Sukhjit Singh, Steve Armentrout, and Robert J. Daverman, Editors**
The Collected Papers of R. H. Bing, Parts 1 and 2
1988

JULIA ROBINSON
December 8, 1918–July 30, 1985

THE COLLECTED WORKS OF
JULIA ROBINSON

THE COLLECTED WORKS OF
JULIA ROBINSON

Solomon Feferman
Editor

American Mathematical Society
Providence, Rhode Island

1991 *Mathematics Subject Classification*. Primary 01A75, 03Dxx.

Library of Congress Cataloging-in-Publication Data

Robinson, Julia, d. 1985.
[Works. 1996]
The collected works of Julia Robinson / Solomon Feferman, editor.
p. cm.
Includes bibliographical references (p. –).
ISBN 0-8218-0575-4
1. Number theory. I. Feferman, Solomon. II. Title.
QA3.R67 1996
512′.7—dc20 96-22866
CIP

A complete list of acknowledgments can be found at
the back of this publication.

∞ The paper used in this book is acid-free and falls within the guidelines
established to ensure permanence and durability.

10 9 8 7 6 5 4 3 2 1 01 00 99 98 97 96

Contents

Foreword

This volume gathers together all the published works of Julia Bowman Robinson, spanning a period of more than thirty years. She was still active mathematically when her career was sadly brought to a close by her premature death at the age of sixty-five due to leukemia. Except for two early publications—one in statistics and one in the theory of games—all these papers constitute important contributions to the theory of effectively calculable functions and especially to its applications. Outstanding among the latter are Robinson's proof of the effective unsolvability of the decision problem for the rational number field (and, in consequence, of that for the first-order theory of all fields), and her work that provided the central step toward the negative solution of Hilbert's Tenth Problem (eventually established by Yuri Matijasevich). These results provide upper bounds for what one can hope to obtain in the way of positive solutions to the decision problem for special classes of fields and for special classes of diophantine equations, respectively. Besides their largely thematic unity, Robinson's papers are distinguished by their clarity of purpose and accessibility of exposition; they should be found rewarding to a large audience of non-specialists as well as specialists.

Following a brief Introduction by Julia Robinson's sister, Constance Reid (the biographer of a number of mathematicians), the reader will find a piece I wrote on Robinson's life and work which appeared in the 1994 volume of the *Biographical Memoirs* of the National Academy of Sciences; I appreciate the willingness of the Academy to have it reprinted here. I wish to thank Constance Reid and Donald Babbitt (publisher of the American Mathematical Society) for encouraging me to take on the title of editor of this volume. The person to whom I am especially indebted in this respect is Christine Thivierge (also of the Society), who carried out all the real work required to organize the necessary materials and see this publication into existence.

Solomon Feferman*

---

*My work as editor of this volume was carried out while spending a year as a fellow at the Center for Advanced Study of the Behavioral Sciences, Stanford, CA, whose facilities and support I have greatly appreciated.

Introduction

Julia Robinson felt that the important thing about a mathematician, and the only thing worth saving, is the mathematics left behind. It would thus be a source of great satisfaction to her that the American Mathematical Society has chosen to publish her collected papers. Such a volume is the biography she would have preferred.

She was a very private person and, until the last months of her life, she turned down all requests for biographical material. Yet she felt strongly, on the basis of her own experience of the inspiration provided by the reading of E. T. Bell's *Men of Mathematics* when she was a college sophomore, that it was valuable for students to know something about the lives of the mathematicians whose names are attached to theorems in textbooks. With her encouragement I wrote a life of David Hilbert, which was followed by biographies of other mathematicians, including Bell himself. Finally, on June 4, 1985, less than two months before her death, she agreed to an interview with me about her life. That interview provided the material for my article, "The Autobiograpy of Julia Robinson," which Professor Feferman cites in his very fine summation of Julia Robinson's life and her mathematical career.

Almost everything that Julia "says" in the "Autobiography"—and this applies in particular to those statements in which she expresses an opinion—is taken verbatim from our interview. She also had me read the finished article aloud to her in the hospital and approved, sometimes after minor modifications, everything I had written there. The "Autobiography" truly represents the account of her life that she was willing to leave behind. There is nothing to which this statement applies more emphatically than to the final words:

> "What I really am is a mathematician. Rather than being remembered as the first woman this or that, I would prefer to be remembered, as a mathematician should, simply for the theorems I have proved and the problems I have solved."

Constance Reid

Publications of Julia Robinson

1948

1. *A note on exact sequential analysis*, Univ. California Publ. Math. (N.S.) **1** (1948), 241–246.

1949

2. *Definability and decision problems in arithmetic*, J. Symbolic Logic **14** (1949), 98–114.

1950

3. *General recursive functions*, Proc. Amer. Math. Soc. **1** (1950), 703–718.

1951

4. *An iterative method of solving a game*, Ann. of Math. **54** (1951), 296–301.

1952

5. *Existential definability in arithmetic*, Trans. Amer. Math. Soc. **72** (1952), 437–449.

1955

6. *A note on primitive recursive functions*, Proc. Amer. Math. Soc. **6** (1955), 667–670.

1959

7. *The undecidability of algebraic rings and fields*, Proc. Amer. Math. Soc. **10** (1959), 950–957.

8. *Problems of number theory arising in metamathematics*, Report of the Institute in the Theory of Numbers (Boulder), 1959, pp. 303–306.

1961

9. *The decision problem for exponential diophantine equations* (with Martin Davis and Hilary Putnam), Ann. of Math. **74** (1961), 425–436.

1962

10. *The undecidability of exponential Diophantine equations*, Logic, Methodology, and Philosophy of Science: Proceedings of the 1960 International Congress (E. Nagel et al., eds.), Stanford Univ. Press, Stanford, 1962, pp. 12–13.

11. *On the decision problem for algebraic rings*, Studies in Mathematical Analysis and Related Topics, (D. Gilbarg et al., eds.), Stanford Univ. Press, Stanford, 1962, pp. 297–304.

1965

12. *The decision problem for fields*, The Theory of Models: Proceedings of the 1963 International Symposium at Berkeley (J. W. Addison et al., eds.), North-Holland, Amsterdam, 1965, pp. 299–311.

1967

13. *An introduction to hyperarithmetical functions*, J. Symbolic Logic **32** (1967), 325–342.

1968

14. *Recursive functions of one variable*, Proc. Amer. Math. Soc. **19** (1968), 815–820.

15. *Finite generation of recursively enumerable sets*, Proc. Amer. Math. Soc. **19** (1968), 1480–1486.

1969

16. *Diophantine decision problems*, Studies in Number Theory (W. J. LeVeque, ed.), MAA Stud. Math., vol. 6, Math. Assoc. America, Washington, DC, 1969, pp. 76–116.

17. *Finitely generated classes of sets of natural numbers*, Proc. Amer. Math. Soc. **21** (1969), 608–614.

18. *Unsolvable diophantine problems*, Proc. Amer. Math. Soc. **22** (1969), 534–538.

19. *Hilbert's tenth problem*, Proceedings of the 1969 Summer Institute in Number Theory (D. J. Lewis, ed.), Proc. Sympos. Pure Math., vol. 20, Amer. Math. Soc., Providence, RI, 1971, pp. 191–194.

1973

20. *Solving diophantine equations*, Proceedings of the Fourth International Congress for Logic, Methodology and Philosophy of Science (P. C. Suppes et al., eds.), North-Holland, Amsterdam, 1973, pp. 63–67.

21. *Axioms for number-theoretic functions*, Selected Questions of Algebra and Logic (A. I. Shirshov et al., eds.), Siberian Academy of Sciences, Novosibirsk, 1973, pp. 253–263.

1974

22. *Two universal three-quantifier representations of recursive enumerable sets* (with Yuri Matijasevich), Theory of Algorithms and Mathematical Logic (B. A. Kushner and N. M. Nagornyi, eds.), Vychisl. Centr. Akad. Nauk SSSR (1974), 112–123, 216.

1975

23. *Reduction of an arbitrary diophantine equation to one in 13 unknowns* (with Yuri Matijasevich), Acta Arith. **27** (1975), 521–553.

1976

24. *Hilbert's tenth problem. Diophantine equations: positive aspects of a negative solution* (with Martin Davis and Yuri Matijasevich), Mathematical Developments Arising from Hilbert Problems (F. E. Browder, ed.), Proc. Sympos. Pure Math., vol. 28, Amer. Math. Soc., Providence, RI, 1976, 323–378.

1980

25. *Between logic and arithmetic*, Lecture notes distributed in conjunction with the Colloquium Lectures given at the Eighty-fourth Summer Meeting of the American Mathematical Society, University of Michigan, Ann Arbor, August 19–22, 1980, pp. 1–11.

JULIA BOWMAN ROBINSON

December 8, 1919–July 30, 1985

BY SOLOMON FEFERMAN

One of my earliest memories is of arranging pebbles in the shadow of a giant saguaro . . . I think I have always had a basic liking for the natural numbers. To me they are the one real thing. We can conceive of a chemistry which is different from ours, or a biology, but we cannot conceive of a different mathematics of numbers. What is proved about numbers will be a fact in any universe.

(From *The Autobiography of Julia Robinson* by Constance Reid)

AS A MATHEMATICIAN, Julia Bowman Robinson will long be remembered for her many important contributions to questions of algorithmic solvability and unsolvability of mathematical problems, in particular for her part in the negative solution of Hilbert's "Tenth Problem." And, despite her expressed wish, she will be remembered as the first woman to be elected to the mathematical section of the National Academy of Sciences, as well as the first woman to be president of the American Mathematical Society. By those who knew her personally she will be remembered for her rare qualities of idealism, integrity, modesty, openness, and generosity, and for her appreciation and encouragement of the work of others.

She was born Julia Bowman on December 8, 1919, in St. Louis, Missouri, the second of two daughters, to Helen Hall Bowman and Ralph Bowers Bowman. Her mother died when she was two years old, and her father retired not long after, having lost interest in his machine tool and equipment business. Ralph Bowman remarried a few years later to Edenia Kridelbaugh, and the family moved first to Arizona and then to San Diego; a third daughter, Billie, joined Constance and Julia a few years later. As a child, Julia was said to be stubborn and slow to talk, but she exhibited a precocious liking for the natural numbers, as evidenced by her earliest reported memories.

At the age of nine, Julia was stricken first with scarlet fever, and then with rheumatic fever, which, after several relapses, forced her to spend a year in bed. As a result, she lost more than two years of school, and there were more serious, lifelong consequences for her health.

Julia returned to school after a year of tutoring, during which she went through the state syllabuses for the fifth, sixth, seventh, and eighth grades. Shy and out of the swim socially, she took to her studies and found herself especially attracted to mathematics. Moving on in high school through the standard courses of algebra, trigonometry, and plane and solid geometry, she stood out as the only girl in the advanced mathematics and physics courses. She graduated from high school with awards in all the sciences (except chemistry, which she had not taken) and a special medal (the Bausch-Lomb) for all-round excellence in mathematics and science.

Following in the footsteps of her older sister Constance, Julia entered San Diego State College (now University) in 1936. It was common for students there to attend State for a couple of years and then transfer to UCLA or the University of California at Berkeley; those who remained generally obtained teaching credentials. Julia majored in mathematics because she liked it and was good at it, but she had no idea that there was such a thing as being a mathematician. When she came across the book *Men of Mathematics* by E. T. Bell, she became very excited at the intellectual vistas that opened up and she was determined to transfer to UCLA or Berkeley to learn some real mathematics. Finances were strained due to the suicide in 1937 of her father, whose retirement savings had been wiped out by the 1929 crash and the subsequent depression. However, with the assistance of an aunt and her older sister, in 1939 she managed finally to transfer to U.C.B. as a senior. It happened to be a particularly fortuitous time to enter the mathematics program at Berkeley, as the noted chairman, Griffith C. Evans, had been building up the department with outstanding researchers and teachers. In her first year at Berkeley, Julia took five mathematics courses, including one in number theory taught by Raphael M. Robinson. She was extremely happy at Berkeley, finding herself among so many students and faculty excited by and talking about mathematics.

Julia received her A.B. degree in 1940 and, at the urging of Raphael Robinson, continued on with graduate studies at Berkeley. She also managed to obtain a part-time position as an assistant to Professor Jerzy Neyman for his work on statistics. During this period, Julia and Raphael developed a personal friendship that blossomed into a courtship, and they married in December 1941. The relationship with Raphael was also to have major significance for Julia's development as a mathematician; he was an excellent teacher, both in class and in one-to-one discussions, and he had a deep knowledge of many parts of classical and modern mathematics. Julia counted Raphael Robinson as one of her two most important scientific influences, the second one being the famous logician Alfred Tarski, who joined the Berkeley faculty in 1942.

Unfortunately, her marriage to a faculty member was to limit Julia's employment opportunities in Berkeley, since there was at that time (and for many years after) a nepotism rule in the U.C. system that prevented members of the same family from working in the same department. This did not affect her at first, since she had already received a teaching assistantship in mathematics for the second year of her graduate studies. Julia had wanted to teach calculus, but Jerzy Neyman pushed her to teach statistics, which she didn't like as much. The connection was useful, though, as Neyman was to employ her in his "Stat Lab" later during the war, and her work there would result in her first published paper.

From the beginning of their marriage, the Robinsons wanted and expected to have a family, and though Julia continued to audit mathematics courses, her preoccupations shifted more toward their home life. She became pregnant but, unfortunately, lost the baby a few months later; then, shortly afterward, on a visit home to San Diego, she contracted viral pneumonia. A doctor there was the first to recognize that there was substantial scar tissue in the mitral valve of the heart, a residue of her early bout with rheumatic fever, and he strongly advised her against becoming pregnant. He also told her mother in private that Julia would be lucky to live to age forty.

Julia was deeply depressed for a long time over not being able to have children, but, encouraged by Raphael, she returned eventually to the rewards of mathematics. She attended the first seminar that Tarski conducted at Berkeley and soon provided an ingenious solution to a definability problem in number theory that had been posed by Tarski's former student Andrzej Mostowski. The question was whether addition can be defined in terms of the successor operation and multiplication. Tarski was impressed by her positive solution to this problem and viewed it as suitable thesis material. It, and other similar work, was eventually to be incorporated as a relatively minor part of her thesis, the major part of which is to be described below.

During the immediate postwar period, Tarski was building a school in mathematical logic at Berkeley. One of the two most important logicians of the 20th century (the other being Kurt Gödel), he was a very systematic yet intense and charismatic lecturer with extensive research programs and a wealth of good problems to match. Tarski's main subjects of interest were metamathematics, set theory, model theory, universal (abstract) algebra, and algebraic theory. During a period of thirty-odd years following the war, he directed the doctoral work of a series of outstanding students, of whom Julia Robinson was one of the first.

Except for two papers, all of Robinson's work is concerned with the *effective solvability* and *unsolvability* of various mathematical problems, as well as with the characterization of various notions of effectiveness. The exceptions were her very first paper (1948), on a topic in sequential analysis, which came out of her work in the U.C.B. Stat Lab during World War II, and a paper (1951) on the theory of games. The latter resulted from her work during a year (1949–50) spent at the RAND Corporation in Santa Monica, where game theory was at the time a topic of intense investigation. Robinson succeeded there in solving in the affirmative a "prize" problem that had been posed by George W. Brown, whether a certain iterative method of play always leads to the value of a two-person zero-sum game; that result has been considered a very significant contribution to the subject.

To return to the main topic of Robinson's work: the informal idea of *effective solvability* (or *decidability*) of a class of mathematical problems of the form—Is $P(a)$ true? (where a ranges over a given set S)—is that provided by an *algorithm*, i.e., a completely determined finitely described procedure which, given any element (or "input"), leads step-by-step to a definite conclusion as to the truth or falsity of the predicate $P(a)$. Similarly an operation (function) f on elements of S to S is said to be *effectively computable* if there is an algorithm which, given any element $a \in S$ as argument, leads to the value (or "output") b of $f(a)$, for which $f(a) = b$. (We here use $\in$ for the membership relation and read "a in S" for "$a \in S$".) The elements of the set S must themselves be presented in finite form as a finite sequence of symbols. Using any one of the standard systems of representation, these notions thus apply to predicates and operations on the set Z of *integers* and its subsets N^+ of *positive integers* and N of *non-negative integers* (*natural numbers*), i.e., to $Z = \{. . . ,-3,-2,-1,0,1,2,3, . . .\}$, $N^+ = \{1,2,3,. . .\}$ and $N = \{0,1,2,3,. . .\}$. The notions extend directly to the set Q of all rational numbers, i.e., all fractions n/m with $n,m \in Z$ and $m \neq 0$. Furthermore, if $A = \{\alpha_1,. . .,\alpha_k\}$ is any finite alphabet, we can apply the notions of effectiveness to predicates and functions on the set $S = W(A)$ of all finite expressions (or "words") $\alpha_{i_1} . . . \alpha_{i_m}$ with letters from A. Finally, if effectiveness is meaningful for predicates and operations on S, the same holds for the set S^n of all n-tuples $(a_1,. . .,a_n)$ where we deal with n-ary relations $P(a_1,. . .,a_n)$ and operations $f(a_1,. . .,a_n) = b$.

It should be noted that the notion of effectiveness for

predicates and relations can be reduced to that for operations, using distinct elements $t_1, t_0 \in S$ (e.g., 1 and 0) to represent *truth* and *falsity*, respectively. By the *characteristic function* of P (an n-ary relation between elements of S), is meant the operation f_P defined on S^n by $f_P(a_1, \ldots, a_n) = t_1$ if $P(a_1, \ldots, a_n)$ is true and $f_P(a_1, \ldots, a_n) = t_0$ otherwise. Then, clearly, P is effectively decidable if and only if f_P is effectively computable.

Since the beginning of abstract mathematics in antiquity, many algorithms have been presented for various specific mathematical problems and operations. In each case it was verified that the algorithm does the required work by following through its application to see that it gives the appropriate answer for an arbitrarily given argument as input. No general notion of effectiveness was required for such verifications, i.e., the informal conception of effectiveness sufficed. However, if one is to show that a problem of the form—Is $P(a)$ true? (for $a \in S$)—is *effectively undecidable*, i.e., that *no possible algorithm* can serve to effectively determine the truth or falsity of $P(a)$ for each $a \in S$, we must have a completely general and precise definition of effective decidability, so that its extent can be sharply delimited. The same applies to the task of showing that a given operation f is *not* effectively computable. By the remark above, once the general notion of effectively computable function is made precise, the same applies to that of effectively decidable predicate or relation.

Several independent proposals for a general, mathematically precise definition of effectively computable function on the natural numbers N (or N^+) were made in the 1930s, and before long all were shown to be equivalent. Among the first was that made by Alonzo Church who advanced, for this, calculability in a certain restricted symbolism for functions called the *λ-calculus*. Independently, and around the same time (using a suggestion of Jacques Herbrand), Kurt Gödel proposed the class of *general recursive functions* as a candidate for the algorithmically computable functions. Before this, in his famous 1931 paper on the incompleteness of *Principia Mathematica* and other formal axiomatic systems concerning the natural numbers, Gödel had made heavy use of a subclass of these, called the *primitive recursive functions*. In 1936, Church's student Stephen C. Kleene proved the equivalence of the λ-definable and general recursive functions; in the same year Church first proposed in print to identify the algorithmically computable functions with either one of these (coextensive) classes, and this proposal was subsequently dubbed *Church's Thesis*. Unaware of the preceding work, Alan Turing in 1937 proposed a notion of mechanically computable functions, where the calculations are carried out by an idealized machine (subsequently dubbed *Turing machines*) without limitation on storage or time. When Turing learned of Church's Thesis, he established the equivalence of the λ-calculable functions with his own mechanically computable functions. All these (and related) contributions bolstered the arguments for Church's Thesis, which has since gained practically universal acceptance; in particular, Turing's notion is considered to be one of the most faithful to the informal concept of a step-by-step finite mechanical procedure. His definition also extends directly to that of effective computability on any set of expressions generated from a finite alphabet.

The systematic development of the theory of effectively computable functions, assuming Church's Thesis, was initially due to Kleene; in particular, working with the Herbrand-Gödel definition, he showed how the general recursive functions can be obtained from the primitive recursive functions, themselves generated by some relatively perspicuous schemata, by the adjunction of one new scheme embodying the construction "do . . . until . . ." It became convenient to work with Kleene's schemata as a means for verifying various properties of the effectively computable functions. As a result of this, the subject of effective computability (both for positive and negative results) came to be called *recursive function theory*. One mildly complicating feature, though, of Kleene's schemata is that they pass through functions of arbitrarily many arguments in order even to obtain the functions of one argument (the unary functions). The problem thus arose of finding a simple set of schemata for generating the unary general recursive functions. An important step in this direction had been made by Raphael Robinson, who obtained a simple set of schemata for the unary primitive recursive functions; he suggested to Julia the problem of extending this to a characterization of the unary general recursive functions. After several years' work, she succeeded in obtaining an elegant solution in 1948, which was published in a paper (1950). She returned to this problem in a later paper (1968, 1) with an even more elegant characterization of the class of unary general recursive functions as the least class containing the zero and successor functions and closed under composition and a new special scheme of general recursion. Related to these papers is her note (1955) on primitive recursive functions. Somewhat in the same spirit, but for other classes of functions or sets, are her papers (1967; 1969, 1; 1973, 2) (in the last, within an axiomatic setting). Of these, one (1967) is noteworthy for its beautifully simple and direct exposition of the theory of hyperarithmetical functions, a class originally obtained by a complicated transfinite iteration of the "jump" process associated with general recursion in a relativized form. These papers comprise one of three main areas of Robinson's contributions to the theory of effective computability.

For an explanation of Julia Robinson's dissertation work with Tarski and her later work on the Diophantine problem, we consider two further notions from recursive function theory. Suppose S is a set to which the notion of effectively computable (*alias* general recursive) function is applicable, either directly or by an effective enumeration of S by N. A subset D of S is said to be *decidable* (or *recursive*) if its characteristic function f_D on S is effectively decidable (or *recurvsive*) if its characteristic function f_D on S is effectively computable; in other words, D is decidable if its membership problem is effectively decidable. A subset E of S is said to be *effectively* (or *recursively*) *enumerable* if it is the range of a recursive function f from N to S, i.e., if E is the set of values $f(0),. . .,f(n),. . .$ for $n \in N$; for simplicity, the empty set is also counted as being effectively enumerable. It is not hard to see that D is decidable if and only if both D and its complement $S - D$ are effectively enumerable.

Of special interest in modern logic are the sets $S = WFF(L)$ of well-formed formulas in a formal Language L, which are special kinds of expressions generated from a finite symbolism. In the first-order predicate calculus these are generated from one or more basic relations $R(x_1,. . .,x_n)$ by closure under $\wedge$("and"), $\vee$ ("or"), $\neg$ ("not"), $\rightarrow$ ("if . . . then . . ."), $\forall x$ ("for all x, . . ."), and $\exists x$ ("there exists x,. . ."). In the following we consider the special case where the basic relations are $x = y$, $x + y = z$, and $x \cdot y = z$. We define $x \neq y$ by $\neg(x = y)$, $x = 0$ by $x + x = x$, $x = 1$ by $(x \cdot x = x) \wedge (x \neq 0)$, etc. In this language we can speak of first-order properties of the structures of the natural numbers

N, the integers Z, as well as of the rational numbers Q, the real numbers R, and the complex numbers C.

A formula A of L, such as $\exists x \ \forall y \ (x \cdot y = y)$, is said to be *closed* if it contains no free variables. For any mathematical structure $(M, R_1, \ldots, R_m)$ in which the basic relations of S are interpreted in a definite way (with $x = y$ interpreted as equality) each closed formula A of S has a definite truth value in M. By the *theory of* M, in symbols $Th(M, R_1, \ldots, R_m)$, is meant the set T of closed well-formed formulas which are true in M under this interpretation. This theory is said to be *decidable* if T is a decidable subset of S, otherwise *undecidable*. One of the basic consequences of the work of Gödel and Church is that the theory of the natural number systems, $Th(N,+,\cdot)$ is undecidable, and the same applies to $Th(Z,+,\cdot)$ since N is definable in the latter theory by: $x \in N$ if and only if $\exists y \ \exists z \ \exists u \ \exists w \ (x = y^2 + z^2 + u^2 + w^2)$ (according to a famous theorem of Lagrange). On the other hand, it was known from earlier work of M. Presburger (1930) that $Th(Z,+)$ and $Th(N,+)$ are decidable; note that the product relation $x \cdot y = z$ is *not* included in these structures.

One of Tarski's main long-term programs, beginning in the 1930s in Warsaw and continuing over into his Berkeley period, was the systematic investigation of the *decision problem* for first-order theories associated with mathematical structures met in practice. Tarski's most famous result in this direction, first announced in 1931 but not published in full until 1948, was his decision procedure for the elementary (first-order) theory of real numbers, i.e., $Th(R,+,\cdot)$; moreover, this was shown to be the same as the theory of any real-closed field $(M,+,\cdot)$, including the countable field of real algebraic numbers, i.e., the real number solutions of polynomial equations over Q. Tarski had also shown (what was easier) that the elementary theory of the complex number field $Th(C,+,\cdot)$ is decidable and that any algebraically closed field $(M,+,\cdot)$ of characteristic zero has the same theory.

This is where things stood when Julia Robinson took up, at Tarski's suggestion (via Raphael Robinson), the question of definability of the set N in $Th(Q,+,\cdot)$. The status of the decision problem for the theory of the field of rational numbers was at that point the big unknown between the undecidable theory of integers $Th(Z,+,\cdot)$ and the decidable theory of real numbers $Th(R,+,\cdot)$. Robinson's exceptional achievement in her thesis work was to show that the set Z (and hence N) *is* first-order definable in $Th(Q,+,\cdot)$, and hence that the latter theory is also undecidable. It was far from obvious how to do this. Her first breakthrough came with the realization that if r is a rational number and r is expressed as a quotient of integers a/b in lowest terms, then b is odd if and only if $(\exists x_1\ \exists x_2\ \exists x_3)(7r^2 + 2 = x_1^2 + x_2^2 + x_3^2)$ is true in Q. Thus one ternary quadratic form can be used to "eliminate" 2 as a factor from the denominator of a rational number. Robinson then found other ternary quadratic forms to eliminate in turn all the other prime numbers from denominators. The final problem that she overcame was to combine these infinitely many different quadratic forms into a finitely described class of forms which would succeed in eliminating all (and only) denominators of fractions in lowest terms and thus in characterizing the integers Z as exactly the rationals satisfying a first-order property in $(Q,+,\cdot)$. Among the important consequences of this deep and difficult work were the result that the axiomatic theory of fields is undecidable; this is a corollary using some general results of Tarski on undecidable theories. Robinson received her Ph.D. in 1948, and the results of the thesis were published soon after in the paper (1949).

In a later paper (1959, 1), Robinson extended her thesis work to algebraic fields $(F,+,\cdot)$ of finite degree over the rationals (subfields of the complex numbers C). Here she showed that N is definable in the ring $(A,+,\cdot)$ of algebraic integers of F and that A is in turn definable in the field F, so that both $Th(A,+,\cdot)$ and $Th(F,+,\cdot)$ are undecidable. The papers (1962, 2) and (1965) contain further related results of interest.

We can now finally return to the third main area of Robinson's mathematical preoccupations, that of existential definability in arithmetic, also called *Diophantine definability*. The background to this problem is as follows.

In the third century A.D., the Greek mathematician Diophantus worked on solving equations with arbitrary integer coefficients, for integer values. The general linear Diophantine equation in two variables is $ax + by = c$ where a,b,c are arbitrarily given integers. An algorithm to determine whether or not there are $x,y \in A$ satisfying $ax + by = c$ makes use of an algorithm, due to Euclid, for finding the greatest common divisor of two integers: provided $a \neq 0$ or $b \neq 0$, the equation has a solution $x,y \in Z$ if and only if the greatest common divisor d of a and b is a divisor of c; then if there is a solution at all, we have a further algorithm for finding all possible solutions. Another classical algorithm, to find all possible $x,y,z \in N^+$ satisfying $x^2 + y^2 = z^2$, connects number theory with geometry; the solutions of this equations such as 3,4,5 and 5,12,13, provide all possible right triangles with each side of integer length. The famous Fermat problem concerns the existence of integer solutions, if any, of $x^m + y^m = z^m$, with $x,y,z \neq 0$ and $m > 2$.

The general Diophantine equation in n variables $x_1, \ldots$

$.,x_n$ is given by an equation $P_1(x_1,\ldots,x_n) = P_2(x_1,\ldots,x_n)$ where P_1, P_2 are polynomials with integer coefficients. By subtraction this can always be brought to the form $P(x_1,\ldots,x_n) = 0$, where P is again such a polynomial, i.e., P is a finite sum of terms of the form $ax_1^{k_1}\ldots x_n^{k_n}$ where a is in Z and the exponents k_i are in N. The statement that (or question whether) there exist integer solutions at all to this equation is symbolized by $(\exists x_1 \ldots x_n \in Z)\ [P(x_1,\ldots,x_n) = 0]$. Closely related questions are whether there exist natural number solutions $(\exists x_1 \ldots x_n \in N)\ [P(x_1,\ldots,x_n) = 0]$ or positive integer solutions $(\exists x_1 \ldots x_n \in N^+)\ [P(x_1,\ldots,x_n) = 0]$.

Interest in Diophantine problems lay mostly dormant until the 17th century, when they were taken up by the outstanding amateur mathematician Pierre de Fermat, who solved many individual Diophantine problems, often challenging others to do the same without revealing his methods. Subsequently the subject was pursued by such great mathematicians as Euler, Lagrange, Gauss, Dirichlet, Dedekind, and others through the 19th century. This work brought to bear many ingenious and difficult arguments, which were organized by various techniques; however, no general theory emerged.

Faced with this situation at the turn of the 20th century, David Hilbert proposed the following decision problem for arbitrary Diophantine equations: *To devise a process according to which it can be determined by a finite number of operations whether the equation is solvable in integers.*[1] This problem was the tenth on his famous list of twenty-three problems in a wide variety of fields, which Hilbert dramatically presented as a challenge in his address to the International Congress of Mathematicians in 1900. Hilbert's optimism about finding a general algorithm for deciding all questions of the form $(\exists x_1,\ldots,x_n \in Z)[P(x_1,\ldots,x_n) = 0]$ with P an integer polynomial was in line with his proclaimed optimism about the eventual solvability of all mathematical problems, but flew in the face of the fact that Diophantine equations were known to be exceptionally resistant to general methods of solution. Indeed, for classifications according to degree, an algorithm was subsequently provided by Carl Ludwig Siegel only for the general problem of Diophantine equations of degree 2 and arbitrarily many unknowns. (The degree of $P(x_1,\ldots,x_n)$ is the maximum of the sums of the exponents $k_1 + \ldots + k_n$ of its terms $ax_1^{k_1} \ldots x_n^{k_n}$; the general polynomial of degree 2 in two variables is thus of the form $ax_1^2 + bx_1x_2 + cx_2^2 + dx_1 + ex_2 + f$.) On the other hand, in classifications according to the number n of variables, some progress has been made in recent years only for $n = 2$ and here only for a wide class of equations of degree higher than 2.[2] Even now, practically nothing systematic is known for nonquadratic equations in more than two variables.

With the emergence in the 1930s of the theory of recursive functions and the demonstrated effective unsolvability of some logical and mathematical problems, it became possible to consider giving a *negative* solution to Hilbert's Tenth Problem. In fact, there were already related negative results due to Gödel in his famous 1931 paper on incompleteness of formal systems of arithmetic. He there associated with each consistent axiom system T containing a sufficient amount of arithmetic, a true statement not provable in T, of the form $\forall x(f(x) \neq 0)$ where f is primitive recursive. Then he succeeded in showing that the relation $f(x) = y$ is definable over N in the form:

$$(1) \qquad f(x) = y \leftrightarrow (Q_1 z_1) \ldots (Q_m z_m)\ [P(x,y,z_1,\ldots,z_m) = 0]$$

where each Q_i is either the universal quantifier "$\forall$" or the existential quantifer "$\exists$", and P is a polynomial with integer coefficients. Somewhat later (1950), Martin Davis succeeded in showing that one can choose in place of the right-hand side of (1), a definition in which all but one of the Q_i are existential, and where the universal quantifier is "bounded," so that f is definable in the form:

$$(2) \qquad f(x) = y \leftrightarrow (\exists u)(\forall w \leq u)(\exists z_1 \ldots z_m) \\ [P(x,y,u,w,z_1,\ldots,z_m) = 0]$$

for a suitable polynomial P. Hence elimination of the bounded universal quantifier would give a negative solution to Hilbert's Tenth Problem. For, by the work of Church, Turing, and Kleene:

(3) (a) Every non-empty recursively enumerable set A of natural numbers is definable in the form

$$x \in A \leftrightarrow \exists y[f(y) = x]$$

with f primitive recursive; and

(b) one can construct a recursively enumerable set A which is not recursive and hence not effectively decidable.

Thus, elimination of the bounded universal quantifier in the form (2) for such A would eventually give

$$(4) \qquad x \in A \leftrightarrow (\exists z_1 \ldots z_m)\ [P(x,z_1,\ldots z_m) = 0]$$

for a suitable integer polynomial P, and this would show that there is no general algorithm for deciding whether a Diophantine equation has any solutions at all in the natural numbers.

A set A of natural numbers is called *Diophantine* if it is definable in the form (4) with the variables "z_i" ranging over N. This is equivalent to its being definable in purely existential form in the structure $(N,+,\cdot)$. Julia Robinson's work here started with the specific question, posed by Tarski, whether the set $\{2,2^2,\ldots,2^n,\ldots\}$ of powers of 2 is Diophantine. At first she tried to establish Tarski's conjecture that the answer would be negative, but failing in that, she began to consider the opposite conjecture, and before long was led to the general problem of the Diophantine definability of arbitrary recursively enumerable sets. In her paper (1952) Robinson reported her first results in this direction. She showed there if R is any binary relation in the natural numbers of *roughly exponential* growth in the sense of (5) next, then the relation x *Pow* y (x is some power of y) is existentially definable in the structure $(N,+,\cdot,R)$, where:

(5) (a) $\exists n\ \forall x\ \forall y\ [R(x,y) \rightarrow y < x^{\cdot^{\cdot^{x}}}\big\}n]$, and
(b) $\neg\exists k\ \forall x\ \forall y\ [R(x,y) \rightarrow y < x^k]$.

Furthermore, she showed that

(6) The relations $x^y = z$, $(x/y) = z$ and $x! = y$ are all existentially definable in terms of *Pow*, hence in terms of any R of roughly exponential growth.

The full significance of Robinson's 1952 results was not to emerge for close to a decade. During that period she continued to work hard on the general problem of Diophantine definability without substantial progress, though, as described above, she was able to obtain very satisfying results in her other two main areas of interest.

Around 1960, Julia received the draft of a paper by Martin Davis and Hilary Putnam in which they showed that if the famous hypothesis that there exist arbitrarily long arithmetic progressions containing only prime numbers were true, then every recursively enumerable (r.e.) set would be exponential Diophantine, i.e., existentially definable in $(N,+,\cdot,Exp)$, where $Exp(x,y) = x^y$. She quickly succeeded in eliminating that hypothesis (still unproved) and also simplified their arguments in many other respects. This led to the joint publication (1961) with Davis and Putnam in which the main result is that every r.e. set A is the set of all x for which a Diophantine equation $P(x,z_1,\ldots,z_m) = 0$ with *variable exponents* has solutions $z_1,\ldots,z_m$ in N. The Davis, Putnam, Robinson paper (1961) also featured the "J. R. Hypothesis," that some relation R of roughly exponential growth is Diophantine. By her 1952 results and this joint paper, the J. R. Hypothesis implies that every r.e. set is Diophantine. But, once more, matters would rest without essential progress for almost a decade.

Julia Robinson's doctor had told her mother in 1941 that she would be lucky to live to age forty. In 1961, when she was forty-one, her heart was functioning so poorly that the only alternative to a life of invalidism was an operation, by a technique then in its pioneering stage, to remove the built-up scar tissue. Her health improved dramatically, and one month after surgery, Julia took up bicycling as a new form of exercise. Over the following years, she bought a half dozen increasingly better bicycles and thereby enjoyed many outings and cycling trips in the U.S. and elsewhere, including The Netherlands. Raphael Robinson sometimes complained that "while other men's wives buy fur coats and diamond bracelets, [my] wife buys bicycles." (*Autobiography*, p. 18.)

Julia continued to work hard on Hilbert's 10th Problem throughout the 1960s. Then in 1970 the news came from the U.S.S.R. that a twenty-two-year old mathematician in Leningrad named Yuri Matijasevič had succeeded in establishing the J. R. Hypothesis. He showed that the sequence of even Fibonacci numbers, known to be of exponential growth, formed a Diophantine relation. This work involved elementary properties of the so-called Pell (Diophantine) equation $x^2 + (a^2 - 1)y^2 = 1$ for $a > 0$. As it happens, Robinson had used the same equation in her 1952 paper and was very familiar with its properties. In retrospect, she realized that she had herself been very close to the solution of Hilbert's 10th Problem. Nevertheless, she was very excited and pleased to see the long effort capped by Matijasevič's achievement, and she immediately sent her generous congratulation: ". . . now I know it is true, it is beautiful, it is wonderful. . . . If you really are 22, I am especially pleased to think that when I first made the conjecture you were a baby and I just had to wait for you to grow up." (*Autobiography*, p. 19.) In 1971, the Robinsons visited Leningrad and had the pleasure of meeting Matijasevič and his wife.

Capitalizing on the breakthrough, Julia Robinson and Yuri Matijasevič subsequently collaborated on three papers, (1974), (1975), and (1976), the last of these jointly with Martin Davis. In the 1975 paper Robinson and Matijasevič succeeded in reducing to 13 the number m of unknowns in the representation (4). In later work (by Matijasevič alone), this number was further reduced to 9. There is still a wide gap between the positive information on algorithmically solvable Diophantine equations in two unknowns mentioned above and the negative results for 9 and more unknowns. The most challenging open problem currently in this direction is whether there is a general algorithm for deciding the solvability of Diophantine equations in three unknowns.

The paper of Davis, Matijasevič, and Robinson (1976) provides a beautifully exposited yet concisely presented survey of these results on Hilbert's 10th Problem, together with a series of new "positive" applications, such as the Diophantine representation (i.e., as the range of positive values of a polynomial) of the set of prime numbers, and the equivalence of many famous problems—including the Riemann Hypothesis—with statements of the form $(\forall z_1 \ldots z_m)[P(z_1,\ldots,z_m) \neq 0]$; the paper also presented a number of further interesting open problems. This work was solicited for the Proceedings of the 1974 *Symposium on Mathematical Developments Arising from Hilbert Problems* (ed. Browder, 1976). Other papers that Robinson wrote herself on the subject of Hilbert's 10th Problem, both before and after the 1970 breakthrough, are (1962, 1), (1969, 1, 3, 4), and (1973, 1).

In recognition of her various outstanding contributions and, in particular, of her central role in the work leading to the solution of Hilbert's 10th Problem, Julia Robinson was elected in 1975 to the mathematical section of the National Academy of Sciences, the first woman to be so honored. In the same year, the University of California at Berkeley offered her a full professorship in the Department of Mathematics with a special arrangement permitting her to serve quarter-time since, although her health had improved greatly as a result of the operation mentioned earlier, she still did not feel that she could take on a full teaching load. In the past she had on occasion taught classes in the Department of Mathematics, but this was her first appointment as a regular member of the faculty at Berkeley since her receipt of the Ph.D. there in 1949.

Other signal honors followed in the next decade. Robinson was chosen as the colloquium lecturer at the 84th Summer Meeting of the American Mathematical Society in 1980; her lectures covered the main areas of her interests between logic and number theory, and the notes for these constitute her last writings, though they were not in final form for publication. In 1982, she was nominated for the presidency of the American Mathematical Society (for the years 1983–84), again the first woman to be so honored. Uncertain whether to take on the position because of her limited energies, she finally decided that, in the words from her *Autobiography* (pp. 20–21): ". . . as a woman and as a mathematician I had no alternative but to accept. I have always tried to do everything I could to encourage talented women to become research mathematicians. I found my service as President of the Society taxing but very, very satisfying." In 1983, Julia Robinson received a MacArthur Fellowship with its substantial award for a five-year period. And, in 1985, she was chosen as a member of the American Academy of Arts and Sciences.

While she was presiding over the summer A.M.S. meeting in Eugene, Oregon, in 1984, it was discovered that she was suffering from leukemia. After prolonged treatment and hospital stays, she enjoyed a remission of several months in the spring of 1985. But then the disease returned, and she died on July 30, 1985 at the age of sixty-five. She is survived by her husband, Raphael Robinson, and two sisters, Constance Reid and Billie Comstock.

One of Julia's last requests was that there be no funeral service and that those wishing to make a gift in her memory contribute to the Alfred Tarski Fund, which she had been instrumental in setting up in honor of her late teacher, friend, and colleague. Modest to the end, she let her character and achievements speak for themselves.

I AM INDEBTED to Constance Reid, as well as to John Addison, Leon Henkin, and Raphael Robinson of the University of California at Berkeley for materials on the life, work, and career of Julia Robinson. From published sources I have drawn most heavily on *The Autobiog-*

raphy of Julia Robinson by Constance Reid, and *Julia Bowman Robinson (1919–1985)* by Constance Reid with Raphael M. Robinson. Other valuable biographical material and personal impressions are listed in the references which follow.

NOTES

1. See Browder (1976), pp. 17–18, for the translation from the original German. I have omitted the adjective "rational" before "integers" to avoid confusion below.

2. For the result on Diophantine equations of degree 2, see Siegel (1972); the work mentioned on Diophantine equations in two variables of degree > 2 is due to Baker (1968).

REFERENCES

Alan Baker, "Contributions to the Theory of Diophantine Equations: I. On the Representation of Integers by Binary Forms, II. The Diophantine Equation $y^2 = x^3 + k$," *Philos. Trans. Roy. Soc. London, Ser. A.* 263(1968):173–208.

Felix E. Browder, ed. *Mathematical Developments Arising from Hilbert Problems. Proc. Symp. Pure Math.* XXVIII (Providence: American Mathematical Society, 1976).

Martin Davis, "Hilbert's Tenth Problem is Unsolvable," *Am. Math. Mon.* 80(1973):233–69.

Lisl Gaal, "Julia Robinson's Thesis," *Assoc. Women Math. Newsl.* 16(3)(1986):6–8.

Kurt Gödel, "Über formal unentscheidbare Sätze der Principia Mathematica und verwandter Systeme I," *Monatsh. Math. Phys.* 38(1932):173–98; reprinted and translated in *Gödel* (1986), pp. 144–95.

Kurt Gödel, *Collected Works, Vol. I. Publications 1929–1936*, ed. by S. Feferman et al. (New York: Oxford University Press, 1986).

Stephen C. Kleene, "Origins of Recursive Function Theory," *Ann. Hist. Computing* 3(1981):52–67.

Constance Reid, *Hilbert* (New York: Springer-Verlag, 1970).

Constance Reid, "The Autobiography of Julia Robinson," *Coll. Math. J.* 17(1986):2–21.

Constance Reid with Raphael M. Robinson, "Julia Bowman Robinson (1919–1985)," *Women of Mathematics, A Bibliographic Source Book,* ed. by Louise S. Grinstein and Paul J. Campbell (Westport, Conn.: Greenwood Press, 1987):182–89.

Carl Ludwig Siegel, "Zur Theorie der quadratischen Formen," *Nach. Akad. Wiss. Göttingen Math. Phys. Kl. II* (1972):21–46.

Craig Smorynski, "Julia Robinson, In Memoriam," *Math. Intelligencer,* 8(1986):77–79.

Alfred Tarski, *A Decision Method for Elementary Algebra and Geometry*, prepared for publication by J. C. C. McKinsey (RAND Report R–109, 1948; 2nd rev. ed. Univeristy of California Press, Berkeley and Los Angeles, 1951).

BIBLIOGRAPHY

1948

A note on exact sequential analysis. *Univ. Calif. Pub. Math. New Ser.* 1:241–46.

1949

Definability and decision problems in arithmetic. *J. Symb. Logic* 14:98–114.

1950

General recursive functions. *Proc. Am. Math. Soc.* 1:703–18.

1951

An iterative method of solving a game. *Ann. Math.* 54:296–301.

1952

Existential definability in arithmetic. *Trans. Am. Math. Soc.* 72:437–49.

1955

A note on primitive recursive functions. *Proc. Am. Math. Soc.* 6:667–70.

1959

The undecidability of algebraic rings and fields. *Proc. Am. Math. Soc.* 10:950–57.

Problems of number theory arising in metamathematics. *Report of the Institute in the Theory of Numbers* (Boulder):303–6.

1961

With Martin Davis and Hilary Putnam. The decision problem for exponential Diophantine equations. *Ann. Math.* 74:425–36.

1962

The undecidability of exponential Diophantine equations. In *Logic, Methodology, and Philosophy of Science: Proceedings of the 1960 International Congress*, ed. E. Nagel et al. (Stanford: Stanford University Press):12–13.

On the decision problem for algebraic rings. In *Studies in Math-*

ematical Analysis and Related Topics (essays in honor of George Pólya). (Stanford: Stanford University Press):297–304.

1965

The decision problem for fields. In *The Theory of Models: Proceedings of the 1963 International Symposium at Berkeley*, ed. by J. W. Addison et al. (Amsterdam: North-Holland):299–311.

1967

An introduction to hyperarithmetical functions. *J. Symb. Logic* 32:325–42.

1968

Recursive functions of one variable. *Proc. Am. Math. Soc.* 19:815–20.

Finite generation of recursively enumerable sets. *Proc. Am. Math. Soc.* 19:1480–86.

1969

Diophantine decision problems. *Studies in Number Theory*, ed. W. J. LeVeque. *MAA Stud. Math.* 6:76–116.

Finitely generated classes of sets of natural numbers. *Proc. Am. Math. Soc.* 21:608–14.

Unsolvable Diophantine problems. *Proc. Am. Math. Soc.* 22:534–38.

Hilbert's tenth problem. In *Proceedings of the 1969 Summer Institute in Number Theory*, ed. D. J. Lewis. *Proc. Symp. Pure Math.* 20, (Providence: AMS):191–94.

1973

Solving Diophantine equations. In *Proceedings of the Fourth International Congress for Logic, Methodology and Philosophy of Science*, ed. P. C. Suppes (Amsterdam: North-Holland):63–67.

Axioms for number-theoretic functions. In *Selected Questions of Algebra and Logic*, ed. A. I. Shirshov et al. (Novosibirsk: Siberian Academy of Sciences):253–63.

1974

With Y. Matijasevič. Two three-quantifier universal representations of recursively enumerable sets. (in Russian). In *Theory of Algorithms and Mathematical Logic* (Moscow):112–23.

1975

With Y. Matijasevič. Reduction of an arbitrary Diophantine equation to one in 13 unknowns. *Acta Arith.* 27:521–53.

1976

With Martin Davis and Y. Matijasevič. Hilbert's 10th problem. Diophantine equations: Positive aspects of a negative solution. *Mathematical Developments Arising from Hilbert Problems, Proc. Symp. Pure Math.* 28, ed. by F. A. Browder. (Providence: AMS):323–78.

1980

Between logic and arithmetic. Lecture notes distributed in conjunction with the Colloquium Lectures given at the Eighty-fourth Summer Meeting of the American Mathematical Society, University of Michigan, Ann Arbor, August 19–22, 1980.

A NOTE ON EXACT SEQUENTIAL ANALYSIS

BY

JULIA ROBINSON

1. THE RESULTS presented below refer to the situation described in detail in the preceding paper by Pólya. Also, the terminology and most of the notation are Pólya's. Consider the strip S defined by

$$bx + k - d < ay < bx + k\,,$$

where a, b, and d are positive integers, with a and b relatively prime and k an integer. The significant cases of the problem will be those with $0 < k < d$, but other values of k will also be used in the solution. The coördinate system is chosen so that the x-axis is directed downward (south) and the y-axis to the right (east). A hiker starts at the origin and walks repeatedly one unit south or one unit east. At each stage the probability is p that he will go south and $q = 1 - p$ that he will go east, where $0 < p < 1$. Assuming that the origin is within the strip, what is the probability that the hiker will first leave the strip by reaching the northern boundary? In order to include the case when the origin is not in S, the question may be thus formulated: what is the probability that the first point of the hiker's path to lie outside of the strip will be to the north of it?

The answer may also be interpreted as the probability of accepting a lot of manufactured goods on the basis of the sequential probability ratio test corresponding to the strip S. Here we make a sequence of tests classifying objects chosen at random from a certain lot as good or bad. After each observation we plot the point (x,y) where y equals the cumulated number of good objects and x is the corresponding number of bad ones. We accept or reject the lot according as the first point which falls outside S is above or below it. If the true proportion of defectives is p, then the probability of accepting the lot is the same as the probability of the random walker first leaving the strip via the northern boundary.

Pólya has shown that this probability has the form Q/R where both Q and R are polynomials in p and q. These polynomials were calculated in a number of cases by the method given in the preceding paper, and an explicit formula was found empirically in the case $a = 1$ where the strip S has the form:

$$bx + k - d < y < bx + k\,.$$

Namely, the probability of accepting is

$$q^k f_{d-k-1}(pq^b)/f_{d-1}(pq^b)\,,$$

where

$$f_n(z) = f_{n,b}(z) = \sum_{0 \leqq r \leqq n/b} \binom{n - rb}{r} (-z)^r\,.$$

In this note I shall give a proof of the formula, and at the end of the paper some numerical results which were calculated by Pólya's method and by the foregoing formula. These will give an idea of the size of the error in the approximate formulas of Wald.

2. Let $L(k) = L_p(k,d,b)$ be the probability that the first point of the hiker's path to lie outside S will be to the north of it. We notice first that if $k \geqq d$, $L(k) = 0$, since then the hiker starts south of the strip. Similarly $L(0) = 1$, since the hiker starts north of S. On the other hand, if $0 < k < d$, the hiker starts within the strip. The probability is p that he will first walk south reaching the point (1,0). At this point, he will find the northern boundary $k + b$ units to the east, so the probability of first reaching it would be $L(k + b)$. In a similar way, the probability is q that he will first go east to the point (0,1); the northern boundary is then $k - 1$ units further east and the probability of first reaching it is then $L(k - 1)$. Combining these results, we see that

$$L(k) = pL(k + b) + qL(k - 1) \qquad \text{for} \qquad 0 < k < d\,.$$

It is easy to see that these equations, together with the boundary conditions $L(0) = 1$ and $L(k) = 0$ for $k > d - 1$, uniquely determine $L(k)$ for $k \geqq 0$. We know that the probability $L(k)$ furnishes one solution. To see that these equations cannot have more than one solution, we proceed as follows: First write the equations in the form

$$L(k - 1) = [L(k) - pL(k + b)]/q \qquad \text{for} \qquad 0 < k < d\,.$$

We may then successively express $L(d - 2)$, $L(d - 3)$, $\cdot \cdot \cdot$, $L(1)$, $L(0)$ as multiples of $L(d - 1)$, utilizing the fact that $L(k) = 0$ for $k > d - 1$. The last equation will have the form

$$L(0) = mL(d - 1)$$

and since we know that $L(0) = 1$, this serves to determine $L(d - 1)$ uniquely and hence also $L(k)$ is uniquely determined for $k \geqq 0$.

Thus, it is sufficient to show that $L^*(k) = q^k f_{d-k-1}(pq^b)/f_{d-1}(pq^b)$ satisfies this system of equations and the boundary conditions in order to prove that $L^*(k) = L(k)$.

LEMMA 1. $f_{n+1}(z) - f_n(z) = -zf_{n-b}(z) \qquad \text{for} \qquad n \geqq 0.$

PROOF: $$f_{n+1}(z) - f_n(z) = \sum_{0 \leqq r \leqq (n+1)/b} \binom{n+1-rb}{r}(-z)^r - \sum_{0 \leqq r \leqq n/b} \binom{n-rb}{r}(-z)^r$$

$$= \sum_{1 \leqq r \leqq n/b} \binom{n-rb}{r-1}(-z)^r = \sum_{0 \leqq r \leqq (n-b)/b} \binom{n-b-rb}{r}(-z)^{r+1}$$

$$= -z f_{n-b}(z)\,.$$

LEMMA 2. $f_{n+1}(pq^b) > q f_n(pq^b) \qquad \text{for} \qquad n \geqq 0.$

PROOF: This is clear for $n = 0$. Suppose it is true for $n < N$. By Lemma 1, we have

$$f_{N+1}(pq^b) = f_N(pq^b) - pq^b f_{N-b}(pq^b) .$$

By the inductive hypothesis, or trivially if $N - b < 0$,

$$q^b f_{N-b}(pq^b) < f_N(pq^b) .$$

Hence

$$f_{N+1}(pq^b) > f_N(pq^b) - pf_N(pq^b) = qf_N(pq^b) ,$$

as was to be shown.

Since $f_0(z) = 1$, it follows from Lemma 2 that

$$f_n(pq^b) > q^n \qquad \text{for} \qquad n > 0 .$$

In particular, the denominator of $L^*(k)$ does not vanish. We can now prove the required formula

$$L^*(k) = pL^*(k+b) + qL^*(k-1) \qquad \text{for} \qquad 0 < k < d .$$

In fact, multiplying through by $f_{d-1}(pq^b)/q^k$, this becomes

$$f_{d-k-1}(pq^b) = pq^b f_{d-k-1-b}(pq^b) + f_{d-k}(pq^b) ,$$

which is Lemma 1 with $n = d - k - 1$. The boundary conditions, $L^*(0) = 1$ and $L^*(k) = 0$ for $k \geqq d$, are clear.

A few further conclusions from the lemmas may be of interest. In the first place, since it was shown that $f_n(pq^b) > 0$ for $n \geqq 0$, it follows that

$$f_{n+1}(pq^b) \leqq f_n(pq^b) \qquad \text{for} \qquad n \geqq 0 .$$

More precisely, we have equality for $0 \leqq n < b$, and strict inequality for $n \geqq b$. Also, by Lemma 2, we have

$$f_{d-1}(pq^b) > q^k f_{d-k-1}(pq^b) , \qquad \text{for} \qquad 0 < k < d ,$$

which furnishes a direct proof that

$$0 < L^*(k) < 1 .$$

This inequality is of course also a consequence of the fact that $L^*(k)$ is the probability $L(k)$ of leaving the strip S by the northern boundary when starting at an interior point.

3. Tables 1 and 2 reproduce samples of the results obtained in the Statistical Laboratory, University of California, by using the exact and the approximate formulae relating to sequential analysis. In both tables the first four columns are sup-

TABLE 1

COMPARISON OF APPROXIMATE AND EXACT FORMULAS RELATING TO SEQUENTIAL ANALYSIS

Stipulated conditions				Inspection plans			Probabilities and expected number of inspections								
α	β	p_1	p_2	s	h_1	h_2		L_{p_1}	L_s	L_{p_2}	$\bar{N}_0$	$\bar{N}_{p_1}$	$\bar{N}_s$	$\bar{N}_{p_2}$	$\bar{N}_1$
.05	.10	.001	.06	.01465	.5418	.6956	Approx.	.95	.56	.10	37	35	26	13	1
				.01471	Same	Same	Exact	.979	.637	.103	37	37.6	37.1	18.1	1
.05	.50	.0025	.03	.01113	.2554	.9163	Approx.	.95	.78	.50	23	23	21	18	1
				.01111	Same	Same	Exact	.975	.828	.513	23	25.8	30.1	27.0	1
.05	.10	.05	.17	.09938	1.6568	2.1271	Approx.	.95	.56	.10	17	30	39	25	3
				.1	Same	Same	Exact	.969	.591	.099	17	31.4	46.8	30.0	3
.05	.50	.05	.17	.09938	.4724	1.6946	Approx.	.95	.78	.50	5	7	9	9	2
				.1	Same	Same	Exact	.965	.798	.493	5	8.3	11.1	11.0	2
.05	.50	.06	.13	.09098	.7546	2.7071	Approx.	.95	.78	.50	9	19	25	25	3
				.09091	Same	Same	Exact	.959	.786	.478	9	21.5	29.8	30.4	3
.05	.10	.07	.22	.1331	1.7042	2.1880	Approx.	.95	.56	.10	13	24	32	21	3
				.1333	Same	Same	Exact	.965	.584	.095	13	25.5	37.8	24.9	3
.05	.10	.12	.30	.1998	1.9660	2.5241	Approx.	.95	.56	.10	10	22	31	21	4
				.2	Same	Same	Exact	.964	.583	.098	10	22.8	35.1	23.5	4
.05	.50	.12	.30	.1998	.5605	2.0108	Approx.	.95	.78	.50	3	5	7	7	3
				.2	Same	Same	Exact	.966	.800	.487	3	6.3	9.0	9.5	3

TABLE 2

COMPARISON OF EFFICIENCY OF INSPECTION PLANS RECOMMENDED BY SRG WITH SOME ALTERNATIVE PLANS

Stipulated conditions				Inspection plans				Exact probabilities and expected number of inspections					
α	β	p_1	p_2		s	h_1	h_2	L_{p_1}	L_{p_2}	$\bar{N}_0$	$\bar{N}_{p_1}$	$\bar{N}_{p_2}$	$\bar{N}_1$
.05	.50	.0025	.03	SRG	.01113	.2554	.9163	.975	.513	23	25.8	27.0	1
				Alternate	.01075	.2473	.7957	.952	.500	23	23.2	18.8	1
.05	.10	.05	.17	SRG	.09938	1.6568	2.1271	.969	.099	17	31.4	30.0	3
				Alternate	.10000	1.7000	1.9000	.954	.097	17	30.5	26.1	3
.05	.50	.05	.17	SRG	.09938	.4724	1.6946	.965	.493	5	8.3	11.0	2
				Alternate	.10000	.5000	1.5000	.952	.486	5	7.9	9.5	2
.05	.10	.07	.22	SRG	.1331	1.7042	2.1880	.965	.095	13	25.5	24.9	3
				Alternate	.1250	1.6250	2.2500	.956	.086	13	23.2	20.0	3

posed to be stipulated by the manufacturer and (or) consumer: if the true proportion of defectives in a lot is p_1 then the probability Lp_1 of accepting the lot must be $1 - \alpha$ or better. If the true proportion of defectives is p_2 then the probability Lp_2 of accepting the lot should be β or less.

For each set of conditions stiuplated in the first four columns there are two lines of figures giving the proposed plan of inspection, the probabilities L_p and, finally, the expectations $\overline{N}_p$ of the number of items to be inspected before a decision is made as to whether to accept or reject the lot. The proposed plan of inspection is characterized by the value of the slope s and by the two intercepts h_1 and h_2 as in formula 1.1 of the article by Professor Pólya. These three constants are exactly those used to characterize the sampling inspection plans published by the Statistical Research Group, Columbia University (Reference [1] in the article by Professor Pólya). The probability of accepting a lot which has a fixed proportion of defectives p, and the expected number of inspections before a decision $\overline{N}_p$ are both functions of s, h_1, and h_2.

Table 1 is intended to illustrate the differences occurring between the values of L_p and $\overline{N}_p$ computed by the approximate formulas of Wald and by the exact ones. It will be seen that these differences vary. Occasionally the probabilities of accepting the lot and the expected number of inspections computed exactly are practically the same as those obtained from the approximate formulas. In other cases, however, the differences are large. For example, the exact value of $\overline{N}_{p_2}$ corresponding to the second set of stipulated conditions exceeds the approximate by 50 per cent. The approximate values of $\overline{N}_p$ are taken directly from tables prepared by the Statistical Research Group just quoted.

Table 2 is intended to give a tentative answer to a question suggested by table 1: Granted that the expected number of inspections computed by the approximate formula is occasionally underestimated, could one think of decreasing the true expected number of inspections by altering the inspection plan?

In other words, table 2 is intended to test the "bestness" of the inspection plans exhibited in the same publication of the Statistical Research Group. Accordingly, the first line of each pair corresponding to fixed α, β, p_1 and p_2 gives the exact values of the L_p and $\overline{N}_p$ which are determined by the s, h_1 and h_2 taken from the report of the S. R. G. The second line gives the exact values of the same functions corresponding to a modified plan, selected by trial and error. It will be seen that occasionally there exists an alternative plan which would result in a substantial saving on sampling inspection. Tables 1 and 2 seem to justify some further work on sequential analysis both theoretical and numerical.

University of California

THE JOURNAL OF SYMBOLIC LOGIC
Volume 14, Number 2, June 1949

DEFINABILITY AND DECISION PROBLEMS IN ARITHMETIC

JULIA ROBINSON

Introduction. In this paper, we are concerned with the arithmetical definability of certain notions of integers and rationals in terms of other notions. The results derived will be applied to obtain a negative solution of corresponding decision problems.[1]

In Section 1, we show that addition of positive integers can be defined arithmetically in terms of multiplication and the unary operation of successor S (where $Sa = a + 1$). Also, it is shown that both addition and multiplication can be defined arithmetically in terms of successor and the relation of divisibility $|$ (where $x|y$ means x divides y). Thus the elementary theory of integers with S and $\cdot$ (or S and $|$) as the only primitive notions, which may seem rather narrow, is sufficient to express every idea or result which can be expressed in elementary number theory, i.e., the arithmetic of integers with $+$ and $\cdot$. An axiomatic problem concerning the system of positive integers with S and $\cdot$ is discussed in Section 2.

In Section 3, we show that the notion of an integer can be defined arithmetically in terms of the notion of a rational number and the operations of $+$ and $\cdot$ on rationals. Hence the arithmetic of rationals is adequate for the formulation of all problems of elementary number theory.

Since the solution of the decision problem is known to be negative for elementary number theory, it follows from our results that the solution of the decision problem is negative for any of the related theories mentioned above. As a further consequence, we see that the solution of the decision problem for the arithmetical theory of arbitrary fields is also negative. These problems will be discussed in Section 4.

The way in which we use the terms "arithmetic," "arithmetical or elementary theory," and "arithmetical definability" calls for some comments. For example, by the arithmetic (or elementary theory) of integers, we mean that part of the general theory of integers which can be developed without using any notions of a set-theoretical nature; that is, the part of the theory which can be formalized within the lower predicate calculus. Thus in formulating statements of the arithmetic of integers, we use only variables representing arbitrary integers; logical constants of the lower predicate calculus: $\wedge$ (and), $\vee$ (or), $\sim$ (not), $\rightarrow$ (if, then), $\leftrightarrow$ (if and only if), $\bigwedge$ (for every), $\bigvee$ (there exists), $=$ (equals); and mathematical constants denoting individual integers such as 0 and 1, relations between integers such as $<$, and operations on integers such as $+$ and $\cdot$.

The notion of arithmetical definability will be explained by means of examples.

Received September 28, 1948.

[1] This paper was submitted as a dissertation in partial fulfillment of the requirements for the degree of Doctor of Philosophy at the University of California. I wish to express my appreciation for the many helpful suggestions of Professor Alfred Tarski, under whose direction the thesis was written. I am particularly indebted for his assistance in preparing Section 4.

By saying for instance that addition of positive integers is arithmetically definable in terms of $\cdot$ and S, we mean that there is a formula ϕ with the following properties:

(i) ϕ is a formula of the elementary theory of positive integers with S and $\cdot$ as the only mathematical constants (thus ϕ contains only variables representing positive integers, the logical constants listed above and the mathematical constants S and $\cdot$).

(ii) ϕ contains three distinct free variables say x, y and z.[2]

(iii) ϕ is satisfied by those and only those ordered triples of integers $\langle x, y, z\rangle$ for which $x + y = z$.[3] A formula ϕ with this property may be referred to as a defining formula for $+$ in terms of $\cdot$ and S. As we shall see in Section 1 such a defining formula actually exists, in fact we can take for ϕ the formula

$$S(x\cdot z)\cdot S(y\cdot z) = S(z\cdot z\cdot S(x\cdot y)).$$

To give another example, the formula

$$\bigvee_{y} x = y + y \tag{1}$$

of the elementary theory of positive integers is satisfied by those and only those positive integers x which are even. Hence this formula shows that the notion of an even integer is arithmetically definable in terms of addition. Now consider the formula

$$\bigwedge_{\mathfrak{A}} ([2 \in \mathfrak{A} \wedge \bigwedge_{y} (y \in \mathfrak{A} \leftrightarrow y + 2 \in \mathfrak{A})] \rightarrow x \in \mathfrak{A}). \tag{2}$$

This formula differs from (1) in that besides variables representing positive integers, it also contains a bound variable representing sets of integers; and in addition to logical constants of the lower predicate calculus it also contains the symbol ϵ denoting the membership relation. On the other hand (2) like (1) contains only one free variable x and is satisfied by those and only those positve integers which are even. If we knew only (2) without knowing (1), we could just say that the notion of an even integer is recursively definable (or recursively arithmetically definable) in terms of addition.

In this example, the explicit arithmetical definition is the simplest and most natural one; in other cases, set-theoretical or in particular recursive definitions may be easy while an explicit arithmetical definition is either impossible or hard to find.

The problem of the axiomatic foundation of the discussion will be disregarded in Sections 1 and 3. Thus when stating that a formula is a theorem, we do not try to derive this formula from any axioms but merely to show that it is true in the intuitive sense and can be so recognized by mathematicians. It will be

---

[2] No attempt is made in this paper to present the metamathematical fragments of the discussion in a rigorous, formal way or to carry through a sharp distinction between metamathematical and mathematical notions. In particular, quotation marks are omitted in most cases since we believe no confusion will arise from this negligence.

[3] For a precise formal definition of the notion of satisfaction and truth, see Tarski [10].

seen that the question of axiomatization will play a rather restricted rôle even in the results established in Section 4.

1. Problems of definability in the arithmetic of integers. In developing the arithmetic of positive integers, we usually consider addition and multiplication as fundamental operations. It is known that a great variety of other arithmetical notions are (explicitly) arithmetically definable in terms of these operations. The problem arises whether one of these operations is definable in terms of the other. Consider first the problem of defining $\cdot$ in terms of $+$. It is obvious that $\cdot$ can be defined recursively in terms of $+$; we first define 1 as the positive integer which is not the sum of two positive integers and then use the familiar recursive definition of multiplication.

The situation changes if we are interested in the arithmetical definability of multiplication in terms of addition. From a result of Presburger [8], we can easily obtain a detailed description of all relations between integers and operations on integers which can be defined arithmetically in terms of $+$; and we do not find $\cdot$ among them. It is easily seen that $+$ is not definable in terms of $\cdot$ in any sense of the word definable. For the system of positive integers has automorphisms with respect to multiplication which permute the primes in an arbitrary way, so that sum is certainly not preserved. This method of proof is due to Padoa [7].

Among the notions which are arithmetically definable in terms of addition, we find the relation $<$ in terms of which the successor operation is arithmetically definable. Conversely, $+$ can be defined in terms of $<$ and $<$ in terms of S by means of recursive definitions. However by an argument exactly analogous to that referred to above to show that $\cdot$ is not arithmetically definable in terms of $+$, it can be shown that $+$ is not arithmetically definable in terms of $<$ and that $<$, hence also $+$, is not arithmetically definable in terms of S.[4]

We can now ask the question whether addition is arithmetically definable in terms of one of these weaker notions combined with multiplication. In this connection we obtain:

THEOREM 1.1. *Addition of positive integers is arithmetically definable in terms of multiplication and the successor operation as well as in terms of multiplication and the relation of less-than.*

Proof. It is easily seen that for any three positive integers, we have $a + b = c$ if and only if a, b, and c satisfy the following formula:

$$S(a\cdot c)\cdot S(b\cdot c) = S[(c\cdot c)\cdot S(a\cdot b)]. \tag{1}$$

To verify this, write formula (1) using ordinary mathematical notation:

$$(1 + ac)\,(1 + bc) = 1 + c^2\,(1 + ab).$$

Multiplying out and cancelling terms appearing on both sides of the equation, we have $(a + b)c = c^2$. Since c is a positive number, this is true if and only if $a + b = c$.

From this argument it follows at once that $+$ is arithmetically definable in

[4] This follows from the results discussed in Hilbert-Bernays [6], vol. I, pp. 234–264.

terms of $\cdot$ and S. We now notice that (1) can be given the following equivalent form:

$$\bigvee_{x,y,z} [S(a\cdot b) = x \wedge S(a\cdot c) = y \wedge S(b\cdot c) = z \wedge S([c\cdot c]\cdot x) = y\cdot z].$$

This clearly remains valid if we eliminate the symbol S by replacing each partial formula of the form

$$Su = v$$

by

$$u < v \wedge \bigwedge_{w} \sim(u < w \wedge w < v).$$

In this way, we see that $+$ is arithmetically definable in terms of $\cdot$ and $<$.

We can now try to improve the results obtained in Theorem 1.1 by replacing multiplication by weaker notions. In the first place, we have in mind the relation of divisibility. This relation is of course arithmetically definable in terms of multiplication while it seems very likely that $\cdot$ is not arithmetically definable in terms of $|$. It can be shown however that $\cdot$ can be defined set-theoretically in terms of $|$. As an improvement of Theorem 1.1, we now obtain:

THEOREM 1.2. *Addition and multiplication of positive integers are arithmetically definable in terms of the successor operation and the relation of divisibility.*

Proof. It is clearly sufficient to show that $\cdot$ is arithmetically definable in terms of $|$ and S. We shall express the fact that two integers a and b are relatively prime by $a \perp b$ and denote the least common multiple of two numbers a and b by $a\circ b$. Then the following equivalence holds in the arithmetic of positive integers:

$$\text{(2)}\qquad a\cdot b = c \leftrightarrow \bigwedge_{x} (a|x \wedge b|x \wedge c|x) \vee \bigwedge_{x,y,m} \{[a \perp x \wedge b \perp y \wedge c \perp x \wedge c \perp y \wedge x \perp y \wedge m|S(a\circ x) \wedge m|S(b\circ y)] \rightarrow \bigvee_{u} [m|u \wedge Su = c\circ(x\circ y)]\}.$$

Suppose first that $a\cdot b = c$ and that x, y and m satisfy the conditions

$$a \perp x \wedge b \perp y \wedge c \perp x \wedge c \perp y \wedge x \perp y \wedge m|S(a\circ x) \wedge m|S(b\circ y).$$

We must show that if a and b are not both 1, then

$$\bigvee_{u} [m|u \wedge Su = (a\cdot b)\circ(x\circ y)].$$

Now because of the conditions of relative primeness, we can replace the symbol $\circ$ by $\cdot$ throughout. Hence we must check that $m|(ax + 1)$ and $m|(by + 1)$ implies that $m|(abxy - 1)$. Writing this out in terms of congruences, we have $ax \equiv -1 \pmod{m}$ and $by \equiv -1 \pmod{m}$ implies $abxy \equiv +1 \pmod{m}$, which certainly holds.

Conversely, suppose the right side of (2) holds and show that $a\cdot b = c$. Notice first that $c = 1$ implies $a = b = 1$, for otherwise we may take x, y, and m all equal to 1 and obtain that there exists a u such that $Su = 1$, which is impossible.

Hence consider the case when c is different from 1. Let m be prime to a and b. Choose x and y prime to a, b, and c and to each other and such that

$$ax \equiv -1 \pmod{m},$$

$$by \equiv -1 \pmod{m}.$$

This is possible since x and y are only restricted to belonging to certain residue classes mod m which are prime to m. Now since $m|S(a \circ x)$ and $m|S(b \circ y)$ and the right side of (2) holds, it follows that there is a u satisfying $m|u \wedge Su = c \circ (x \circ y)$ or in terms of congruences $cxy \equiv +1 \pmod{m}$. Hence $c \equiv ab$ for arbitrarily large values of m and thus $c = ab$.

The relation $\perp$ and the operation $\circ$ are both arithmetically definable in terms of divisibility. In fact, the following equivalences can be used as definitions:

$$a \perp b \leftrightarrow \bigwedge_x (x|a \wedge x|b \rightarrow \bigwedge_y x|y)$$

and

$$c = a \circ b \leftrightarrow \bigwedge_x (a|x \wedge b|x \leftrightarrow c|x).$$

Hence we can eliminate the symbols $\perp$ and $\circ$ from (2) and obtain a defining formula for $\cdot$ in terms of S and $|$.

We might also try to improve Theorem 1.2 by replacing divisibility by the relation of relative primeness. However I have not been able to determine whether $\cdot$ is arithmetically definable in terms of $\perp$ and S or even in terms of $\perp$ and $+$.

It is easy to see that Theorem 1.1 can be extended to the arithmetic of arbitrary (not only positive) integers and furthermore to the arithmetic of arbitrary integral domains with unity, i.e., for every integral domain addition can be defined arithmetically in terms of multiplication and the successor operation S with $Sa = a + 1$.[5] The definition of $+$ in terms of $\cdot$ and S used in the proof of Theorem 1.1 can easily be modified in order to make it apply to arbitrary integers or to the elements of arbitrary integral domains.

On the other hand, I have been unable to extend Theorem 1.2 to arbitrary integers; the difficulty in this case reduces to that of defining the notion of a positive integer in terms of $|$ and S.

2. An axiomatic problem regarding the arithmetic of positive integers. Whenever addition and multiplication can be defined in terms of other notions of the arithmetic of positive integers, it is clearly possible to formulate an axiom system for arithmetic involving these other notions as the only primitive ones. This can be done in a mechanical way by simply eliminating $+$ and $\cdot$ from any of the familiar axiom systems involving only these two notions. The axioms thus obtained are usually complicated and artificial, and the problem of finding a simple and elegant axiom system of the kind desired may present considerable difficulty. We shall consider here only one problem of this kind.

---

[5] This improves an analogous result for the theory of abstract fields recently published by B. A. Bernstein [1].

We start here with the familiar axiom system due to Peano. It contains four primitive notions 1, S, $\cdot$, and $+$ and consists of the following axioms:

A1. $\bigwedge_{a} \sim Sa = 1$

A2. $\bigwedge_{a,b} (Sa = Sb \rightarrow a = b)$

A3. $(\mathcal{P}(1) \wedge \bigwedge_{a} [\mathcal{P}(a) \rightarrow \mathcal{P}(Sa)]) \rightarrow \bigwedge_{a} \mathcal{P}(a)$

A4. $\bigwedge_{a} a + 1 = Sa$

A5. $\bigwedge_{a,b} a + Sb = S(a + b)$

A6. $\bigwedge_{a} a \cdot 1 = a$

A7. $\bigwedge_{a,b} a \cdot Sb = (a \cdot b) + a.$

Here the induction principle A3 is an axiom-schema, that is an infinite system of axioms. It is understood that the symbol $\mathcal{P}(a)$ may be replaced by any arithmetical formula containing just one free variable a, if suitable replacements of $\mathcal{P}(1)$ and $\mathcal{P}(Sa)$ are also made.[6]

We put the induction principle in this form if we decide to formalize the arithmetic of positive integers entirely within the lower predicate calculus. If on the other hand we want to use set theory in developing the arithmetic of positive integers, we replace the axiom-schema A3 by the following axiom:

A3′. $\bigwedge_{\mathfrak{A}} ([1 \in \mathfrak{A} \wedge \bigwedge_{a} (a \in \mathfrak{A} \rightarrow Sa \in \mathfrak{A})] \rightarrow \bigwedge_{a} a \in \mathfrak{A}).$

In such a case, we can also replace A4—A7 by explicit set-theoretical definitions of $+$ and $\cdot$.

By our Theorem 1.1, it follows that the symbol $+$ may be eliminated from Peano's axioms. This can be done mechanically by transforming A4, A5, and A7 with the help of the formula defining $+$ in terms of S and $\cdot$ in the proof of Theorem 1.1. We thus obtained the formulations:

A4′. $\bigwedge_{a} S(a \cdot Sa) \cdot S(1 \cdot Sa) = S([Sa \cdot Sa] \cdot Sa)$

A5′. $\bigwedge_{a,b,c} [S(a \cdot c) \cdot S(b \cdot c) = S[(c \cdot c) \cdot S(a \cdot b)] \rightarrow$
$S(a \cdot Sc) \cdot S(Sb \cdot Sc) = S([Sc \cdot Sc] \cdot S(a \cdot Sb))]$

A7′. $\bigwedge_{a,b} S[(a \cdot b) \cdot (a \cdot Sb)] \cdot S[a \cdot (a \cdot Sb)] = S([(a \cdot Sb) \cdot (a \cdot Sb)] \cdot S[(a \cdot b) \cdot a]).$

[6] We could consider a more general axiom schema in which $\mathcal{P}(a)$ could be replaced by any arithmetical formula containing in addition to a also other free variables b, c, $\cdots$ (in which case the whole schema would have to be preceded by the quantifiers $\bigwedge_{b,c,\ldots}$). It is shown in Hilbert-Bernays [6], vol. I, p. 343, that this more general form of the induction principle is actually no stronger than A3.

In view of the character of the defining formula which is used, we have to add one more axiom to guarantee the operational character of addition, i.e., the uniqueness of the result of addition. This new axiom is:

$$\text{A8.} \quad \bigwedge_{a,b,c,d} \{[S(a\cdot c)\cdot S(b\cdot c) = S([c\cdot c]\cdot S(a\cdot b)) \qquad S(a\cdot d)\cdot S(b\cdot d) = S([d\cdot d]\cdot S(a\cdot b))] \rightarrow c = d\}.$$

A moment's reflection suffices to see that the axiom system consisting of A1 –A3, A4′, A5′, A6, A7′, and A8 is equivalent to the original axiom system (when supplemented by the definition of $+$ in terms of $\cdot$ and S) and is adequate for the development of the arithmetic of positive integers within the lower predicate calculus. We shall refer to this axiom system as System $\mathfrak{S}$; and shall use the symbol $\mathfrak{S}'$ to denote the axiom system obtained from $\mathfrak{S}$ by replacing A3 by A3′. We do not know whether the axioms of $\mathfrak{S}$ or $\mathfrak{S}'$ are mutually independent.

We now want to consider another simpler axiom system with the same mathematical constants as $\mathfrak{S}$ and which is also possibly an adequate basis for the development of the arithmetic of positive integers. This system consists of the old axioms A1, A2, A3, A6 and the following three new axioms:

$$\text{A4}''. \quad \bigwedge_{a,b} a\cdot b = b\cdot a$$

$$\text{A5}''. \quad \bigwedge_{a,b,c} a\cdot (b\cdot c) = (a\cdot b)\cdot c$$

$$\text{A7}''. \quad \bigwedge_{a,b} Sa\cdot S(a\cdot b) = S(a\cdot S(b\cdot Sa)).$$

A4″ and A5″ are of course well-known consequences of the Peano axioms. Axiom A7″ is unfamiliar, but expresses the simple identity

$$(1 + a)\ (1 + ab) = 1 + a\ (1 + b\ (1 + a)).$$

The new axiom system will be referred to as System $\mathfrak{T}$; and by $\mathfrak{T}'$ we shall denote the system obtained from $\mathfrak{T}$ by replacing A3 by A3′. We are interested in whether Systems $\mathfrak{S}$ and $\mathfrak{T}$ are equivalent (mutually derivable) and whether consequently $\mathfrak{T}$ can actually serve as a basis for the arithmetic of integers. The question can be answered affirmatively if we replace $\mathfrak{S}$ and $\mathfrak{T}$ by $\mathfrak{S}'$ and $\mathfrak{T}'$ and permit outselves to use set-theoretical devices in derivations. In fact, we have:

THEOREM 2.1. *Systems $\mathfrak{S}'$ and $\mathfrak{T}'$ are equivalent.*

Proof. The derivation of the axioms of $\mathfrak{T}'$ from those of $\mathfrak{S}'$ presents no difficulty. By defining addition in an appropriate way, we can easily derive from $\mathfrak{S}'$ all of Peano's axioms and hence we can obtain the axioms of $\mathfrak{T}'$ by means of the ordinary recursive procedure.

To derive the axioms of $\mathfrak{S}'$ from those of $\mathfrak{T}'$ we proceed in the following way. Since A1, A2, and A3′ (i.e., "proper" Peano axioms) are available in $\mathfrak{T}'$, we introduce ordinary addition and multiplication by means of the usual recursive definitions. Then we show by induction that the multiplication operation thus defined satisfies the remaining axioms of $\mathfrak{S}'$. The only thing that we do not yet know is whether this recursive multiplication coincides with the operation $\cdot$

of the axioms of $\mathfrak{T}'$. Hence the proof will be completed if we establish the following:

LEMMA. *The only binary operation $\cdot$ defined on the positive integers and satisfying the following formulas for all positive integers*

(i) $a \cdot 1 = a$

(ii) $a \cdot b = b \cdot a$

(iii) $a \cdot (b \cdot c) = (a \cdot b) \cdot c$

(iv) $Sa \cdot S(a \cdot b) = S(a \cdot S(b \cdot Sa))$

is ordinary multiplication.

Proof. Let $\times$ denote ordinary multiplication, that is multiplication introduced recursively. We need to show that for every a and b, $a \cdot b = a \times b$. We proceed by induction. Clearly $a \cdot 1 = 1 \cdot a = a = 1 \times a$. Suppose that it has been shown that $m \cdot n = m \times n$ for $m \leqq a$ and for all n. We need to show that $Sa \cdot n = Sa \times n$ for all n. Let $\mathfrak{N}$ be the set of all n such that $Sa \cdot n = Sa \times n$.

We show first that the inductive hypothesis together with (i)–(iv) implies that if $b \in \mathfrak{N}$ then $S(a \times b) \in \mathfrak{N}$. Using the inductive hypothesis, we have from (iv) that

$$Sa \cdot b = Sa \times b \rightarrow Sa \cdot S(a \times b) = S(a \times S(b \times Sa)).$$

But since $\times$ also satisfies (iv), we have

$$Sa \times S(a \times b) = S(a \times S(b \times Sa)).$$

Therefore,

$$Sa \cdot b = Sa \times b \rightarrow Sa \cdot S(a \times b) = Sa \times S(a \times b),$$

and if $b \in \mathfrak{N}$ then $S(a \times b) \in \mathfrak{N}$.

Next we shall show that if $m \leqq a$ and $b \in \mathfrak{N}$, then $m \times b \in \mathfrak{N}$. Suppose $Sa \cdot b = Sa \times b$, then

$$Sa \cdot (m \times b) = Sa \cdot (m \cdot b) = (Sa \cdot b) \cdot m = (Sa \times b) \times m = Sa \times (m \times b).$$

Hence if $b \in N$, $m \times b \in \mathfrak{N}$. Here we used (ii) and (iii) as well as the inductive hypothesis.

Also for $m \leqq a$, if $m \times b \in \mathfrak{N}$, then $b \in \mathfrak{N}$. Suppose $m \leqq a$ and $Sa \cdot (m \times b) = Sa \times (m \times b)$. Then

$$(Sa \cdot b) \times m = Sa \cdot b \cdot m = Sa \cdot (m \times b) = Sa \times (m \times b) = (Sa \times b) \times m.$$

Now using the cancellation law for ordinary multiplication we have $Sa \cdot b = Sa \times b$. Therefore if $m \leqq a$ and $m \times b \in \mathfrak{N}$ then $b \in \mathfrak{N}$.

We are now ready to show by induction in b that $b \in \mathfrak{N}$. Suppose that for all $b < n$, $b \in \mathfrak{N}$, and show that $n \in \mathfrak{N}$. It is sufficient to consider $n > 1$, since $Sa \cdot 1 = Sa \times 1 = Sa$ by (i).

CASE I. n and a relatively prime. Then there are r and s such that

$$S(a \times r) = n \times s \wedge r < n \wedge s \leqq a.$$

By the inductive hypothesis $r \in \mathfrak{N}$ since $r < n$. Hence $S(a \times r) \in \mathfrak{N}$, that is $n \times s \in \mathfrak{N}$. But $s \leqq a$, so $n \in \mathfrak{N}$.

Case II. n and a have a common factor $k > 1$. Say $n = k \times t$ with $k \leqq a$ and $t < n$. Hence by the inductive hypothesis, $t \in \mathfrak{N}$, and since $k \leqq a$ this implies $k \times t \in \mathfrak{N}$. But $k \times t$ is just n, so $n \in \mathfrak{N}$.

We have just shown that $a \cdot b = a \times b$ for every a and b. In the proof, we have used various theorems from elementary number theory but these can all be derived from Peano's axioms. The proof of our lemma and hence also of Theorem 2.1 is thus complete.

We now return to the problem of equivalence of $\mathfrak{S}$ and $\mathfrak{T}$ on the basis of the lower predicate calculus. This problem seems to be more difficult and is still open. As in the proof of Theorem 2.1 we can easily deduce the axioms of $\mathfrak{T}$ from those of $\mathfrak{S}$. Going in the opposite direction, we can easily derive A4′ and A7′ from the axioms of $\mathfrak{T}$. However I have not been able to derive from $\mathfrak{T}$ either A5′ or A8. In fact, I have not even been able to show that either the cancellation law

$$\bigwedge_{a,b,c} (a \cdot c = b \cdot c \rightarrow a = b) \tag{1}$$

or all true equalities of the form

$$\alpha \cdot \beta = \gamma \tag{2}$$

where α, β, and γ stand for arbitrary constant terms of the form 1, $S1$, $SS1$, $\cdots$ are derivable from the axioms of Systems $\mathfrak{T}$. The only result which we know in this connection is that all equalities (2) become derivable if we enrich $\mathfrak{T}$ by including the cancellation law (1). This can be shown by analyzing the proof of Theorem 2.1, more specifically the lemma formulated there.

3. Problems of definability in the arithmetic of rationals. Given any statement of the arithmetic of rational numbers with $+$ and $\cdot$, an equivalent statement of integral arithmetic with the same fundamental operations can be found by replacing each rational variable by the ratio of two integer variables, clearing of fractions, and adjoining the condition that the denominators are not zero. It is much less simple to determine whether with every statement of the arithmetic of integers we can correlate an equivalent statement in the arithmetic of rationals and whether consequently the arithmetic of rationals is adequate to express all problems of elementary number theory.

The answer to this question is an affirmative one. To show this, it clearly suffices to prove that the notion of an integer is (explicitly) arithmetically definable in the arithmetic of rationals in terms of the operations $+$ and $\cdot$.

In establishing this result, we need not refrain from using any known theorem concerning integers or rationals since, as was stated in the introduction, we are disregarding here the problem of axiomatic foundations of our discussion. In order to make the argument more easily readable, we shall use in it besides the variables capital A, B, C, $\cdots$ representing arbitrary rationals also small a, b, c, $\cdots$ which are assumed to be integers.

Theorem 3.1. (i) *For a rational number N to be an integer it is necessary and sufficient that it satisfy the following formula*

$$\bigwedge_{A,B} (\{(\bigvee_{X,Y,Z} 2 + BZ^2 = X^2 + AY^2) \wedge \bigwedge_{M} [(\bigvee_{X,Y,Z} 2 + ABM^2 + BZ^2$$
$$= X^2 + AY^2) \rightarrow (\bigvee_{X,Y,Z} 2 + AB(M+1)^2 + BZ^2 = X^2 + AY^2)]\}$$
$$\rightarrow (\bigvee_{X,Y,Z} 2 + ABN^2 + BZ^2 = X^2 + AY^2)).$$

(ii) *Hence the notion of an integer and that of a positive integer is arithmetically definable in terms of the notion of a rational and the operations of addition and multiplication on rationals.*

Proof. (i) Let $\phi(A, B, K)$ stand for

$$\bigvee_{X,Y,Z} 2 + ABK^2 + BZ^2 = X^2 + AY^2.$$

Then we can rewrite the formula given in (i) as follows:

(1) $$\bigwedge_{A,B} \{(\phi(A, B, 0) \wedge \bigwedge_{M} [\phi(A, B, M) \rightarrow \phi(A, B, M + 1)]) \rightarrow \phi(A, B, N)\}.$$

It may be noticed that our formula is closely related to the formula

$$\bigwedge_{\mathfrak{A}} \{[0 \in \mathfrak{A} \wedge \bigwedge_{M} (M \in \mathfrak{A} \rightarrow M + 1 \in \mathfrak{A})] \rightarrow N \in \mathfrak{A}\},$$

which defines the notion of a non-negative integer set-theoretically.

We first show that any given integer N satisfies (1). In fact, suppose A and B satisfy the hypothesis of (1),

(2) $$\phi(A, B, 0) \wedge \bigwedge_{M} [\phi(A, B, M) \rightarrow \phi(A, B, M + 1)].$$

If now N is a positive integer, then by induction it satisfies the conclusion of (1), i.e., $\phi(A, B, N)$. On the other hand, since N occurs in $\phi(A, B, N)$ only to an even power, it is clear that

$$\phi(A, B, -N) \leftrightarrow \phi(A, B, N).$$

Thus if N is any integer and A and B satisfy (2), then $\phi(A, B, N)$ holds; in other words, N satisfies (1).

It is much harder to show that every rational number N which satisfies (1) is an integer. We shall need several lemmas. The first two are special cases of a general theorem of Hasse [5] which gives necessary and sufficient conditions for rational representation of a rational number by a given quadratic form in any number of variables.[7] The symbol (k/p) is the familiar Legendre symbol giving the quadratic character of $k \pmod p$.

LEMMA 1. *If p is a prime $\equiv 3 \pmod 4$ then $X^2 + Y^2 - pZ^2$ represents a non-zero rational number M, if and only if M is not of the form*

$$p \cdot k \cdot S^2 \quad \textit{with } (k/p) = 1$$
$$\textit{or} \quad k \cdot S^2 \quad \textit{with } k \equiv p \pmod 8.$$

[7] A forthcoming book by Gordon Pall, ***The arithmetical theory of quadratic forms***, will give an elementary account of the theory.

LEMMA 2. *If p and q are odd primes with $p \equiv 1 \pmod 4$ and $(q/p) = -1$, then $X^2 + qY^2 - pZ^2$ represents a non-zero rational number M if and only if M is not of the form*

$$p \cdot k \cdot S^2 \quad \text{with } (k/p) = -1$$

$$\text{or} \quad q \cdot k \cdot S^2 \quad \text{with } (k/q) = -1.$$

LEMMA 3. *If p is a prime $\equiv 3 \pmod 4$ then*

$$2 + pM^2 + pZ^2 = X^2 + Y^2$$

has a solution for X, Y, and Z if and only if the denominator of M in lowest terms is odd and prime to p.

Proof. Let $M = n/d$ in lowest terms. Put $m = 2d^2 + pn^2$. It is sufficient to determine when m can be represented rationally by the form $X^2 + Y^2 - pZ^2$. Suppose that d is odd and prime to p. Then m is prime to p, so m is not of the form $p \cdot k \cdot S^2$. Also $m \equiv 1, 2 \pmod 4$, so it is not of the form $k \cdot S^2$ with $k \equiv p \pmod 8$ since $p \equiv 3 \pmod 4$. Hence by Lemma 1, m can be represented.

Suppose $p|d$ and put $d = p \cdot r$. Then $m = p \cdot k$ where $k = 2pr^2 + n^2$ is prime to p, since we assumed that n/d is in lowest terms, so $(k/p) = (n^2/p) = 1$. Thus, m is of the form $p \cdot k \cdot S^2$ with $(k/p) = 1$. Therefore by Lemma 1, m is not represented by $X^2 + Y^2 - pZ^2$.

Suppose $2|d$ and put $d = 2s$. Then $m = 8s^2 + pn^2$. But n is odd, so $m \equiv p \pmod 8$ and hence is not represented by the forms $X^2 + Y^2 - pZ^2$.

LEMMA 4. *If p and q are odd primes with $p \equiv 1 \pmod 4$ and such that $(q/p) = -1$, then*

$$2 + pqM^2 + pZ^2 = X^2 + qY^2$$

has a solution for X, Y, Z if and only if the denominator of M in lowest terms is prime to p and q.

Proof. Let $M = n/d$ in lowest terms and put $m = 2d^2 + pqn^2$. It is sufficient to check that m is represented by the form $X^2 + qY^2 - pZ^2$ if and only if d is prime to p and q.

Suppose that d is prime to p and q, then m is also prime to p and q. Hence by Lemma 2, m is represented by the form $X^2 + qY^2 - pZ^2$.

Suppose that $p|d$ and put $d = p \cdot r$, then $m = p \cdot k$ where $k = 2pr^2 + qn^2$. Since qn^2 is prime to p, the quadratic character $(k/p) = (qn^2/p) = (q/p) = -1$. Hence m is not represented by $X^2 + qY^2 - pZ^2$.

Similarly, suppose $q|d$ and put $d = q \cdot s$. Then $m = qk$ where $k = 2qr^2 + pn^2$ and $(k/q) = (pn^2/q) = (p/q) = (q/p) = -1$. Hence m is not represented by the given form.

LEMMA 5. *If p is a prime $\equiv 1 \pmod 4$, there is an odd prime q such that $(q/p) = -1$.*

Proof. Let s be any non-residue of p. Then either s or $s + p$ is odd and an odd non-residue of p must have an odd prime factor which is also a non-residue of p. Let this be q.

We are now ready to proceed with the proof of (i). We wish to show that if N satisfies (1), then N is an integer.

Consider first the case when $A = 1$ and $B = p$ where p is any prime $\equiv 3 \pmod 4$. Lemma 3 states that $\phi(1, p, M)$ if and only if the denominator of M in lowest terms is not divisible by 2 or p. Hence $A = 1$ and $B = p$ satisfy (2) since M and $M + 1$ have the same denominator and the denominator of 0 in lowest terms is ± 1. Therefore, if N satisfies (1), we must have $\phi(1, p, N)$. Hence the denominator of N is not divisible by 2 or p. But this holds for any prime $\equiv 3 \pmod 4$, hence the denominator of N is not divisible by 2 or by any prime $\equiv 3 \pmod 4$.

Next let p be any prime $\equiv 1 \pmod 4$. Choose q by Lemma 5 to be an odd prime such that $(q/p) = -1$. Then, Lemma 4 states that $\phi(q, p, M)$ if and only if the denominator of M in lowest terms is not divisible by p or q. Hence, as before, $A = q$ and $B = p$ satisfy the condition (2) so that if N satisfies (1) we must have $\phi(q, p, N)$. Hence the denominator of N in lowest terms is not divisible by any prime $\equiv 1 \pmod 4$.

Combining these results, we see that the denominator of N is not divisible by any prime, and therefore must be ± 1. Hence N is an integer.

(ii) From the formula stated in (i) we can easily eliminate the symbols 1 and 2 and replace X^2 by $X \cdot X$ (cf. the proof of the second part of Theorem 1.1). Hence the definability of integers in terms of the notion of a rational and the operations of addition and multiplication follows at once.

To extend this result to the notion of a positive integer it is sufficient to recall that the latter notion is definable in terms of that of an arbitrary integer and the operations + and $\cdot$; in fact, using the symbolic expression Int(A) to express the fact that A is an arbitrary integer, the following formula holds if and only if A is a positive integer:

$$\mathrm{Int}(A) \wedge \sim A = 0 \wedge \bigvee_{X,Y,Z,W} A = X^2 + Y^2 + Z^2 + W^2.$$

Another defining formula with fewer quantifiers which may serve the same purpose was suggested by R. M. Robinson:

$$\mathrm{Int}(A) \wedge \sim A = 0 \wedge \bigvee_{X,Y} [\mathrm{Int}(X) \wedge \sim X = 0 \wedge (A = X^2 \vee X^2 = 1 + AY^2)].$$

As an obvious consequence of Theorem 3.1, we obtain:

THEOREM 3.2. *Let n be any fixed positive integer and let R be an n-ary relation between rational numbers. Let R' be the relation which holds between $2n$ integers, $p_1, p_2, \cdots, p_n; q_1, \cdots, q_n$ if and only if $q_1 \neq 0, \cdots, q_n \neq 0$, and R holds between $p_1/q_1, \cdots, p_n/q_n$. Then R is arithmetically definable in terms of $+$ and $\cdot$ on rationals if (and only if) R' is arithmetically definable in terms of $+$ and $\cdot$ on integers.*

This consequence of our discussion is interesting because of a result of Gödel [4] which shows that the variety of relations between integers (and operations on integers) which are arithmetically definable in terms of addition and multiplication of integers is very great. For instance from Theorem 3.2 and Gödel's result, we can conclude that the relation which holds between three rationals A, B, and N if and only if N is a positive integer and $A = B^N$ is definable in the arithmetic of rationals.

4. Applications to the decision problem in arithmetic. From the results obtained in Sections 1 and 3 various consequences can be derived concerning the decision problem in the arithmetic of integers and rationals. In deriving these consequences, we make use of some unpublished results of Tarski and Mostowski. Hence we start with a brief account of these results.

The following remarks apply to mathematical theories which are assumed to be formalized within the lower predicate calculus (with identity and without variable predicates). We assume that in all these theories the same logical symbols, the same logical axioms and axiom-schemata, and the same rules of inference are used. Thus two such theories differ from each other only by the range of their variables, their specific mathematical constants (primitive symbols) and their specific mathematical axioms.

In each theory, we define in the usual way what we understand by a formula; in particular, all axioms turn out to be formulas and any application of rules of inference to given formulas yields a new formula. A formula which can be obtained from logical and mathematical axioms by applying rules of inference is referred to as a provable formula. A theory $\mathfrak{T}$ is called *consistent* if not every formula in it is provable; this amounts to saying that no two formulas one of which is the negation of the other are both provable.

A theory $\mathfrak{T}'$ is called an *extension* of the theory $\mathfrak{T}$ if every provable formula in $\mathfrak{T}$ is also provable in $\mathfrak{T}'$. A set of formulas is called *decidable* if a method exists which permits us in each particular case to decide in a finite number of steps if a given formula belongs to the set;[8] otherwise, it is called *undecidable.* A theory is decidable if the set of its provable formulas is decidable. A theory $\mathfrak{T}$ is said to be *essentially undecidable*[9] if it is consistent and if no consistent theory $\mathfrak{T}'$ which is an extension of $\mathfrak{T}$ is decidable.

As an example of the theories discussed, we may consider the arithmetic of positive integers with $+$ and $\cdot$ as the only mathematical constants; however, for our purpose it is convenient to introduce in this theory the symbol Pos to denote the notion of a positive integer. The mathematical axioms are essentially Peano's axioms listed in Section 2. We modify them slightly in order to eliminate the symbols S and 1; in formulating these axioms, we use the symbolic expression $\mathfrak{U}(c)$ (to mean $c = 1$) as an abbreviation of the formula

$$\mathrm{Pos}(c) \wedge \bigwedge_{x,y} [\mathrm{Pos}(x) \wedge \mathrm{Pos}(y) \rightarrow \sim(x + y = c)].$$

The resulting axioms are:

B1. $\bigvee_{c} \mathfrak{U}(c)$

B2. $\bigwedge_{a,b,c} ([\mathrm{Pos}(a) \wedge \mathrm{Pos}(b) \wedge \mathfrak{U}(c) \wedge a + c = b + c] \rightarrow a = b)$

B3. $\bigwedge_{a,b,c} \{[\mathrm{Pos}(a) \wedge \mathrm{Pos}(b) \wedge \mathfrak{U}(c)] \rightarrow a + (b + c) = (a + b) + c\}$

[8] Using more technical terminology, a set of formulas is decidable if the set of integers correlated with the formulas is general recursive.

[9] This notion was introduced by Tarski.

B4. $\bigwedge_{a,c} [\{\mathrm{Pos}(a) \wedge \mathfrak{U}(c)\} \to a \cdot c = a]$

B5. $\bigwedge_{a,b,c} \{[\mathrm{Pos}(a) \wedge \mathrm{Pos}(b) \wedge \mathfrak{U}(c)] \to a \cdot (b + c) = (a \cdot b) + a\}$

B6. $\bigwedge_{c} \{[\mathfrak{U}(c) \wedge \mathcal{P}(c) \wedge \bigwedge_{a} ([\mathrm{Pos}(a) \wedge \mathcal{P}(a)] \to \mathcal{P}(a + c))]$
$$\to \bigwedge_{a} (\mathrm{Pos}(a) \to \mathcal{P}(a))\}.$$

We shall call the theory based on these axioms $\mathfrak{P}$. The same remarks that were made in Section 2 about A6 apply to the induction principle B6.

It was shown by Church [2] using essentially the method of Gödel [4] that the theory $\mathfrak{P}$ based upon the above axioms is undecidable. Rosser [9] showed that this theory is even essentially undecidable.

It is well known that the following formulas are provable in Theory $\mathfrak{P}$:

B7. $\bigwedge_{a,b,c} [(\mathrm{Pos}(a) \wedge \mathrm{Pos}(b) \wedge \mathrm{Pos}(c) \wedge a + c = b + c) \to a = b]$

B8. $\bigwedge_{a,b} [(\mathrm{Pos}(a) \wedge \mathrm{Pos}(b)) \to a + b = b + a]$

B9. $\bigwedge_{a,b,c} [(\mathrm{Pos}(a) \wedge \mathrm{Pos}(b) \wedge \mathrm{Pos}(c)) \to a + (b + c) = (a + b) + c]$

B10. $\bigwedge_{a,b} [(\mathrm{Pos}(a) \wedge \mathrm{Pos}(b) \wedge \sim a = b) \to \bigvee_{c} (\mathrm{Pos}(c) \wedge [a = b + c \vee b = a + c])]$

B11. $\bigwedge_{a,b,c} [(\mathrm{Pos}(a) \wedge \mathrm{Pos}(b) \wedge \mathrm{Pos}(c)) \to a \cdot (b + c) = (a \cdot b) + (a \cdot c)].$

It is also clear that if we replace B2, B3, and B5 by B7–B11 in the axiom system of the theory $\mathfrak{P}$ we obtain an equivalent axiom system. Hence the results of Church and Rosser previously mentioned apply equally well to the new axiom system. In this connection, however, Mostowski and Tarski have recently obtained a stronger result:

I. *The theory $\mathfrak{P}'$ with primitive mathematical terms* $+$, $\cdot$, *and* Pos *and with the mathematical axioms* B1, B4, B7–B11 *is essentially undecidable.*

The method of proof is similar to that used by earlier authors.

As is easily seen, an important difference between $\mathfrak{P}$ and $\mathfrak{P}'$ is that $\mathfrak{P}'$ does not have the induction principle and hence has only a finite number of mathematical axioms. In consequence, as we shall see, Theorem I has a much wider range of applications than the older results in this direction, i.e., it enables us to establish the undecidability of a much greater variety of mathematical theories.

The applications of Theorem I are based on the following general method developed by Tarski. Let $\mathfrak{T}_1$ be a theory (of the type considered) with the (specific) mathematical constants $O_1, \cdots, O_n$ and $\mathfrak{T}_2$ be another theory with the mathematical constants $Q_1, \cdots, Q_m$. We shall speak of a possible definition of one of the symbols O_i in terms of $Q_1, \cdots, Q_m$. To fix the idea, assume for example that O_i is the symbol for a binary operation. Then by a possible definition of O_i in terms of $Q_1, \cdots, Q_m$ we understand any equivalence of the form

$$\bigwedge_{a,b,c} [aO_ib = c \leftrightarrow \phi]$$

where ϕ stands for any formula containing only three free variables a, b, and c and containing no mathematical constants other than $Q_1, \cdots, Q_m$. We now say that Theory $\mathfrak{T}_1$ is *consistently interpretable* in Theory $\mathfrak{T}_2$ if there is a consistent theory $\mathfrak{T}$ satisfying the following conditions: (i) $\mathfrak{T}$ is an extension of both $\mathfrak{T}_1$ and $\mathfrak{T}_2$, i.e., both $O_1, \cdots, O_n$ and $Q_1, \cdots, Q_m$ are mathematical constants of $\mathfrak{T}$ and all axioms of both $\mathfrak{T}_1$ and $\mathfrak{T}_2$ are axioms (or at least provable formulas) of $\mathfrak{T}$. (ii) For each of the symbols O_i ($i = 1, \cdots, n$) which does not occur in the sequence $Q_1, \cdots, Q_m$, there is an axiom (or provable formula) in $\mathfrak{T}$ which is a possible definition of O_i in terms $Q_1, \cdots, Q_m$.

Using this terminology Tarski stated the following general principle:

II. *If a theory $\mathfrak{T}_1$ has only a finite number of mathematical axioms and is essentially undecidable, then every theory $\mathfrak{T}_2$ in which $\mathfrak{T}_1$ is consistently interpretable is undecidable (although not necessarily essentially undecidable).*

By applying II to two theories $\mathfrak{T}_1$ and $\mathfrak{T}_2$ with the same mathematical constants, we obtain the following formulation:

III. *Let $\mathfrak{T}_1$ and $\mathfrak{T}_2$ be two theories with the same mathematical constants. If $\mathfrak{T}_1$ has only a finite number of mathematical axioms and is essentially undecidable, and if the theory $\mathfrak{T}$ which has the same mathematical constants as $\mathfrak{T}_1$ and $\mathfrak{T}_2$ and whose axiom system is the union of those of $\mathfrak{T}_1$ and $\mathfrak{T}_2$ is consistent, then $\mathfrak{T}_2$ is undecidable.*

From I–III various general results regarding the decision problem have been derived. Thus, I and III imply at once the following corollary:

IV. *Every theory $\mathfrak{T}$ with mathematical symbols* Pos, $+$, *and* $\cdot$ *is undecidable provided only that all the axioms of $\mathfrak{T}$ are true formulas of the arithmetic of positive integers; or at least that all the axioms of $\mathfrak{T}$ are compatible with those of Theory $\mathfrak{P}'$ of Theorem I (i.e., the theory based on the union of the two axiom systems is consistent).*

This result comprehends two particular cases which are known from the literature: the case when all true sentences are regarded as axioms of $\mathfrak{T}$; and the case when $\mathfrak{T}$ has no mathematical axioms at all, so that $\mathfrak{T}$ reduces to a purely logical theory whose undecidability amounts to the undecidability of the lower predicate calculus which was established by Church [3].

By using II instead of III, Theorem IV can be extended to the arithmetic of arbitrary integers. Since among the true formulas of the arithmetic of arbitrary integers involving $+$ and $\cdot$, we find those which serve as postulates defining the notion of an abstract ring, this result implies the following interesting corollary:

V. *The general (arithmetical) theory of abstract rings is undecidable.*

Finally, by means of I and II Tarski has established the undecidability of the general theories of lattices and groups.

New theorems in this domain can be obtained by combining I–III with the results of the earlier sections of this paper. In fact we obtain:

THEOREM 4.1. *Every theory $\mathfrak{T}$ whose mathematical constants are* Pos (*or* Int), S, *and* $\cdot$ *is undecidable provided only that the axioms are true formulas in the arithmetic of positive integers (or of arbitrary integers). The result (in its application* *to positive integers) remains valid if the multiplication symbol · is replaced by the divisibility symbol* |.[10]

THEOREM 4.2. *Every theory* $\mathfrak{T}$ *whose mathematical constants are* Rt *(to denote the notion of a rational number),* + *, and* · *is undecidable provided only that the axioms of* $\mathfrak{T}$ *are true formulas in the arithmetic of rationals.*

The proofs of these two theorems are entirely analogous. We want to outline briefly the proof 4.2 in order to clarify the use of I–III.

Given the theories $\mathfrak{P}'$ of Theorem I and the theory $\mathfrak{T}$ satisfying the hypothesis of 4.2, we construct a new theory $\mathfrak{T}'$ in the following way. The mathematical constants of $\mathfrak{T}'$ are Rt, Pos, + , and ·. The axiom system of $\mathfrak{T}'$ consists: (i) of all axioms of $\mathfrak{P}'$ listed in I; (ii) of all axioms of $\mathfrak{T}$; and (iii) of an equivalence which is a possible definition of Pos in terms of Rt, + , and · and which results from Theorem 3.1. The theory $\mathfrak{T}'$ thus obtained is obviously consistent since all its axioms are true formulas in the arithmetic of rational numbers. Hence according to the definition of consistent interpretability, $\mathfrak{P}'$ is consistently interpretable in $\mathfrak{T}$; and therefore, by I and II, $\mathfrak{T}$ is undecidable.

It may be noticed that the symbol + in 4.2 can be replaced by S; for the addition of rationals, like that of integers, is definable in terms of S and ·.

Since among the formulas involving + and · which are true in the arithmetic of rationals, we find those which are used as postulates to define the notion of an abstract field, we obtain as a particular case of 4.2:

THEOREM 4.3. *The general (arithmetical) theory of abstract fields is undecidable.*[11, 12]

It should be emphasized that the general theory of abstract fields is by no means essentially undecidable, for extensions of this theory are known which are decidable. In fact, as it was stated by Tarski [11], the arithmetical theory of real closed fields and that of algebraically closed fields (thus in particular the arithmetic of algebraic numbers, real algebraic numbers, real numbers, and that of complex numbers with + and ·) are decidable. Hence many further problems are suggested which are still open.

In particular, we can ask whether the arithmetical theory of various algebraic extensions of the field of rational numbers are decidable. This applies to both finite and infinite extensions.

A specially interesting case, in view of its application to the decision problem for geometry, is that of the field of all constructible numbers, i.e., numbers which can be obtained from 1 by means of rational operations and extracting square roots.

We can also look for purely mathematical characterization of those fields

[10] The problem of the undecidability of the system of integers with multiplication and less-than was proposed by Mostowski.

[11] The decision problem for fields was proposed by Tarski.

[12] Using the results formulated in theorems 4.1–4.3, Tarski has recently shown that the general (arithmetical) theory of modular lattices and that of projective geometry are also undecidable. This improves his earlier result regarding the undecidability of the general lattice theory.

whose theory is decidable or undecidable. Regarding those fields with characteristic 0, parallel problems arise if, instead of fields whose theories are undecidable, we consider those in which the notion of an integer is arithmetically definable. We do not know at present whether these two series of problems are equivalent; if the notion of integer is arithmetically definable in a field then the arithmetical theory of this field is undecidable, but the problem of whether the converse holds (for fields of characteristic zero) is still open.

References

[1] BERNSTEIN, B. A., *Weak definitions of a field*, ***Duke mathematical journal***, vol. 14 (1947), pp. 475–482.

[2] CHURCH, ALONZO, *An unsolvable problem of elementary number theory*, ***American journal of mathematics***, vol. 58(1936), pp. 345–363.

[3] CHURCH, ALONZO, *A note on the Entscheidungsproblem*, this JOURNAL, vol. 1(1936), pp. 40–41, 101–102.

[4] GÖDEL, KURT, *Über formal unentscheidbare Sätze der Principia Mathematica und verwandter Systeme I*, ***Monatshefte für Mathematik und Physik***, vol. 38(1931), pp. 173–198.

[5] HASSE, H., *Über die Darstellbarkeit von Zahlen durch quadratische Formen im Körper der rationalen Zahlen*, ***Journal für die reine und angewandte Mathematik***, vol. 152 (1923), pp. 129–148.

[6] HILBERT, D., and BERNAYS, P., ***Grundlagen der Mathematik***.

[7] PADOA, A., *Un nouveau système irréductible de postulats pour l'algèbre*, ***Comptes rendus du 2-e Congrès International des Mathématiciens***, 1902, pp. 249–256.

[8] PRESBURGER, M., *Über die Vollstandigkeit eines gewissen Systems der Arithmetik*, ***Comptes rendus du I Congrés des Pays Slaves***, Warsaw 1929.

[9] ROSSER, B., *Extensions of some theorems of Gödel and Church*, this JOURNAL, vol. 1(1936), pp. 87–91.

[10] TARSKI, A., *Der Wahrheitsbegriff in den formalisierten Sprachen*, ***Studia philosophica***, vol. 1(1935), pp. 261–405.

[11] TARSKI, A., *New investigations on the completeness of deductive theories* (abstract), this JOURNAL, vol. 4(1939), p. 176. [Added in proof: A detailed account has since appeared in A. TARSKI, ***A decision method for elementary algebra and geometry***, Project RAND, publication R-109, August 1948.]

UNIVERSITY OF CALIFORNIA

GENERAL RECURSIVE FUNCTIONS

JULIA ROBINSON

1. **Introduction.** A primitive recursive function is one which can be obtained from the initial functions I_{nk} $(1 \leqq k \leqq n)$, O_n $(n \geqq 0)$, and S, by repeated substitution and recursion. Here

$$I_{nk}(x_1, \cdots, x_n) = x_k, \qquad O_n(x_1, \cdots, x_n) = 0,$$

S denotes the successor function, and the recursion scheme has the form

$$F(\mathfrak{x}, 0) = A\mathfrak{x}, \qquad F(\mathfrak{x}, Sy) = B(\mathfrak{x}, y, F(\mathfrak{x}, y)),$$

where we have put $\mathfrak{x} = (x_1, \cdots, x_n)$, $n \geqq 0$.[1]

The class of general recursive functions is obtained if we allow one additional scheme for defining functions, namely

$$F\mathfrak{x} = \mu y\{A(\mathfrak{x}, y) = 0\},$$

where the symbol on the right denotes the smallest y such that $A(\mathfrak{x}, y) = 0$, under the assumption that there is such a y for each $\mathfrak{x}$.

Kleene showed that this definition of general recursive function is equivalent to Herbrand-Gödel metamathematical definition.[2] In this paper we shall be concerned with the mathematical (as opposed to metamathematical) aspects of the theory of general recursive functions. Starting from the definition stated, we shall investigate the possible restrictions on the defining schemes. Part of the results obtained are already known from the work of Kleene, but we go further in this direction than Kleene did. No previous knowledge of general recursive functions is assumed in this paper.

It will be convenient to have a logical symbolism to express the conditions that appear in applications of the μ-rule. We shall use: $\bigwedge$ (for every), $\bigvee$ (there exists), $\wedge$ (and), $\vee$ (or), $\sim$ (not), $\rightarrow$ (if – then), $\leftrightarrow$ (if and only if).

The following equivalences will be useful:

Presented to the Society, November 27, 1948; received by the editors August 22, 1949.

[1] Further discussion of this definition may be found, for example, in R. M. Robinson, *Primitive recursive functions*, Bull. Amer. Math. Soc. vol. 53 (1947) pp. 925–942. This paper will be referred to as PRF. The notation of the present paper is the same as in PRF.

[2] See S. C. Kleene, *General recursive functions of natural numbers*, Math. Ann. vol. 112 (1936) pp. 727–766 and *Recursive predicates and quantifiers*, Trans. Amer. Math. Soc. vol. 53 (1943) pp. 41–73.

$$a = 0 \wedge b = 0 \leftrightarrow a + b = 0, \qquad a = 0 \vee b = 0 \leftrightarrow ab = 0,$$
$$a \neq 0 \leftrightarrow 0^a = 0.$$

In connection with the last formula, we recall that $0^0 = 1$. Notice also that

$$a = b \leftrightarrow (a - b) + (b - a) = 0,$$
$$a \geqq b \leftrightarrow [(a - b) + b] - a = 0$$

where an arbitrary difference function $x - y$ is used; that is, $x - y$ has the usual meaning when $x \geqq y$, but a completely arbitrary meaning for $x < y$.

At various places, we shall need pairing functions, that is functions J, K, and L such that

$$KJ(x, y) = x \quad \text{and} \quad LJ(x, y) = y.$$

It will also be convenient to have a one-to-one correspondence between pairs of numbers (x, y) and numbers z. This is expressed by the identity

$$J(Kz, Lz) = z.$$

Furthermore, sometimes we shall want $J(x, y)$ to be a monotone increasing function of x and y. All of these conditions are satisfied by Cantor's diagonal enumeration. This one to one correspondence between pairs of numbers (x, y) and numbers z is established by the equation

$$2z = (x + y)^2 + 3x + y.$$

Notice that

$$8z + 1 = (2x + 2y + 1)^2 + 8x$$

and hence $[(8z+1)^{1/2}]$ is either $2x+2y+1$ or $2x+2y+2$. It follows easily that J, K, L can all be obtained by substitution alone, if we adjoin

$$x^2, \quad x + y, \quad x - y, \quad [x/2], \quad [x^{1/2}]$$

to the initial functions.

2. **Various forms of the μ-rule.** A more general form of the μ-rule is

$$F\mathfrak{x} = \mu y\{A_1(\mathfrak{x}, y) = A_2(\mathfrak{x}, y)\}.$$

This can be reduced to the standard form

$$F\mathfrak{x} = \mu y\{A(\mathfrak{x}, y) = 0\}$$

by putting $A(\mathfrak{x}, y)=|A_1(\mathfrak{x}, y)-A_2(\mathfrak{x}, y)|$. Consider next the more special form

$$F\mathfrak{x} = \iota y\{B(\mathfrak{x}, y) = 0\}$$

where the right side denotes the only y such that $B(\mathfrak{x}, y)=0$, under the assumption that there is just one such y for each $\mathfrak{x}$. This is readily seen to be equivalent to the preceding form if we take

$$B(\mathfrak{x}, y) = A(\mathfrak{x}, y) + \sum_{z<y} 0^{A(\mathfrak{x},z)}.$$

Indeed, $B(\mathfrak{x}, y)=0$ is equivalent to $A(\mathfrak{x}, y)=0$ and $A(\mathfrak{x}, z)\neq 0$ for $z<y$. Notice that $B(\mathfrak{x}, y)$ is primitive recursive if $A(\mathfrak{x}, y)$ is.

By successive applications of pairing functions, we can reduce the number of components of $\mathfrak{x}=(x_1, \cdots, x_n)$ which occur in applications of the μ-rule to one. Consider for example the reduction from two components to one. We are given

$$F(x_1, x_2) = \mu y\{A(x_1, x_2, y) = 0\}.$$

Putting

$$B(x, y) = A(Kx, Lx, y)$$

and

$$Gx = \mu y\{B(x, y) = 0\},$$

we see that

$$F(x_1, x_2) = GJ(x_1, x_2).$$

The same reduction is also valid with ι in place of μ.

We shall next discuss the relation between the following two forms of the μ-rule:

$$Fx = \mu y\{A(x, y) = 0\},$$

$$Gx = \mu z\{Bz = x\}.$$

It is easily seen that

$$Fx = L\mu z\{A(Kz, Lz) = 0 \wedge Kz = x\},$$

where J, K, and L are Cantor's pairing functions. Indeed, the equation $z=J(x, y)$, with x fixed, establishes a one-to-one correspondence between values of y satisfying $A(x, y)=0$ and values of z such that $A(Kz, Lz)=0\wedge Kz=x$. Also, the smallest y corresponds to the smallest z. It follows that

$$L\mu z\{0^{A(Kz,Lz)}Kz = x\} = \begin{cases} Fx & \text{if } x \neq 0, \\ 0 & \text{if } x = 0. \end{cases}$$

Thus, putting

$$Bz = 0^{A(Kz,Lz)}Kz$$

and defining Gx as above, we see that

$$Fx = LGx + 0^x F0.$$

We may also put $Fx = HGx$, where

$$Hy = \begin{cases} Ly & \text{if } y \neq G0, \\ F0 & \text{if } y = G0. \end{cases}$$

(H is a primitive recursive function.)

The situation is somewhat different for the ι-rule. A more elaborate method is needed for reducing

$$Fx = \iota y\{A(x, y) = 0\}$$

to the use of

$$Gx = \iota z\{Bz = x\}$$

and certain elementary substitutions. As before, we have

$$Fx = L\iota z\{A(Kz, Lz) = 0 \wedge Kz = x\}$$

or putting

$$Cz = \operatorname{sgn} A(Kz, Lz),$$

also

$$Fx = L\iota z\{Cz = 0 \wedge Kz = x\}.$$

For each x there is a unique z such that $Kz = x$ and $Cz = 0$. It is these values of z in which we are interested. We wish to construct a function B from which these values can be read in a simple way, but also so that $\iota z\{Bz = x\}$ is a legitimate operation, that is, so that for every x there is exactly one z with $Bz = x$, even though the value of z may be immaterial for our purposes.

We shall construct a function B such that

$$B(2z) = 2x \leftrightarrow Cz = 0 \wedge Kz = x.$$

The function B will assume all distinct odd values for odd arguments and for even arguments $2z$ for which $Cz = 1$. This will insure the legitimacy of the operation $\iota z\{Bz = x\}$. Such a function B may be defined as follows:

$$B(2z) = \begin{cases} 2Kz & \text{if } Cz = 0, \\ 2z - 1 + 2\sum_{u \leqq z} Cu & \text{if } Cz = 1, \end{cases}$$

$$B(2z + 1) = 2z + 1 + 2\sum_{u \leqq z} Cu.$$

Notice that B can be obtained from A by substitutions and primitive recursions, and hence is primitive recursive if A is. Finally, we see that if G is defined as above, we have

$$Fx = L[G(2x)/2].$$

REMARK. Thus, if substitution and primitive recursion are available, the strongest and weakest forms of the μ-rule are equivalent.

3. Elimination of recursion. In this section we shall show that all general recursive functions can be obtained without the use of the recursion scheme, that is, by using only substitution and the μ-rule, if we adjoin suitable functions to the initial functions. Indeed, we shall see that it is sufficient to adjoin $x+y$ and Q if we use the μ-rule in the form $F\mathfrak{x} = \mu y\{A(\mathfrak{x}, y) = 0\}$,[3] and to adjoin $x+y$ and E if we use the more special form of the μ-rule $Fx = \mu y\{Ay = x\}$. Here $Ex = x - [x^{1/2}]^2$ is the excess of x over a square, and $Qx = 0^{Ex}$ is the characteristic function of squares. In the latter case, in place of E we may use either K or L. I do not know whether the recursion scheme can be omitted if the ι-rule is substituted for the μ-rule and a finite number of primitive recursive functions are adjoined to the initial functions.

We first recall a device due to Gödel.[4] Any finite sequence of numbers $a_0, a_1, a_2, \cdots, a_y$ can be obtained as remainders when a fixed number is divided by $1+v(z+1)$, for $z = 0, 1, 2, \cdots, y$. Denoting by $R(u, t)$ the remainder when u is divided by t, the required conditions are

$$R(u, 1 + v(z + 1)) = a_z \qquad \text{for } 0 \leqq z \leqq y.$$

These conditions are equivalent to

[3] A similar result was proved by Kleene, adjoining the functions $x+y$, xy, and $0^{|x-y|}$. The method of eliminating recursion is essentially the same as in the present paper. See Kleene, *A note on recursive functions*, Bull. Amer. Math. Soc. vol. 42 (1936) pp. 544–546.

[4] See Kurt Gödel, *Über formal unentscheidbare Sätze der Principia Mathematica und verwandter Systeme* I, Monatshefte für Mathematik und Physik vol. 38 (1931) pp. 173–198.

$$u \equiv a_z \pmod{1 + v(z+1)}, \qquad a_z < 1 + v(z+1).$$

We need only to choose v sufficiently large and so that the moduli are relatively prime, and then choose u satisfying the congruences.

If we put $w = J(u, v)$, then the sequence is determined from a single number w. Putting for brevity

$$T_z w = R(Kw, 1 + Lw \cdot Sz),$$

we have

$$a_z = T_z w \qquad \text{for } z = 0, 1, 2, \cdots y.$$

Thus $T_z w$ denotes the zth term in the sequence determined by w. (Here J can be taken to be any pairing function with suitably associated functions K and L.)

We are now ready to consider the elimination of recursive definitions. Suppose that F is defined recursively from A and B by the equations

$$F(\mathfrak{x}, 0) = A\mathfrak{x}, \qquad F(\mathfrak{x}, Sy) = B(\mathfrak{x}, y, F(\mathfrak{x}, y)).$$

The sequence of numbers

$$F(\mathfrak{x}, 0), F(\mathfrak{x}, 1), \cdots, F(\mathfrak{x}, y)$$

may be represented by a single number w in the manner described above. Choosing the smallest possible value for w, we see that

$$F(\mathfrak{x}, y) = T_y \mu w \{ T_0 w = A\mathfrak{x} \wedge \bigwedge_z [z < y \rightarrow T_{Sz} w = B(\mathfrak{x}, z, T_z w)] \}.$$

This may also be written in the form

$$F(\mathfrak{x}, y) = T_y \mu w \{ T_0 w = A\mathfrak{x} \wedge \mu z \{ T_{Sz} w \neq B(\mathfrak{x}, z, T_z w) \vee z = y \} = y \}.$$

Thus we see that $F(x, y)$ has been obtained from the given functions A and B, and the function T by two applications of the μ-rule. These are not in standard form, but can be reduced to standard form by elementary transformations, making use of the functions $x+y$, $x-y$, xy, and 0^x. The function T is obtained by substitution from K, L, R, and S. Thus to make possible the elimination of recursion, it is sufficient to adjoin $x+y$, $x-y$, xy, 0^x, K, L, and R to the usual initial functions. All general recursive functions can then be obtained using substitution and the μ-rule in the standard form.

We shall show that actually it is sufficient to adjoin $x+y$ and Q to the initial functions.

We note first that $0^x = QSS(Qx + QSx)$. We shall next define a function $x^{1/2}$. This symbol, in distinction to $[x^{1/2}]$, will be used to denote any function which gives the square root of x when x is a square, but whose values for other arguments are immaterial. A suitable definition is

$$x^{1/2} = \mu y\{Q(x + 2y + 1) = 1\}.$$

The condition in braces may also be written in the standard form $0^{QS(x+y+y)} = 0$.

It was proved by Euler that it is impossible to find four distinct squares forming an arithmetic progression. Using this fact, we may define x^2 as follows:

$$\begin{aligned} x^2 = \mu y\{Qy &= 1 \wedge Q(y + 3y^{1/2} + x + 4) \\ &= 1 \wedge Q(y + 2y^{1/2} + 2x + 4) \\ &= 1 \wedge Q(y + y^{1/2} + 3x + 4) = 1\}. \end{aligned}$$

In the first place, $y = x^2$ satisfies the condition. Conversely, if y satisfies the condition, then y is a square, so that $y^{1/2}$ indeed represents the square root of y. Thus $y + 4y^{1/2} + 4$ is also a square so that the condition requires four squares in arithmetic progression. They must all be equal, hence $y^{1/2} = x$ or $y = x^2$. Thus x^2 is the only y satisfying the condition in braces. The condition may be reduced to standard form by using

$$a \neq 0 \wedge b \neq 0 \wedge c \neq 0 \wedge d \neq 0 \leftrightarrow 0^a + 0^b + 0^c + 0^d = 0.$$

The following definitions may now be made:

$$\begin{aligned} x - y &= \mu z\{Q((x + y)^2 + x + 3y + 1 + z) = 1\}, \\ [x/2] &= \mu y\{Sy + Sy > x\}, \\ [x^{1/2}] &= \mu y\{(Sy)^2 > x\}, \\ x \cdot y &= [(((x + y)^2 - x^2) - y^2)/2], \\ [x/y] &= \mu z\{y \cdot Sz > x \vee y = 0\}, \\ R(x, y) &= x - y[x/y]. \end{aligned}$$

The conditions involving inequalities may be reduced to standard form using the difference function. Finally, K and L may also be obtained by substitution. Thus it is indeed sufficient to adjoin $x + y$ and Q to the initial functions in order to obtain all general recursive functions by substitution and use of the μ-rule.

We shall now prove that all general recursive functions can be obtained by substitution and inversion $A^{-1}x = \mu y\{Ay = x\}$ if we adjoing $x + y$ and E to the initial functions. The reduction of the general form of the μ-rule to this form required use of the functions J,

K, L, $x+y$, xy, and 0^x. We needed $x+y$ and Q in the general case. All of these functions can be obtained from

$$x+y, \quad x-y, \quad x^2, \quad [x/2], \quad [x^{1/2}], \quad 0^x$$

by subtitution. Thus we need only to show that these functions can be obtained from the usual initial functions and $x+y$ and E by substitution and inversion.

In the first place, we see that $E^{-1}(2x)=x^2+2x$. It follows easily that

$$E(E^{-1}(2x+2y)+3x+y+4) = x-y \qquad \text{if } x \geqq y.$$

Now we define

$$x^2 = E^{-1}(2x) - 2x.$$

We may then put

$$\text{sgn } x = ES(x^2), \qquad O^x = 1 - \text{sgn } x.$$

It is easily seen that the predecessor function P can defined by

$$Px = E\mu y\{ESy = x\}.$$

Now observing that $ESSE^{-1}$ is the characteristic function of even numbers, and that E assumes all values for even arguments, we find that

$$[x/2] = E\mu y\{2Ey + ESSE^{-1}y = x\}.$$

Clearly,

$$x^{1/2} = [EPx/2] + \text{sgn } x$$

if x is a square, and

$$[x^{1/2}] = (x - Ex)^{1/2}.$$

We have thus defined all the required functions.

Finally, we observe that instead of E we may equally well adjoin either K or L to the initial functions (the sum $x+y$ being adjoined in any case). The pattern of values assumed by K and L is indicated in the following table:

$$\begin{aligned} x &= 0, 1, 2, 3, 4, 5, 6, 7, 8, 9, \cdots, \\ Kx &= 0, 0, 1, 0, 1, 2, 0, 1, 2, 3, \cdots, \\ Lx &= 0, 1, 0, 2, 1, 0, 3, 2, 1, 0, \cdots. \end{aligned}$$

In particular, Kx is the excess of x over a triangular number. We notice that $K^{-1}x = x(x+3)/2$. The following formulas are then easily verified:

$$K\,(K^{-1}(x + y) + 2x + 3) = x - y \qquad \text{for } x \geqq y,$$

$$x^2 = 2K^{-1}x - 3x, \qquad Px = K\mu y\{KSy = x\},$$

$$Kx + Lx = KP(x - Kx) + KS(x - Kx).$$

Thus Lx can be obtained by subtraction. Noting that of all representations of x in the form u^2+v, the minimum value of $u+v$ is attained only when $u=[x^{1/2}]$ and $v=Ex$, we see that $J(u, v)$ is smallest in this case. Hence

$$Ex = L\mu y\{(Ky)^2 + Ly = x\}.$$

Thus K may be used as an initial function in place of E. To see that L may equally well be used, we need only note the equation

$$Kx = L(x + 2Lx + 2).$$

4. **General recursive functions of one variable.** The results of this section, as far as Theorem 3, are exactly parallel to those of PRF, §7. The omitted proofs can be copied from there practically verbatim.

It is easy to see that all general recursive functions of more than one variable can be obtained from general recursive functions of one variable, the function $x+y$, and the various identity functions, by substitution alone. Furthermore, as we shall show, all general recursive functions of one variable can be obtained without using functions of more variables in the process.

THEOREM 1. *The functions of one variable which are obtainable from certain initial functions by substitution and inversion are also obtainable if substitution is not allowed except for defining functions of one variable.*

THEOREM 2. *If the initial functions include the identity and zero functions, and the function $x+y$, but no other function of more than one variable, then all functions of one variable obtainable by substitution and inversion are also obtainable by repeated use of the three formulas*

$$Fx = Ax + Bx, \qquad Fx = BAx, \qquad Fx = B^{-1}x$$

to define F when A and B are known functions of one variable. The third formula is used only when B assumes all values.

THEOREM 3. *All general recursive functions of one variable can be obtained by starting with the two functions S and E, and repeatedly using any of the formulas*

$$Fx = Ax + Bx, \qquad Fx = BAx, \qquad Fx = B^{-1}x$$

to construct a new function from the known functions A and B. The third formula is used only when B assumes all values.

PROOF. The identity and zero functions of one variable, $I = I_{11}$ and $O = O_1$, can be defined by

$$Ix = EE^{-1}x, \qquad Ox = ESE^{-1}(x + x).$$

Thus by Theorem 2, we can obtain all the functions of one variable which we could get by substitution and inversion from all the identity and zero functions, $x+y$, S, and E. But by §3, all general recursive functions can be so obtained.

REMARK. In place of E in this theorem, we may also use either K or L. In virtue of the results of §3, we need only note the definitions of I and O,

$$Ix = KK^{-1}x, \qquad Ox = KSK^{-1}x,$$

$$Ix = LL^{-1}x, \qquad Ox = L(Ix + L^{-1}x).$$

Furthermore, in place of S we may also use SO. For without using S we can define I, and then we put $Sx = Ix + SOx$.

REMARK. Comparing this theorem with the corresponding theorem of PRF, §7, we see that the only difference is the substitution of the scheme $Fx = B^{-1}x$ for the scheme $Fx = B^xO$, the result of this change being that we obtain all general recursive functions instead of just primitive recursive functions.

We shall now make a further reduction, eliminating the scheme $Fx = Ax + Bx$ from the preceding theorem, a suitable change in the initial functions being made. It will be convenient to introduce the following notation. If $Fx = Ax + Bx$ and $Gx = C(Ax, Bx)$, we shall denote F and G by $A + B$ and $C(A, B)$.

THEOREM 4. *All general recursive functions of one variable can be obtained by starting with a certain two primitive recursive functions and repeatedly using the formulas*

$$Fx = BAx, \qquad Fx = B^{-1}x$$

to construct a new function from known functions A and B. The latter formula is used only when B assumes all values.

PROOF. We first make the observation that any finite set of initial functions can be replaced by two initial functions. Indeed, the functions $F_1, F_2, \cdots, F_n$ are evidently obtainable by substitution from the two functions, K and

$$J(L, J(F_1, J(F_2, \cdots, J(F_{n-1}, F_n) \cdots))).$$

Thus we need only show that a finite set of primitive recursive initial functions is possible.

To each general recursive function F, we make correspond another function F^* by the formula

$$F^* = J(FK, L),$$

where J, K, and L are the Cantor pairing functions. We know that F can be generated from SO and K by addition, substitution, and inversion. Each step of this construction can be mirrored for the star functions in such a way that addition is unnecessary. Since

$$KF^*J(I, I) = KJ(FK, L)J(I, I) = KJ(F, I) = F,$$

we see that the function F itself can then be recovered.

We notice first that

$$B^*A^* = J(BK, L)J(AK, L) = J(BAK, L) = (BA)^*.$$

If B assumes all values, then B^* does also, and we have

$$\begin{aligned}(B^*)^{-1}x &= \mu z\{B^*z = x\} = \mu z\{J(BKz, Lz) = x\}\\ &= \mu z\{BKz = Kx \wedge Lz = Lx\} = J(\mu y\{By = Kx\}, Lx)\\ &= J(B^{-1}Kx, Lx) = (B^{-1})^*x.\end{aligned}$$

We have used the fact that J is a one-to-one monotone pairing function. Finally, we shall show that $(A+B)^*$ is obtained from A^* and B^* by substitution, making use of the special functions

$$G_1 = J(K, I), \qquad G_2 = J(KL, I), \qquad G_3 = J(K + KL, LLL).$$

Indeed we have

$$\begin{aligned}A^*G_1 &= J(AK, L)J(K, I) = J(AK, I),\\ B^*G_2 &= J(BK, L)J(KL, I) = J(BKL, I),\\ G_3B^*G_2 &= J(K + KL, LLL)J(BKL, I) = J(BKL + K, LL),\\ G_3B^*G_2A^*G_1 &= J(BKL + K, LL)J(AK, I) = J(BK + AK, L),\end{aligned}$$

that is,

$$(A + B)^* = G_3B^*G_2A^*G_1.$$

Thus if F is any general recursive function, the corresponding function F^* can be obtained from the two initial functions $(SO)^* = J(SO, L)$ and $K^* = J(KK, L)$ by use of substitution and inversion only, making use of the three auxiliary functions G_1, G_2, and G_3.

Finally, F itself is obtained by substitution from F^*, K, and $J(I, I)$. Thus all general recursive functions of one variable may be obtained by substitution and inversion from the seven functions,

$$K,\ J(I, I),\ J(SO, L),\ J(KK, L),\ J(K, I),\ J(KL, I),\ J(K + KL, LLL).$$

As already remarked, these seven functions can be replaced by two suitable functions, so the proof of the theorem is complete.

We shall however investigate the possibility of replacing these seven functions by fewer functions of a simpler nature than one of the two obtained by the above process. We start by proving a lemma valid for arbitrary pairing functions.

LEMMA. *If M and N can be obtained by substitution from I, K, and L, then $J(M, N)$ can be obtained by substitution from $J(LK, KL)$ and $J(L, I)$.*

PROOF. We notice first that

$$J(AK, BL)J(Cx, Dx) = J(AKJ(Cx, Dx), BLJ(Cx, Dx)) = J(ACx, BDx).$$

Thus

$$J(AK, BL)J(C, D) = J(AC, BD).$$

This and similar reductions will be used freely below. In particular, we observe the identities

$$J(L, K)J(A, B) = J(B, A),$$
$$J(KK, L)J(A, B) = J(KA, B),$$
$$J(LK, L)J(A, B) = J(LA, B).$$

From these it is clear that all required functions $J(M, N)$ can be obtained by substitution from the four functions

$$J(I, I), \qquad J(L, K), \qquad J(KK, L), \qquad J(LK, L).$$

It remains to show that these four functions can in turn be obtained from the two functions

$$G = J(LK, KL), \qquad H = J(L, I).$$

The following computation shows that this is possible.

$$GH = J(LK, KL)J(L, I) = J(LL, K),$$
$$GH^2 = J(LL, K)J(L, I) = J(L, L),$$
$$GH^3 = J(L, L)J(L, I) = J(I, I),$$
$$G^2H^3 = J(LK, KL)J(I, I) = J(L, K),$$

$$G^2H^4 = J(L, K)J(L, I) = J(I, L),$$
$$G^3H^4 = J(LK, KL)J(I, L) = J(L, KL),$$
$$G^3H^5 = J(L, KL)J(L, I) = J(I, K),$$
$$G^4H^5 = J(LK, KL)J(I, K) = J(L, KK),$$
$$G^2H^3G^4H^5 = J(L, K)J(L, KK) = J(KK, L),$$
$$GHG^2H^3 = J(LL, K)J(L, K) = J(LK, L).$$

THEOREM 5. *All general recursive functions of one variable can be obtained by substitution and inversion from the three functions* $J(LK, KL)$, $J(I, SO)$, $J(K+KL, I)$.

PROOF. It is sufficient to show that we can define the seven functions listed at the end of the proof of Theorem 4. In the first place,

$$J(K + KL, I)J(I, SO) = J(I + KSO, J(I, SO)) = J(I, J(I, SO)),$$
$$J(LK, KL)J(I, J(I, SO)) = J(L, I).$$

Hence by the lemma we can obtain all functions $J(M, N)$ where M and N are obtained from I, K, L by composition. Since the pairing function J is assumed one-to-one, we obtain $K = J(KK, LK)$. The remaining functions are obtained in virtue of the equations

$$J(K + KL, LLL) = J(K, LLLL)J(K + KL, I),$$
$$J(SO, L) = J(L, LK)J(I, SO).$$

We shall now show that it is impossible to reduce the two initial functions in Theorem 4 to a single function of any sort.

THEOREM 6. *There is no function A from which all general recursive functions can be obtained by substitution and inversion.*

PROOF. Such a function A would have to assume all values. For otherwise no function assuming all values could be obtained by substitution, and inversion would be impossible. If A were a one-to-one mapping of the integers onto themselves, then only such permutations would be definable. Hence A must assume some value more than once.

We shall show that the only functions assuming all values which are obtainable from A are $A^0 = I$, A, A^2, A^3, $\cdots$, and hence inversion is possible only for powers of A. Clearly, it is sufficient to show that

$$F = A^k(A^l)^{-1}A^m(A^n)^{-1} \cdots A^p(A^q)^{-1}$$

assumes all values only if it is a power of A. If $k < l$, then F does not assume all values, since $A^k(A^l)^{-1}$ does not. Indeed, we know that

A^{l-k} assumes some value at least twice; hence if $A^k(A^l)^{-1}$ assumed all values, then $I=A^{l-k}A^k(A^l)^{-1}$ would have to assume some value twice. Hence $k \geqq l$, and then $A^k(A^l)^{-1}=A^{k-l}A^l(A^l)^{-1}=A^{k-l}$. Thus F can be simplified. Repeating the argument, we find that if F assumes all values, it must be a power of A. Since no permutations are obtained except the identity, we do not obtain all general recursive functions.

5. **Representation of general recursive functions in terms of primitive recursive functions.** Starting from the Herbrand-Gödel metamathematical definition of general recursive function, Kleene showed that every such function can be expressed in the form

$$F\mathfrak{x} = A\mu y\{B(\mathfrak{x}, y) = 0\},$$

where A and B are primitive recursive functions and the required y exists for each x. Starting from the definition of this paper, we shall prove the possibility of such a representation, without using metamathematical arguments. We may also replace μ by ι in this formula.

THEOREM. *Every general recursive function F can be expressed in the form*

$$F\mathfrak{x} = A\iota y\{B(\mathfrak{x}, y) = 0\}$$

where A and B are primitive recursive functions, and for every $\mathfrak{x}$ *there is a unique y such that* $B(\mathfrak{x}, y)=0$.

PROOF. By successive applications of the pairing functions, we can reduce the theorem to the special case $\mathfrak{x}=x$. We can then apply the results of §4 concerning general recursive functions of one variable. We notice first that if F is primitive recursive, the representation

$$Fx = F\iota y\{\,|\,x - y\,| = 0\}$$

may be used. Thus according to Theorem 4 of §4 we need only check that the set of functions which can be represented in the form stated in the theorem is closed under substitution and inversion.

Suppose that

$$Fx = A\iota y\{B(x, y) = 0\}, \qquad Gx = C\iota y\{D(x, y) = 0\},$$

where A, B, C, D are primitive recursive. Then

$$\begin{aligned} FGx &= A\iota z\{B(Gx, z) = 0\} \\ &= A\iota z\{B(C\iota y\{D(x, y) = 0\}, z) = 0\}. \end{aligned}$$

In this formula, y and z are uniquely determined by x. Putting $w = J(y, z)$ (J is taken to be Cantor's pairing function), we see that

$$FGx = AL\iota w\{B(CKw, Lw) - 0 \wedge D(x, Kw) = 0\},$$

which reduces to the required form. Thus the set of functions considered is closed under substitution.

Now consider inversion. Again suppose that

$$Fx = A\iota y\{B(x, y) = 0\}.$$

Let us assume first $B(x, y) = 0$ determines y as a monotone non-decreasing function of x. The inverse function is determined by the formula

$$F^{-1}z = \mu x\{Fx = z\} = \mu x\{A\iota y\{B(x, y) = 0\} = z\}.$$

Now y is determined uniquely by x, and increases with x, hence the same is true of $w = J(x, y)$. Since the smallest possible w corresponds to the smallest possible x, we see that

$$F^{-1}z = K\mu w\{ALw = z \wedge B(Kw, Lw) = 0\}.$$

This can be reduced to the required form by the method of §2.

If B does not determine y as a monotone function of x, we can find a new representation of F with this property. Notice first that for a given x, we can choose u so that Ku, KLu, $\cdots$, KL^xu all have preassigned values, and the smallest possible u increases with each of them. Hence

$$Fx = AKL^x\mu u\{B(0, Ku) = 0 \wedge B(1, KLu) = 0 \wedge \cdots \wedge B(x, KL^xu) = 0\},$$

or

$$Fx = AKL^x\mu u\{G(x, u)] = 0\}$$

where

$$G(x, u) = \sum_{n=0}^{x} B(n, KL^nu).$$

Clearly, the smallest u such that $G(x, u) = 0$ increases with x, since more conditions are imposed on it. Putting $v = J(2^xu, L^xu)$, and noticing that we always have $J(2z, 0) \geqq J(z, z)$, we see that v also increases with x. It follows that

$$Fx = AKL\mu v\{G(x, [Kv/2^x]) = 0 \wedge Lv = L^x[Kv/2^x]\}.$$

The functions occurring in the braces are all primitive recursive functions of v and x, and the smallest v satisfying the condition increases with x. We may now reduce to the desired form as in §2.

COROLLARY. *Any general recursive function of one variable can be expressed in the form* AB^{-1}, *where* A *and* B *are primitive recursive and* B *assumes all values.*

PROOF. The reduction to this form is made as in §2.

COROLLARY. *A function* $F\mathfrak{x}$ *(of* n *variables) is general recursive if and only if it has the form* $F\mathfrak{x}=AC\mathfrak{x}$, *where* Ay *and* $0^{|y-C\mathfrak{x}|}$ *are primitive recursive functions (of one and* $n+1$ *variables, respectively).*[5]

PROOF. If F is general recursive, it has the form

$$F\mathfrak{x} = A\iota y\{B(\mathfrak{x}, y) = 0\},$$

where A and B are primitive recursive. Putting

$$C\mathfrak{x} = \iota y\{B(\mathfrak{x}, y) = 0\},$$

we see that

$$0^{|y-C\mathfrak{x}|} = 0^{B(\mathfrak{x},y)}$$

is primitive recursive. Conversely, suppose $F\mathfrak{x}=AC\mathfrak{x}$, where Ay and $0^{|y-C\mathfrak{x}|}$ are primitive recursive. Then

$$F\mathfrak{x} = A\iota y\{0^{|y-C\mathfrak{x}|} = 1\}$$

is general recursive.

BERKELEY, CALIF.

[5] Stated by A. Tarski in a seminar at the University of California in the summer of 1943. Also see Th. Skolem, *Some remarks on recursive arithmetic*, Norske Videnskabers Selskabs, Forhandlinger vol. 17 (1944) pp. 103–106.

ANNALS OF MATHEMATICS
Vol. 54, No. 2, September, 1951

AN ITERATIVE METHOD OF SOLVING A GAME

BY JULIA ROBINSON

(July 28, 1950)

A two-person game[1] can be represented by its pay-off matrix $A = (a_{ij})$. The first player chooses one of the m rows and the second player simultaneously chooses one of the n columns. If the ith row and the jth column are chosen, then the second player pays the first player a_{ij}.

If the first player plays the ith row with probability x_i and the second player plays the jth column with probability y_j, where $x_i \geqq 0$, $\sum x_i = 1$, $y_j \geqq 0$, and $\sum y_j = 1$, then the expectation of the first player is $\sum\sum a_{ij}x_iy_j$. Furthermore,

$$\min_j \sum_i a_{ij} x_i \leqq \max_i \sum_j a_{ij} y_j, \tag{1}$$

since

$$\min_j \sum_i a_{ij} x_i \leqq \sum_i \sum_j a_{ij} x_i y_j \leqq \max_i \sum_j a_{ij} y_j.$$

The minimax theorem of game theory (see [1] page 153) asserts that for some set of probabilities $X = (x_1, \cdots, x_m)$ and $Y = (y_1, \cdots, y_n)$ the equality holds in (1). Such a pair (X, Y) is called a solution of the game. The value v of the game is defined by

$$v = \min_j \sum_i a_{ij} x_i = \max_i \sum_j a_{ij} y_j,$$

where (X, Y) is a solution of the game.

In this paper, we shall show the validity of an iterative procedure suggested by George W. Brown [2]. This method corresponds to each player choosing in turn the best pure strategy against the accumulated mixed strategy of his opponent up to then.

Let $A = (a_{ij})$ be an $m \times n$ matrix. $A_{i.}$ will denote the ith row of A and $A_{.j}$, the jth column. Similarly, if $V(t)$ is a vector, then $v_j(t)$ is the jth component. Let $\max V(t) = \max_j v_j(t)$ and $\min V(t) = \min_j v_j(t)$. In this notation, (1) can be rewritten as follows:

$$\min \sum_i A_{i.} x_i \leqq \max \sum_j A_{.j} y_j, \tag{2}$$

whenever $x_i \geqq 0$, $\sum x_i = 1$, $y_j \geqq 0$, and $\sum y_j = 1$.

DEFINITION 1. A system (U, V) consisting of a sequence of n-dimensional vectors $U(0), U(1), \cdots$ and a sequence of m-dimensional vectors $V(0), V(1), \cdots$ is called a *vector system* for A provided that

$$\min U(0) = \max V(0),$$

[1] More technically, a finite two-person zero-sum game. See [1] in the bibliography at the end of the paper.

and

$$U(t+1) = U(t) + A_{i\cdot}, \qquad V(t+1) = V(t) + A_{\cdot j},$$

where i and j satisfy the conditions

$$v_i(t) = \max V(t), \qquad u_j(t) = \min U(t).$$

Thus a vector system for A can be formed recursively from a given $U(0)$ and $V(0)$. At each step, the row added to U is determined by a maximum component of V and the column added to V is determined by a minimum component of U.

An alternate notion of vector system is obtained if the condition on j in Definition 1 is replaced by

$$u_j(t+1) = \min U(t+1).$$

A vector system of this new type can also be built up recursively. The only difference is that here successive U and V are determined alternately while in the other definition U and V could be obtained simultaneously. In all the following proofs and theorems, either definition may be used.

In the special case $U(0) = 0$ and $V(0) = 0$, we see that $U(t)/t$ is a weighted average of the rows of A and $V(t)/t$ is a weighted average of the columns. Hence for every t and t',

$$\frac{\min U(t)}{t} \leqq v \leqq \frac{\max V(t')}{t'}.$$

If for some t and t', these two bounds are equal, we have a solution of the game. Unfortunately, this is not always the case. However George Brown [2] conjectured that as t and t' tend to ∞, the two bounds approach v. The main result of this paper is to prove this for any vector system. In numerical examples, vector systems of the second kind appear to converge more rapidly than the first.

THEOREM.[2] *If (U, V) is a vector system for A, then*

$$\lim_{t\to\infty} \frac{\min U(t)}{t} = \lim_{t\to\infty} \frac{\max V(t)}{t} = v.$$

The proof will be divided into four lemmas.

LEMMA 1. *If (U, V) is a vector system for a matrix A, then*

$$\liminf_{t\to\infty} \frac{\max V(t) - \min U(t)}{t} \geqq 0.$$

PROOF. For each t,

$$V(t) = V(0) + t\sum_j y_j A_{\cdot j} \text{ where } y_j \geqq 0, \sum y_j = 1,$$

[2] The solution to Problem 5 in the RAND Mathematical Problem Series II is contained as a special case of this theorem.

and

$$U(t) = U(0) + t\sum_i x_i A_{i\cdot} \text{ where } x_i \geqq 0, \sum x_i = 1.$$

Hence

$$\max V(t) \geqq \min V(0) + t \max \sum_j y_j A_{\cdot j} \geqq \min V(0) + tv,$$

and

$$\min U(t) \leqq \max U(0) + t \min \sum_i x_i A_{i\cdot} \leqq \max U(0) + tv.$$

Therefore,

$$\liminf_{t\to\infty} \frac{\max V(t) - \min U(t)}{t} \geqq 0.$$

DEFINITION 2. If (U, V) is a vector system for A, then we say that the ith row is eligible in the interval (t, t') provided that there exists t_1 with

$$t \leqq t_1 \leqq t'$$

and

$$v_i(t_1) = \max V(t_1).$$

Similarly, the jth column is eligible in the interval (t, t') if there exists t_2 with

$$t \leqq t_2 \leqq t'$$

and

$$u_j(t_2) = \min U(t_2).$$

LEMMA 2. *Given a vector system (U, V) for A, then if all the rows and columns of A are eligible in the interval $(s, s + t)$,*

$$\max U(s + t) - \min U(s + t) \leqq 2at$$

and

$$\max V(s + t) - \min V(s + t) \leqq 2at,$$

where

$$a = \max_{i,j} | a_{ij} | .$$

PROOF. Let j be such that

$$u_j(s + t) = \max U(s + t).$$

Choose t' with $s \leqq t' \leqq s + t$ so that

$$u_j(t') = \min U(t').$$

Then

$$u_j(s + t) \leqq u_j(t') + at = \min U(t') + at,$$

since the change in the ith component in t steps is not more than at. But

$$\min U(s + t) \geqq \min U(t') - at,$$

so that

$$\max U(s + t) - \min U(s + t) \leqq 2at.$$

Similarly,

$$\max V(s + t) - \min V(s + t) \leqq 2at.$$

LEMMA 3. *If all the rows and columns of A are eligible in $(s, s + t)$ for a given vector system (U, V), then*

$$\max V(s + t) - \min U(s + t) \leqq 4at.$$

PROOF. By Lemma 2,

$$\max V(s + t) - \min U(s + t) \leqq 4at + \min V(s + t) - \max U(s + t).$$

Hence it is sufficient to show that $\min V(s + t) \leqq \max U(s + t)$. Now applying (2) to the transpose of A, we have

$$\min \sum_j A_{\cdot j} y_j \leqq \max \sum_i A_{i \cdot} x_i,$$

whenever $x_i \geqq 0$, $\sum x_i = 1$, $y_j \geqq 0$, and $\sum y_j = 1$. In particular, choose x_i and y_j satisfying

$$U(s + t) = U(0) + (s + t) \sum A_{i \cdot} x_i,$$
$$V(s + t) = V(0) + (s + t) \sum A_{\cdot j} y_j.$$

Then

$$\begin{aligned} \min V(s + t) &\leqq \max V(0) + (s + t) \min \sum A_{\cdot j} y_j \\ &\leqq \min U(0) + (s + t) \max \sum A_{i \cdot} x_i \\ &\leqq \max U(s + t). \end{aligned}$$

LEMMA 4. *To every matrix A and $\varepsilon > 0$, there exists t_0 such that for any vector system (U, V),*

$$\max V(t) - \min U(t) < \varepsilon t \qquad \textit{for } t \geqq t_0.$$

PROOF. The theorem holds for matrices of order 1 since $U(t) = V(t)$ for all t. Assume the theorem holds for all submatrices of A, then we will show by induction that it holds for A. Choose t^* so that for any vector system (U', V') corresponding to a submatrix A' of A, we have

$$\max V'(t) - \min U'(t) < \tfrac{1}{2}\varepsilon t \qquad \text{whenever } t \geqq t^*.$$

We shall prove that if in the given vector system (U, V) for A, some row or column is not eligible in the interval $(s, s + t^*)$, then

$$\max V(s + t^*) - \min U(s + t^*) < \max V(s) - \min U(s) + \tfrac{1}{2}\varepsilon t^*. \tag{3}$$

Suppose, for example, that the kth row is not eligible in the interval $(s, s + t^*)$. Then we can construct a vector system (U', V') for the matrix A' obtained by deleting the kth row of A, in the following way:

$$U'(t) = U(s + t) + C,$$
$$V'(t) = \text{Proj}_k V(s + t) \qquad \text{for } t = 0, 1, \cdots, t^*,$$

where C is the n-dimensional vector all of whose components are equal to $\max V(s) - \min U(s)$ and $\text{Proj}_k V$ is the vector obtained from V by omitting the kth component. The rows of A' will be numbered $1, 2, \cdots, k-1, k+1, \cdots, m$. Notice first that $\min U'(0) = \max V'(0)$. Furthermore, if

$$U(s + t + 1) = U(s + t) + A_{i\cdot}, \qquad V(s + t + 1) = V(s + t) + A_{\cdot j},$$

then

$$U'(t + 1) = U'(t) + A'_{i\cdot}, \qquad V'(t + 1) = V'(t) + A'_{\cdot j}.$$

Also $v_i(s + t) = \max V(s + t)$ if and only if $v'_i(t) = \max V'(t)$ and $u_j(s + t) = \min U(s + t)$ if and only if $u'_j(t) = \min U'(t)$ for $0 \leqq t \leqq t^*$. Hence we see that U' and V' must satisfy the recursive restrictions of the definition of a vector system for $0 \leqq t \leqq t^*$, since U and V do. Naturally, we may continue U' and V' indefinitely to form a vector system for A'.

Now by the choice of t^*, we know that

$$\max V'(t^*) - \min U'(t^*) < \tfrac{1}{2}\varepsilon t^*.$$

Hence

$$\begin{aligned} \max V(s + t^*) &- \min U(s + t^*) \\ &= \max V'(t^*) - \min U'(t^*) + \max V(s) - \min U(s) \\ &< \max V(s) - \min U(s) + \tfrac{1}{2}\varepsilon t^*. \end{aligned}$$

We can now show that given any vector system (U, V) for A,

$$\max V(t) - \min U(t) < \varepsilon t \qquad \text{for } t \geqq 8at^*/\varepsilon.$$

Consider $t > t^*$. Let θ with $0 \leqq \theta < 1$ and q a positive integer be so chosen that $t = (\theta + q)t^*$.

CASE 1. Suppose there is a positive integer $s \leqq q$ so that all rows and columns of A are eligible in the interval $((\theta + s - 1)t^*, (\theta + s)t^*)$. Take the largest such s, then

$$\begin{aligned} (4) \quad \max V(t) &- \min U(t) \\ &\leqq \max V((\theta + s)t^*) - \min U((\theta + s)t^*) + \tfrac{1}{2}\varepsilon(q - s)t^*. \end{aligned}$$

We obtain this inequality by repeated application of (3), since in each of the intervals

$$((\theta + r - 1)t^*, (\theta + r)t^*) \qquad \text{for } r = s + 1, \cdots, q,$$

some row or column of A is not eligible. From Lemma 3 and the choice of s, we have

$$\max V((\theta + s)t^*) - \min U((\theta + s)t^*) \leqq 4at^* . \tag{5}$$

From (4) and (5), we obtain

$$\max V(t) - \min U(t) \leqq 4at^* + \tfrac{1}{2}\varepsilon(q - s)t^* < (4a + \tfrac{1}{2}\varepsilon q)t^* .$$

CASE 2. If there is no such s, then in each interval $((\theta + r - 1)t^* , (\theta + r)t^*)$ some row or column of A is not eligible. Hence

$$\max V(t) - \min U(t) < \max V(\theta t^*) - \min U(\theta t^*) + \tfrac{1}{2}\varepsilon q t^* \leqq 2a\theta t^* + \tfrac{1}{2}\varepsilon q t^*.$$

Therefore, in either case,

$$\max V(t) - \min U(t) < (4a + \tfrac{1}{2}\varepsilon q)t^* \leqq 4at^* + \tfrac{1}{2}\varepsilon t < \varepsilon t \qquad \text{for } t \geqq 8at^*/\varepsilon.$$

From Lemmas 1 and 4, we see that

$$\lim_{t\to\infty} \frac{\max V(t) - \min U(t)}{t} = 0.$$

But from (1),

$$\limsup_{t\to\infty} \frac{\min U(t)}{t} \leqq v,$$

$$\liminf_{t\to\infty} \frac{\max V(t)}{t} \geqq v.$$

Hence

$$\lim_{t\to\infty} \frac{\min V(t)}{t} = \lim_{t\to\infty} \frac{\max V(t)}{t} = v,$$

which completes the proof of the theorem.

THE RAND CORPORATION,
SANTA MONICA, CALIFORNIA

BIBLIOGRAPHY

[1] J. VON NEUMANN and O. MORGENSTERN, Theory of Games and Economic Behavior, Princeton University Press.

[2] G. W. BROWN, *Some notes on computation of Games Solutions*, RAND Report P-78, April 1949, The RAND Corporation, Santa Monica, California.

EXISTENTIAL DEFINABILITY IN ARITHMETIC

BY

JULIA ROBINSON

1. **Introduction.** A relation $\rho(x_1, \cdots, x_n)$ among natural numbers is said to be *arithmetically definable* if there is a formula containing the free variables $x_1, \cdots, x_n$, any number of bound variables, and symbols for particular natural numbers, involving only the mathematical symbols $+$ and $\cdot$, and the logical symbols: $\bigwedge$ (for every), $\bigvee$ (there exists), $\wedge$ (and), $\vee$ (or), $\sim$ (not), and $=$ (equals), which holds if and only if the relation $\rho(x_1, \cdots, x_n)$ is satisfied.

A relation $\rho(x_1, \cdots, x_n)$ is said to be *existentially definable* if there is a formula of the type described above which does not contain either of the logical symbols $\bigwedge$ or $\sim$. We shall consider the existential definability of certain relations, e.g. $x = y^z$, $x = y!$, and "x is a prime," which are known to be arithmetically definable by a general theorem of Gödel (see §6)(1).

It can easily be seen that a relation $\rho(x_1, \cdots, x_n)$ is existentially definable if and only if there is a polynomial $P(x_1, \cdots, x_n, u_1, \cdots, u_k)$ with (positive or negative) integer coefficients such that

$$\rho(x_1, \cdots, x_n) \leftrightarrow \bigvee_{u_1, \cdots, u_k} P(x_1, \cdots, x_n, u_1, \cdots, u_k) = 0.$$

Thus, a set of natural numbers is existentially definable if and only if it is the set of values of a parameter for which a certain diophantine equation is solvable in natural numbers.

Solving the diophantine equation $P(x) = 0$ in integers is equivalent to solving $P(x)P(-x) = 0$ in natural numbers. Also solving $P(x) = 0$ in natural numbers is equivalent to solving $P(u^2 + v^2 + w^2 + z^2) = 0$ in integers. Hence we see that sets of natural numbers which are existentially definable are also existentially definable if we allow the variables to range over the integers.

A simple example of an existentially definable set is the set of perfect squares. The complementary set is also existentially definable, since

$$\sim \bigvee_{y} x = y^2 \leftrightarrow \bigvee_{u,v,w} [(u^2 + v + 1 = x \wedge (u + 1)^2 = x + w + 1)].$$

The tenth problem in Hilbert's famous list is to find an effective method for deciding if a given diophantine equation is solvable. Since there are many classical diophantine equations with one parameter for which no effective

Presented to the International Congress of Mathematicians, September 4, 1950; received by the editors March 30, 1951.

(1) Professor Alfred Tarski suggested the study of existentially definable relations to me.

method is known to determine the solvability for an arbitrary value of the parameter, it is very unlikely that a decision procedure can be found. For example, no way is known to determine the values of a for which the diophantine system,

$$x^2 + ay^2 = s^2, \qquad x^2 - ay^2 = t^2,$$

is solvable. (This problem was first studied by the Arabs in the Middle Ages.) If there is some non-recursive set which is existentially definable, then we could conclude that there is no effective method for determining the solvability of an arbitrary diophantine equation.

A relation $\rho(x_1, \cdots, x_n)$ is said to be existentially definable in terms of the relations $\phi_1, \cdots, \phi_k$ if there is a defining formula of the type described in the definition of existential definability except that it may also contain the relations $\phi_1, \cdots, \phi_k$. Clearly, if ρ is existentially definable in terms of certain existentially definable relations, then ρ itself is existentially definable.

The main result of this paper is that the relation $x = y^z$ is existentially definable in terms of any relation of roughly exponential growth. Let the "nth super power of x," $x * n$, be defined recursively by

$$x * 0 = 1, \quad x * (n + 1) = x^{(x*n)}.$$

We shall prove that the relation $x = y^z$ is existentententially definable in terms of any relation $\phi(u, v)$ such that

$$\bigvee_n \bigwedge_{u,v} [\phi(u, v) \rightarrow v < u * n], \tag{1.1}$$

and

$$\sim \bigvee_n \bigwedge_{u,v} [\phi(u, v) \rightarrow v < u^n]. \tag{1.2}$$

It is not known whether any ϕ satisfying these conditions is existentially definable. At present, very little is known about the size of solutions of diophantine equations with a finite number of solutions. However, it seems to me very likely that some such relation ϕ is existentially definable. If this is the case, then we could conclude that $x = y^z$ is existentially definable. In particular, this would mean that there is a polynomial $P(x, y, z, n, u_1, \cdots, u_k)$ with integer coefficients such that

$$x^n + y^n = z^n \leftrightarrow \bigvee_{u_1, \cdots, u_k} P(x, y, z, n, u_1, \cdots, u_k) = 0.$$

In other words, we could reduce Fermat's equation (or any exponential equation) to a polynomial equation in more variables. On the other hand, if it should be shown that $x = y^z$ is not existentially definable, then we could conclude that no relation satisfying (1.1) and (1.2) is existentially definable.

Various properties of solutions of the Pell equation $x^2 - (a^2 - 1)y^2 = 1$ are listed in §2. These are needed in §§3 and 4. §3 contains the main result of the paper. In §4, we show that $x=y^z$ is existentially definable in terms of the relation x Pow y (x is a power of y). It is not known whether Pow is existentially definable or not. However we show that the relation $\sim x$ Pow y is existentially definable. In §5, we investigate existential definability in terms of exponentiation. The relations $x=y!$ and "x is a prime" are shown to be existentially definable in terms of exponentiation. In §6, there is a discussion of the relation of recursive enumerability and existential definability.

2. **Pell's equation.** In this section, we give various lemmas concerning solutions of the equation $x^2-(a^2-1)y^2=1$ in natural numbers. The first lemma is a special case of the well known theorem on Pell's equation, $x^2-ay^2=1$. [See Landau, *Vorlesungen über Zahlentheorie*, vol. 1, pp. 57–64.] The remaining lemmas can be proved easily. In all of the lemmas, a is any natural number greater than 1 and the other free variables are any natural numbers.

LEMMA 1.

$$x^2-(a^2-1)y^2=1 \leftrightarrow \bigvee_n x+y(a^2-1)^{1/2}=(a+(a^2-1)^{1/2})^n.$$

LEMMA 2. *If a_n and a_n' are defined by the equation*

$$a_n + a_n'(a^2-1)^{1/2} = (a+(a^2-1)^{1/2})^n,$$

then

$$\begin{aligned} a_0 &= 1, & a_0' &= 0, \\ a_1 &= a, & a_1' &= 1, \\ a_{n+2} &= 2a\cdot a_{n+1} - a_n, & a_{n+2}' &= 2a\cdot a_{n+1}' - a_n'. \end{aligned}$$

LEMMA 3. $a_{n+1}=a\cdot a_n+a_n'(a^2-1)$.

LEMMA 4. $a^n \leqq a_n \leqq (2a)^n$. *The equality holds on the right only for $n=0$.*

LEMMA 5. $a_n-a_n'(a-y)\equiv y^n \pmod{2ay-y^2-1}$.

This follows easily by induction from Lemma 2.

LEMMA 6.

$$a_n = \sum_{k=0,\ k \text{ even}}^{n} \binom{n}{k} a^{n-k}(a^2-1)^{k/2}$$

and

$$a_n' = \sum_{k=1,\ k \text{ odd}}^{n} \binom{n}{k} a^{n-k}(a^2-1)^{(k-1)/2}.$$

LEMMA 7.

$$a_n' \equiv na^{n-1} \pmod{a^2 - 1},$$

hence

$$a_n' \equiv n \pmod{a - 1}.$$

This follows immediately from Lemma 6.

LEMMA 8. *Let $\psi(a, u)$ be the relation defined by the following equivalence,*

$$\psi(a, u) \leftrightarrow \bigvee_{x,y} [x^2 - (a^2 - 1)(a - 1)^2 y^2 = 1 \wedge x > 1 \wedge a > 1 \wedge u = ax].$$

Then

$$\psi(a, u) \rightarrow u \geqq a^a$$

and

$$a > 1 \rightarrow \bigvee_u (\psi(a, u) \wedge u < a^{2a}).$$

Proof. By Lemma 7, $u = ax \geqq a \cdot a_{a-1}$. Hence we obtain the first conclusion by Lemma 4. On the other hand, $u = a \cdot a_{a-1}$ satisfies $\psi(a, u)$ and by Lemma 4, $a \cdot a_{a-1} \leqq a \cdot (2a)^{a-1} < a^{2a}$.

LEMMA 9. *If $x > 1$ and $a > x^n$, then*

$$a_n x^n \leqq (ax)_n < a_n(x^n | 1).$$

Proof. By Lemma 6,

$$a_n x^n = \sum_{k=0,\ k \text{ even}}^{n} \binom{n}{k} (ax)^{n-k} (a^2x^2 - x^2)^{k/2}$$

and

$$(ax)_n = \sum_{k=0,\ k \text{ even}}^{n} \binom{n}{k} (ax)^{n-k} (a^2x^2 - 1)^{k/2}.$$

Hence $a_n x^n \leqq (ax)_n$. Also,

$$(ax)_n/(a_n x^n) \leqq \left\{(a^2x^2 - 1)/(a^2x^2 - x^2)\right\}^{n/2} \leqq (1 - 1/a^2)^{-n},$$

and

$$(1 - 1/a^2)^n > 1 - n/a^2 > 1 - 1/a \geqq 1 - 1/(x^n + 1).$$

Therefore,

$$a_n x^n \leqq (ax)_n < a_n(x^n + 1).$$

LEMMA 10. *If $x > 1$ and $a > x^n$, then*

$$a_n \leqq (ax)_m \leqq a \cdot a_n \leftrightarrow m = n.$$

Proof. From Lemma 9 and $a > x^n$, we have

$$a_n \leqq (ax)_n \leqq a \cdot a_n.$$

Since $(ax)_m$ is an increasing function of m, it is sufficient to show that

$$(ax)_{n-1} < a_n$$

and

$$(ax)_{n+1} > a \cdot a_n.$$

Applying Lemma 9 with n replaced by $n-1$, we obtain

$$(ax)_{n-1} < a \cdot a_{n-1} \leqq a_n.$$

Also,

$$(ax)_{n+1} > ax \cdot (ax)_n > a \cdot a_n.$$

3. **Exponentiation.** In this section, we shall show that the relation $x = y^z$ is existentially definable in terms of any relation of roughly exponential growth.

LEMMA. *There is an existential definition of a relation $\rho(x, y)$ satisfying*

$$\rho(x, y) \rightarrow y < x^x \tag{3.1}$$

and

$$\sim \bigvee_n \bigwedge_{x,y} (\rho(x, y) \rightarrow y < x^n), \tag{3.2}$$

in terms of any relation $\phi(u, v)$ satisfying (1.1) *and* (1.2).

Proof. Case 1. There is a k such that $\phi(u, v) \rightarrow v < u^{ku}$. Clearly, if we let ρ be defined by the equivalence:

$$\rho(x, y) \leftrightarrow \bigvee_v (\phi(x, v) \wedge y^k \leqq v),$$

then $\rho(x, y)$ satisfies (3.1) and (3.2).

Case 2. $\phi(u, v) \rightarrow v < u * n$ and there is no k such that $\phi(u, v) \rightarrow v < u^{ku}$. In this case, we shall show that we can define existentially in terms of ϕ, a relation $\phi_1(u, v)$ such that

$$\phi_1(u, v) \rightarrow v < u * (n-1) \tag{3.3}$$

and

$$\sim \bigvee_m (\phi_1(u, v) \rightarrow v < u^m). \tag{3.4}$$

Then, either ϕ_1 satisfies the condition of Case 1 or, by Case 2, we can define ϕ_2 satisfying

$$\phi_2(u, v) \to v < u * (n - 2)$$

and

$$\sim \bigvee_m (\phi_2(u, v) \to v < u^m).$$

By continuing this process, we must eventually obtain a relation ϕ_r satisfying the condition of Case 1. Hence a relation ρ satisfying (3.1) and (3.2) can be existentially defined in terms of a relation ϕ satisfying (1.1) and (1.2).

It remains to prove that given ϕ satisfying the conditions of Case 2, we can define existentially in terms of ϕ a relation ϕ_1 satisfying (3.3) and (3.4). Let ϕ_1 be defined by the equivalence

$$\phi_1(u, v) \leftrightarrow \bigvee_a [\psi(a, u) \wedge \phi(a, v)],$$

where ψ is the relation of Lemma 8. Clearly, $\phi_1(u, v) \to v < u * (n-1)$, since $\psi(a, u) \to u \geqq a^a$ and $\phi(a, v) \to v < a * n < (a^a) * (n-1) \leqq u * (n-1)$. Also, there is no m with $\phi_1(u, v) \to v < u^m$, since for each m, there exist a, u, and v such that

$$\phi(a, v) \wedge v \geqq a^{2ma}$$

and

$$\psi(a, u) \wedge u < a^{2a},$$

so that $v \geqq (a^{2a})^m > u^m$. This completes the proof of the lemma.

We now turn to the proof of our main result. We wish to show that the relation $x = y^z$ can be defined existentially in terms of any relation ϕ satisfying (1.1) and (1.2). In view of the lemma, it is sufficient to show that $x = y^z$ can be defined existentially in terms of any relation ρ satisfying (3.1) and (3.2).

By Lemma 9, we have

$$(3.5) \quad (y > 1 \wedge a > y^z) \to [x = y^z \leftrightarrow a_z x \leqq (ay)_z < a_z(x + 1)].$$

Furthermore, by Lemma 10,

$$(3.6) \quad \begin{aligned} &(y > 1 \wedge a > y^z) \\ &\quad \to [u = (ay)_z \leftrightarrow \bigvee_v (u^2 - (a^2y^2 - 1)v^2 = 1 \wedge a_z \leqq u \leqq a \cdot a_z)]. \end{aligned}$$

Hence,

$$(3.7) \quad \begin{aligned} &(1 < x < a \wedge y > 1 \wedge a > y^z) \\ &\quad \to [x = y^z \leftrightarrow \bigvee_{u,v} \{u^2 - (a^2y^2 - 1)v^2 = 1 \wedge a_z x \leqq u < a_z(x + 1)\}]. \end{aligned}$$

We can replace the requirement $a > y^z$ in the hypothesis of (3.7) by the stronger condition

$$\bigvee_{b,c} [b > y \wedge b > z \wedge a > c \wedge \psi(b, c)],$$

where ψ is the existentially definable relation of Lemma 8. Hence from (3.7) we can obtain an existential definition of $x = y^z$ in terms of the relation $r = a_z$, namely,

$$\begin{aligned} x = y^z \leftrightarrow & [(z = 0 \wedge x = 1) \vee (z > 0 \wedge y < 2 \wedge x = y) \\ & \vee (y > 1 \wedge x > 1 \wedge z > 0 \wedge \bigvee_{a,b,c,u,v} \{x < a \wedge b > y \wedge b > z \\ & \wedge a > c \wedge \psi(b, c) \wedge u^2 - (a^2y^2 - 1)v^2 = 1 \wedge a_z x \leqq u < a_z(x + 1)\})]. \end{aligned} \tag{3.8}$$

Consequently, it would be sufficient to give an existential definition of the relation $r = a_z$ in order to obtain an existential definition of $x = y^z$. But at present I am unable to do exactly this. However, by Lemma 7,

$$\begin{aligned} & (1 < r < a_a \wedge 0 < z < a) \\ & \qquad \rightarrow [r = a_z \leftrightarrow \bigvee_s r^2 - (a^2 - 1)(z + s(a - 1))^2 = 1]. \end{aligned} \tag{3.9}$$

Here we see how we can make use of ρ. In order that (3.9) can be used to define $r = a_z$, we must limit the size of r. This can be done by means of ρ (for some values of a). Namely,

$$[r \leqq d \wedge \rho(a, d)] \rightarrow r < a^a < a_a.$$

On the other hand, there are infinitely many values of a for which there is a d satisfying $\rho(a, d) \wedge d > a_z$. Fortunately, in (3.8), we need only define a_z for a sufficiently large value of a. Hence we obtain

$$\begin{aligned} & (x > 1 \wedge y > 1 \wedge z > 0) \rightarrow \\ & [x = y^z \leftrightarrow \bigvee_{a,b,c,d,u,v,r,s} \{x < a \wedge b > y \wedge b > z \wedge \psi(b, c) \\ & \wedge r > 1 \wedge a > c \wedge r \leqq d \wedge \rho(a, d) \wedge u^2 - (a^2y^2 - 1)v^2 = 1 \\ & \wedge rx \leqq u < r(x + 1) \wedge r^2 - (a^2 - 1)(z + s(a - 1))^2 = 1\}]. \end{aligned}$$

Therefore, $x = y^z$ is existentially definable in terms of any relation ρ satisfying (3.1) and (3.2). Consequently, it is existentially definable in terms of any relation ϕ satisfying (1.1) and (1.2).

4. **Powers and non-powers.** We shall show first that exponentiation can be defined in terms of the relation x Pow y. Let $R(a, b, c, d)$ be the relation which holds if and only if $b > 1$, $d > 1$, and there is an n, such that $a = b^n$ and $c = d^n$. We shall show that R is existentially definable in terms of Pow and, then, that $x = y^z$ is existentially definable in terms of R. The following equivalence gives an existential definition of R in terms of Pow:

$$
\begin{aligned}
R(a, b, c, d) \leftrightarrow \{&b > 1 \wedge d > 1 \wedge a \operatorname{Pow} b \wedge c \operatorname{Pow} d \\
&\wedge \bigvee_e [e \operatorname{Pow}(bd + 1) \wedge ae \operatorname{Pow} b(bd + 1) \\
&\wedge ce \operatorname{Pow} d(bd + 1)]\}.
\end{aligned} \tag{4.1}
$$

If $b>1$, $d>1$, $a=b^n$, and $c=d^n$, then $e=(bd+1)^n$ satisfies the right side of the equivalence. If the right side holds, then there are k, l, m, s, and t so that $a=b^k$, $c=d^l$, $e=(bd+1)^m$, $ae=\{b(bd+1)\}^s$, and $ce=\{d(bd+1)\}^t$. Hence

$$
\begin{aligned}
ae &= b^k(bd + 1)^m = b^s(bd + 1)^s, \\
ce &= d^l(bd + 1)^m = d^t(bd + 1)^t.
\end{aligned}
$$

Therefore, all the exponents are equal and $R(a, b, c, d)$ holds.

Now, notice that if $u=2^z$, then

$$
u + z \leqq u(u + 1)^z/u^z \leqq u + z + (2^z - 1)/2^z < u + z + 1. \tag{4.2}
$$

This inequality provides a way to determine the exponent corresponding to a given power of y. Thus,

$$
\begin{aligned}
x = y^z \leftrightarrow \{&(z = 0 \wedge x = 1) \vee (y = 0 \wedge z > 0 \wedge x = 0) \\
&\vee (y = 1 \wedge x = 1) \vee (y > 1 \wedge \bigvee_{u,v,w} [R(u, 2, x, y) \\
&\wedge R(v, u, x, y) \wedge R(w, u + 1, x, y) \\
&\wedge v(u + z) \leqq uw < v(u + z + 1)]\}.
\end{aligned} \tag{4.3}
$$

Clearly, if $x=y^z$, the right side is satisfied by (4.2). Suppose that the right side holds, then x Pow y, say $x=y^t$. It remains to show that $t=z$. But by (4.2), $v(u+t)\leqq uw<v(u+t+1)$. By hypothesis,

$$
v(u + z) \leqq uw < v(u + z + 1),
$$

hence $t=z$. Since R is existentially definable in terms of Pow, we have shown that $x=y^z$ is existentially definable in terms of Pow.

I have been unable to find an existential definition of the set of powers of 2. However some sets of logarithmic density can be defined existentially. For example, let T be the set of values of x for which the equation $x^2-3y^2=1$ is solvable. Then, by Lemma 1, elements of T are roughly powers of $(2+3^{1/2})$.

There is a simple existential definition of the set of natural numbers which are not power of 2, namely,

$$
\sim x \operatorname{Pow} 2 \leftrightarrow \bigvee_{u,v} x = (2u + 3)v.
$$

It is not so easy to give an existential definition of the non-powers of 6. However, we shall now give an existential definition of the relation $\sim x$ Pow y.

Let x Pow y (mod z) be the relation which holds if and only if there exists an n such that $x \equiv y^n \pmod{z}$. We first check the following formula,

$$\begin{aligned}(4.4)\quad &(x > 1 \wedge y > 1 \wedge y \mid x)\\ &\rightarrow \{\sim x \text{ Pow } y \leftrightarrow \bigvee_{u,v} (u \text{ Pow } y \ (\text{mod } x-1) \wedge v \text{ Pow } y \ (\text{mod } x-1)\\ &\wedge 0 < u < v < yu \wedge v < x)\}.\end{aligned}$$

Here $y \mid x$ (y divides x) is existentially definable, since

$$y \mid x \leftrightarrow \bigvee_{u} x = yu.$$

Suppose $x > 1 \wedge y > 1 \wedge y \mid x \wedge \sim x$ Pow y. Choose n so that $y^n < x < y^{n+1}$. Then there are v and k so that $y^{n+1} = v + k(x-1)$ with $0 \leqq v < x-1$ and $0 < k \leqq y$. By using divisibility arguments, we see that actually $1 < v < x-1$, $0 < k < y$, $y \nmid v$. Hence v is not a power of y, so there is an l with $y^l < v < y^{l+1}$. Let $u = y^l$. Then for this u and v, the right side of the equivalence in (4.4) is satisfied.

Suppose $x > 2 \wedge y > 1 \wedge y \mid x \wedge x$ Pow y. If $z < x$, then z Pow y (mod $x-1$) if and only if z Pow y. Hence the right side of the equivalence in (4.4) does not hold. The case $x = 2$ and $y = 2$ can be treated separately. Thus (4.4) is verified.

We shall now show that the relation u Pow v (mod m) is existentially definable provided v and m are relatively prime. From Lemma 5, we have for $a > 1$,

$$\begin{aligned}&u \text{ Pow } v \ (\text{mod } 2av - v^2 - 1) \leftrightarrow\\ &\quad \bigvee_{x,y,r} [x^2 - (a^2-1)y^2 = 1 \wedge u = x - y(a-v) \pm r(2av - v^2 - 1)].\end{aligned}$$

Now if $(m, v) = 1$, i.e. m and v are relatively prime, then there is an $a > 1$ such that $m \mid (2av - v^2 - 1)$. Hence

$$\begin{aligned}&(u \text{ Pow } v \ (\text{mod } m) \wedge (m, v) = 1) \leftrightarrow\\ &\quad \bigvee_{a,w,k} [a > 1 \wedge w \text{ Pow } v \ (\text{mod } 2av - v^2 - 1)\\ &\qquad\qquad \wedge m \mid (2av - v^2 - 1) \wedge u = w + km].\end{aligned}$$

Finally, we give an existential definition of the relation $\sim x$ Pow y,

$$\begin{aligned}(4.5)\quad \sim x \text{ Pow } y \leftrightarrow &[(x = 0 \wedge y > 0) \vee (x > 1 \wedge y \nmid x)\\ &\vee (x > 1 \wedge y = 1)\\ &\vee (x > 1 \wedge y > 1 \wedge y \mid x \wedge \sim x \text{ Pow } y)].\end{aligned}$$

The relation $x \nmid y$ is existentially definable since

$$x \nmid y \leftrightarrow \bigvee_{u,v} (y = ux + v \wedge 0 < v < x).$$

Hence we see that each of the alternatives in (4.5) is existentially definable, so the relation $\sim x$ Pow y is existentially definable.

5. **Existential definability in terms of exponentiation.** We shall show that the relations $r=\binom{n}{k}$, $y=x!$, and "p is a prime" are all existentially definable in terms of exponentiation.

In order to see that $r=\binom{n}{k}$ is existentially definable in terms of exponentiation, we first establish the following identity:

$$\binom{n}{k} = [2^{nk}(1+2^{-n})^n] - 2^n[2^{n(k-1)}(1+2^{-n})^n] \qquad \text{for } n,\ k > 0. \tag{5.1}$$

If $0<k\leqq n$, then

$$2^{nk}(1+2^{-n})^n = 2^{nk}\sum_{\nu=0}^{n}\binom{n}{\nu}2^{-n\nu} < \sum_{\nu=0}^{k}\binom{n}{\nu}2^{n(k-\nu)} + (2^n-1)/2^n.$$

Hence

$$[2^{nk}(1+2^{-n})^n] = \sum_{\nu=0}^{k}\binom{n}{\nu}2^{n(k-\nu)}.$$

Similarly,

$$2^n[2^{n(k-1)}(1+2^{-n})^n] = \sum_{\nu=0}^{k-1}\binom{n}{\nu}2^{n(k-\nu)}.$$

Thus, we see that (5.1) holds provided that $0<k\leqq n$. It can be readily verified in the remaining cases.

From (5.1), we obtain the following existential definition of the binomial coefficient in terms of exponentiation (and less than):

$$\begin{aligned} r = \binom{n}{k} \leftrightarrow \{&(n = 0 \wedge k > 0 \wedge r = 0) \vee (k = 0 \wedge r = 1) \\ &\vee (n > 0 \wedge k > 0 \wedge \bigvee_{u,v} (r = u - 2^n v \\ &\wedge 2^{n^2}u \leqq 2^{nk}(1+2^n)^n < 2^{n^2}(1+u) \\ &\wedge 2^{n^2}v \leqq 2^{n(k-1)}(1+2^n)^n < 2^{n^2}v)\}. \end{aligned} \tag{5.2}$$

Next we show that the relation $y=x!$ is existentially definable in terms of exponentiation and the binomial coefficient. In fact,

$$x! = \left[r^x\Big/\binom{r}{x}\right] \qquad \text{for any } r > (2x)^{x+1}. \tag{5.3}$$

The following chain of inequalities will establish (5.3) for $x>0$. In deriving them, we use the elementary inequalities,

$$1/(1-\theta) < 1 + 2\theta \qquad \text{for } 0 < \theta < 1/2,$$

$$(1+\theta)^x < 1 + 2^x\theta \qquad \text{for } 0 < \theta < 1.$$

$$r^x \Big/ \binom{r}{x} = x! \Big/ \left\{\left(1 - \frac{1}{r}\right)\left(1 - \frac{2}{r}\right) \cdots \left(1 - \frac{x-1}{r}\right)\right\}$$

$$< x! \Big/ \left(1 - \frac{x}{r}\right)^x < x!\left(1 + \frac{2x}{r}\right)^x$$

$$< x!(1 + 2^x \cdot 2x/r) < x! + 1 \qquad \text{for } r > (2x)^{x+1}.$$

The existential definition of the relation $y = x!$ in terms of exponentiation can be written down easily from (5.2) and (5.3).

Finally, to see that the set of primes is existentially definable in terms of exponentiation, we use the fact that p is a prime if and only if $p > 1$ and p is relatively prime to $(p-1)!$. Hence

$$\text{(5.4)} \qquad \begin{aligned} &\sim \bigvee_{x,y} [p = (x+2)(y+2) \vee p = 0 \vee p = 1] \\ &\leftrightarrow \bigvee_{u,v} [pu + (p-1)!v = 1 \wedge p > 1]. \end{aligned}$$

6. **Recursively enumerable relations.** In this section, we shall discuss some results concerning recursively enumerable relations. (It is assumed that the reader is familiar with this concept. For a good account of the theory together with references to earlier papers, see Kleene [3].)

Gödel [2, p. 191] proved that every primitive recursive relation is arithmetically definable. Hence all recursively enumerable relations are arithmetically definable. All of the relations which we have considered are primitive recursive relations and, so, arithmetically definable.

Clearly, every existentially definable relation is recursively enumerable. It is not known whether the converse is true. Davis [1] has shown that every recursively enumerable relation $\phi(x_1, \cdots, x_n)$ can be put in the form

$$\text{(6.1)} \qquad \bigvee_{u} \bigwedge_{v \leqq u} \rho(x_1, \cdots, x_n, u, v)$$

where ρ is an existentially definable relation. (Here the inequality under the quantifier restricts the variable, thus

$$\bigwedge_{v \leqq u} \phi \leftrightarrow \bigwedge_{v} [v \leqq u \rightarrow \phi], \qquad \bigvee_{v \leqq u} \phi \leftrightarrow \bigvee_{v} [v \leqq u \wedge \phi].)$$

We shall show that ρ can also be taken to be primitive recursive. Let ϕ be written in the form (6.1),

$$\phi(x_1, \cdots, x_n) \leftrightarrow \bigvee_{u} \bigwedge_{v \leqq u} \bigvee_{y_1, \cdots, y_k} P(x_1, \cdots, x_n, u, v, y_1, \cdots, y_k) = 0.$$

Then

$$\phi(x_1, \cdots, x_n) \leftrightarrow \bigvee_{u,s} \bigwedge_{v \leqq u} \bigvee_{y_1, \cdots, y_k \leqq s} P = 0.$$

The relation

$$\bigvee_{y_1, \cdots, y_k \leqq s} P = 0$$

is primitive recursive as well as existentially definable, since Gödel [2, p. 180] proved that if, in an arithmetical definition of a given relation, all of the bound variables are bounded by primitive recursive functions, then the relation is primitive recursive. Also,

$$\bigvee_{u,s} \bigwedge_{v \leqq u} \bigvee_{y_1, \cdots, y_k \leqq s} P = 0 \leftrightarrow \bigvee_{z} \bigwedge_{v \leqq z} \Big\{ \bigvee_{u,s \leqq z} \bigvee_{y_1, \cdots, y_k \leqq s} [z = (u + s)^2 + u \wedge (P = 0 \vee v > u)] \Big\}.$$

Here we use the fact that the function $(u+s)^2+u$ is univalent. The expression in braces is both primitive recursive and existentially definable. Hence every recursively enumerable relation can be expressed in the form (6.1) where ρ is an existentially definable primitive recursive relation.

The complement of an existentially definable relation is not necessarily existentially definable, for otherwise every arithmetically definable relation would also be existentially definable and hence recursively enumerable, which is not the case(2). However if the complement of every existentially definable primitive recursive relation is existentially definable, then every recursively enumerable relation is existentially definable. To see this, let ϕ be put in the form (6.1) with ρ a primitive recursive existentially definable relation. Then,

$$\phi(x_1, \cdots, x_n) \leftrightarrow \bigvee_{u} \sim \bigvee_{v \leqq u} \sim \rho(x_1, \cdots, x_n, u, v).$$

By hypothesis $\sim\rho$ is existentially definable and primitive recursive. Hence

$$\bigvee_{v \leqq u} \sim \rho$$

is existentially definable and primitive recursive. Consequently,

$$\sim \bigvee_{v \leqq u} \sim \rho$$

is existentially definable. Therefore ϕ is existentially definable.

Several years ago, Tarski proved that the complement of every set of the form

$$\bigvee_{u} P(x, u) = 0$$

(2) I am indebted to Martin Davis for pointing this out to me.

is existentially definable. (Here P is a polynomial with integer coefficients. In this theorem, it will be more convenient to let the variables range over the integers instead of the natural numbers.) Since this result has not been published, we shall sketch the proof.

THEOREM (TARSKI). *Let* $P(x)=a_nx^n+\cdots+a_0$ *be a polynomial and let* $\phi(a_0, \cdots, a_n)$ *be the relation which holds if and only if* $P(x)=0$ *has an integral root, i.e.*

$$\phi(a_0, \cdots, a_n) \leftrightarrow \bigvee_x a_nx^n + \cdots + a_0 = 0.$$

Then the relation $\sim\phi$ *is also existentially definable.*

Proof. From Tarski's generalization of Sturm's theorem (see [4, p. 51]), it is possible to give a system $\mathfrak{S}(a_0, \cdots, a_n, b, c)$ of inequalities and equations with integer coefficients such that there is no real root of $P=0$ in the closed interval $\langle b, c\rangle$, if and only if $\mathfrak{S}(a_0, \cdots, a_n, b, c)$. We see that for integral $a_0, \cdots, a_n, b$, and c, $\mathfrak{S}(a_0, \cdots, a_n, b, c)$ is an existentially definable relation among integers. If $b=-\infty$ or $c=+\infty$, then the corresponding $\mathfrak{S}$ is still existentially definable. The following equivalence gives an existential definition of the relation $\sim\phi$,

$$\begin{aligned}\sim\phi(a_0, \cdots, a_n) \leftrightarrow \bigvee_{r_1,\cdots,r_n} & [\mathfrak{S}(a_0, \cdots, a_n, -\infty, r_1) \\ & \wedge \mathfrak{S}(a_0, \cdots, a_n, r_1+1, r_2) \wedge \cdots \\ & \wedge \mathfrak{S}(a_0, \cdots, a_n r_{n-1}+1, r_n) \\ & \wedge \mathfrak{S}(a_0, \cdots, a_n, r_n+1, +\infty)].\end{aligned}$$

To see that this is a valid equivalence, notice that:

(1) Every integer is included in one of the closed intervals in which there are no real solutions.

(2) Since there are not more than n real roots, these can be isolated in the n open intervals, (r_i, r_{i+1}) with $i=1, \cdots, n$.

BIBLIOGRAPHY

1. M. Davis, *Arithmetical problems and recursively enumerable predicates* (abstract), Journal of Symbolic Logic vol. 15 (1950) pp. 77–78.

2. K. Gödel, *Über formal unentscheidbare Sätze der Principia Mathematica und verwandter Systeme* I, Monatshefte für Mathematik und Physik vol. 38 (1931) pp. 173–198.

3. S. C. Kleene, *Recursive predicates and quantifiers*, Trans. Amer. Math. Soc. vol. 53 (1943) pp. 41–73.

4. A. Tarski, *A decision method for elementary algebra and geometry*, Rand Corporation, Santa Monica, Calif., 1948. Reprinted by the University of California Press, Berkeley, Calif., 1951.

BERKELEY, CALIF.

A NOTE ON PRIMITIVE RECURSIVE FUNCTIONS

JULIA ROBINSON

The purpose of this note is to prove three theorems on primitive recursive functions of one variable which are analogous to three earlier theorems on general recursive functions. Theorems 1, 2, and 3 parallel closely Theorems 4, 5, and 6 of GRF §4.[1] We shall use the same notation as before, calling special attention to the fact that J, K, and L are the Cantor pairing functions.

THEOREM 1. *All primitive recursive functions of one variable can be obtained by starting with a certain two primitive recursive functions and repeatedly using the formulas*

$$Fx = BAx, \qquad Fx = B^x0$$

to construct a new function from known functions A and B.

PROOF. As before, we need only show that a finite set of primitive recursive functions is sufficient, since these can then be reduced to two functions by pairing (GRF p. 712). Again, to every function F, we let correspond a function $F^* = J(FK, L)$. We shall show that if F is defined by any of the three formulas

$$Fx = Ax + Bx, \qquad Fx = BAx, \qquad Fx = B^x0,$$

then F^* can be obtained from A^*, B^*, and several specific functions by substitution and recursion alone. Now, by a theorem of R. M. Robinson,[2] all primitive recursive functions of one variable can be obtained from S and K by successive additions, substitutions, and recursions of the form $Fx = B^x0$. Hence if F is a primitive recursive function, then F^* can be obtained from a suitable set of initial functions (including S^* and K^*) by substitution and recursion. We also have $F = KF^*J(I, I)$. Therefore, if K and $J(I, I)$ are included in the initial set of functions, we can obtain not only all the star functions corresponding to primitive recursive functions but also all the primitive functions themselves.

We still need to show how F^* can be obtained from A^* and B^*. Referring to GRF, p. 713, we see that $(BA)^* = B^*A^*$ and $(A+B)^*$

Received by the editors November 12, 1954.

[1] Julia Robinson, *General recursive functions*, Proc. Amer. Math. Soc. vol. 1 (1950) pp. 703–718.

[2] R. M. Robinson, *Primitive recursive functions*. II, Proc. Amer. Math. Soc. vol. 6 (1955) pp. 663–666.

$=G_3B^*G_2A^*G_1$ where $G_1=J(K, I)$, $G_2=J(KL, I)$, and $G_3=J(K+KL, LLL)$. Let F be defined by $Fx=B^x0$. Then $F^*x=J(B^{Kx}0, Lx)$. F^* cannot in general be defined by recursion, since F^* is unbounded but is not necessarily univalent. We turn instead to a related function $J(FK, I)$ which is univalent and from which we can obtain F^* by substitution, since $F^*=J(K, LL)J(FK, I)$. Now any univalent function which vanishes at zero can be expressed as M^x0 for a properly chosen function M. We first determine M so that $J(FKx, x)=M^x0$. We shall then show that $M=HB^*$ where H is a fixed function independent of B. We need to express $J(FKSx, Sx)$ in terms of $J(FKx, x)$ $=J(B^{Kx}0, x)$, making use of the fact that KSx is either SKx or 0. Hence,

$$J(FKSx, Sx) = J(\operatorname{sgn} KSx \cdot BB^{Kx}0, Sx).$$

However, $Sx=SLJ(FKx, x)$ and $B^{Kx}0=KJ(FKx, x)$, so Mu $=J(\operatorname{sgn} KSLu \cdot BKu, SLu)$. We still must show that $M=HB^*$ where H is independent of B. This can easily be done, since $BKu=KB^*u$ and $Lu=LB^*u$ and $Mu=J(\operatorname{sgn} KSLB^*u \cdot KB^*u, SLB^*u)=HB^*u$ where $H=J(\operatorname{sgn} KSL\cdot K, SL)$.[3] This completes the proof, for we have shown that any set of initial functions which is sufficient to yield S^*, K^*, K, $J(I, I)$, G_1, G_2, G_3, $J(K, LL)$ and H by substitution and recursion will generate all primitive recursive functions.

We now turn to the problem of replacing these nine functions by fewer functions of a simpler nature than the complicated function obtained from the nine by pairing.

THEOREM 2. *All primitive recursive functions of one variable can be obtained by substitution and recursion of the form $Fx=B^x0$ from the four functions $J(LK, KL)$, $J(I, SO)$, $J(K+KL, I)$, and $J(\operatorname{sgn} KL\cdot K, L)$.*

PROOF. It is sufficient to show that we can define the nine functions listed at the end of the proof of Theorem 1. First of all,

$$J(LK, KL)J(K + KL, I)J(I, SO) = J(LK, KL)J(I, J(I, SO)) = J(L, I).$$

Hence by the lemma on page 714 of GRF, we can obtain all functions $J(M, N)$ where M and N are obtained from I, K, and L by substitution. Since the Cantor pairing functions are one-to-one, we also obtain $K=J(KK, LK)$.

We still have three functions to define which we can obtain by substitution in the following way:

---

[3] Here, of course, we let $KL\cdot K$ denote the function G defined by $Gx=KLx\cdot Kx$

$$J(K, LLLL)J(K + KL, I) = J(K + KL, LLL),$$

$$J(K, LLL)J(K + KL, I)J(L, K)J(I, SO) = J(K + KL, LL)J(SO, I) = J(SK, L),$$

$$\begin{aligned} J(\operatorname{sgn} KL\cdot K, L)J(L, K)J(SK, L)J(L, K) &= J(\operatorname{sgn} KL\cdot K, L)J(L, K)J(SL, K) \\ &= J(\operatorname{sgn} KL\cdot K, L)J(K, SL) = J(\operatorname{sgn} KSL\cdot K, SL). \end{aligned}$$

We shall now show that it is impossible to reduce the two initial functions of Theorem 1 to a single function of any sort, even if we allow the more general form of recursion, $Fx = B^x a$.

LEMMA. *Consider the class of functions generated from a single function A by repeated substitution and recursion of the form $Fx = B^x a$, where various values of a may be used. If F is any function of this class, then one of the following two conditions is satisfied:*

(1) *$F = A^k$ for some positive integer k.*

(2) *There exist a number a and a finite exceptional set α such that for every x, either $Fx = A^n a$ for some positive n or $Fx \in \alpha$.*

PROOF. To show this, we need only check that if G and H satisfy either condition, then GH and $G^x b$ also satisfy one of the alternatives. First consider GH. Clearly, if G and H both satisfy (1), so also does GH. If G satisfies (2), then GH satisfies (2), since the range of GH is included in the range of G. Finally, suppose G satisfies (1) and H satisfies (2), i.e. there exist k, a, and a finite set α, so that $G = A^k$ and for every x, either there is an n such that $Hx = A^n a$ or $Hx \in \alpha$. It follows that for every x, $GH = A^{k+n} a$ or GHx belongs to a finite set, namely the map of α by G. Hence GH satisfies (2).

Now consider $Fx = G^x b$. If G satisfies (1), then $Fx = A^{kx} b$, so F satisfies (2). If G satisfies (2), so also does F since every value of F is a value of G except possibly for $F0 = b$.

THEOREM 3. *There is no single function A from which all primitive recursive functions can be obtained by substitution and recursion of the form $Fx = B^x a$.*

PROOF. *Case* 1. A assumes only a finite number of values more than once. We shall show that every function obtained from A has this property. Functions obtained by recursion are either bounded or univalent and hence do have the stated property. Finally, if G and H both have this property, we must show that GH does also. Clearly, if $GHx_1 = GHx_2$, then either $Hx_1 = Hx_2$ or if we put $y_1 = Hx_1$ and $y_2 = Hx_2$, we have $y_1 \neq y_2$ and $Gy_1 = Gy_2$. Hence the set of possible values z for which $z = GHx$ has more than one solution must be finite. Thus in Case 1, a function such as K cannot be defined and hence not all primitive recursive functions can be obtained from A.

Case 2. A assumes infinitely many values more than once. In this case, we use the lemma. Notice that if H is not univalent, then GH is not univalent. Hence no power of A is univalent. On the other hand, $A^x a$ omits infinitely many values. For if $Au = Av$, and $A^x a$ assumes both of the values u and v, then it is ultimately periodic and hence bounded. If $A^x a$ is not bounded, then it must omit one of each pair of values u, v for which $Au = Av$. Hence any function satisfying the condition (2) of the lemma must omit infinitely many values. Therefore every univalent function which is obtained from A by substitution and recursion must omit infinitely many values. Thus, we cannot define S.

We have shown that there is no function A from which we can obtain both K and S by substitution and recursion. Hence there is no function A which can generate all primitive recursive functions of one variable.

BERKELEY, CALIF.

THE UNDECIDABILITY OF ALGEBRAIC RINGS AND FIELDS

JULIA ROBINSON

1. **Introduction.** Our main result is the solution of the decision problem for algebraic fields of finite degree over the rationals. We will give a definition of the natural numbers within the arithmetic of any such field, thereby showing that the field is undecidable.

By the *arithmetic* of a field F, we mean the mathematical theory whose statements are constructed from the logical symbols $\wedge$ (and), $\vee$ (or), $\neg$ (not), $\rightarrow$ (if $\cdots$ then $\cdots$), $\leftrightarrow$ (if and only if), $\bigwedge$ (for every), $\bigvee$ (there exists), and $=$ (equals); the mathematical symbols $+$ and $\cdot$; and variables whose range is F. Similarly, the arithmetic of a ring $\boldsymbol{R}$ or of the natural numbers $\boldsymbol{N}$ is defined by restricting the variables either to $\boldsymbol{R}$ or to $\boldsymbol{N}$.

An *arithmetical definition* of an n-ary relation ρ within a particular mathematical theory is an equivalence with $\rho(u_1, \cdots, u_n)$ on the left and with an expression of the theory, having $u_1, \cdots, u_n$ as its only free variables, on the right.

We say a mathematical theory is *decidable* if there is an effective method of determining the validity of each statement of the theory. If there is no such method, the theory is *undecidable*. It is clear that if there is a mechanical way of transforming each statement of an undecidable theory into an equivalent statement of another theory, the second theory is also undecidable. This principle, together with the fact that the arithmetic of natural numbers is undecidable, enables us to solve the decision problem for fields of finite degree over the rationals.[1]

The relation $n \in \boldsymbol{N}$ can be defined arithmetically in the ring $\boldsymbol{I}$ of rational integers as follows:

$$n \in \boldsymbol{N} \leftrightarrow \bigvee_{x,y,z,w} n = x^2 + y^2 + z^2 + w^2.$$

Hence it is possible to transform routinely any statement of the arithmetic of $\boldsymbol{N}$ into an equivalent statement of the arithmetic of $\boldsymbol{I}$.

Presented to the Society, January 22, 1959; received by the editors January 29, 1959.

[1] For a complete account of much more powerful arguments to deduce the undecidability of one theory from that of another, see Tarski, Mostowski, and Robinson [9]. This monograph also contains a proof of the fundamental theorem that the arithmetic of the natural numbers is undecidable. This was first proved by Rosser extending the work of Church and Gödel.

Applying the principle stated above, we see that the ring of rational integers is undecidable. In §2, we give a definition of the natural numbers within the arithmetic of the ring R of the algebraic integers of a field of finite degree over the rationals. As before, it follows that R is undecidable.[2]

Let R be any integral domain and F its quotient field. Every statement of the arithmetic of F can be transformed into an equivalent statement of the arithmetic of R by replacing each variable whose range is F by the ratio of two variables with range R and adjoining the condition that the denominator is not 0, then clearing the resulting equations of fractions. The undecidability of F thus implies the undecidability of R. It is not known whether the converse always holds. If R can be defined arithmetically in F, we can transform mechanically any sentence of the arithmetic of R into an equivalent sentence of the arithmetic of F. Hence in this case the converse holds.

The first field shown to be undecidable was the rational field [6]. This was done by using the theory of ternary quadratic forms to define the ring of integers arithmetically in the rational field, and thus reducing the decision problem for the rational field to that of the ring of integers, which was known to be undecidable. In §3, we apply a similar method to fields of finite degree over the rationals. We no longer obtain an arithmetical definition of the rational integers but instead an arithmetical definition of the algebraic integers of the field. Combining the results of §§2 and 3, we see that the natural numbers can be defined arithmetically in any field of finite degree over the rationals, and therefore such a field is undecidable.

2. **The definability of the natural numbers in R.** In this section, we will define arithmetically the relation $n \in N$, i.e. n is a natural number, in the ring R of the algebraic integers of a field of finite degree over the rationals.

LEMMA 1. *There are only finitely many numbers a of R such that*

$$a + 1 \mid f \wedge \cdots \wedge a + l \mid f$$

where l is the degree of R and f is any number of R different from 0.

PROOF. For any a in R, put $P_a(x) = (x + a^{(1)}) \cdots (x + a^{(l)})$ where $a^{(1)}, \cdots, a^{(l)}$ are the l conjugates of a. If k is a rational integer, the

[2] I first proved the undecidability of R by an entirely different method. The definition of the natural numbers given here was discovered jointly with R. M. Robinson. He had shown earlier that the ring of the algebraic integers of a field of finite degree over the rationals which has at most one fundamental unit is undecidable [7].

norm $N(a+k)$ is given by $P_a(k)$. Since $N(f)$ is not 0, it has only a finite number of rational integer divisors. If a satisfies the conditions of the lemma, $N(a+k)$ divides $N(f)$ for $k=1, \cdots, l$. Only a finite number of values can be assumed by $N(a+k)=P_a(k)$ for those a's which satisfy the lemma. Since P_a is of degree l and has leading coefficient 1, it is determined uniquely by $P_a(1), \cdots, P_a(l)$. There can be only a finite number of such polynomials and, since $P_a(-a)=0$, only a finite number of a's satisfying the conditions of the lemma.

THEOREM. *Let* $\boldsymbol{R}$ *be the ring of the algebraic integers of a field of degree* l *over the rationals. The set* $\boldsymbol{N}$ *of natural numbers is arithmetically definable in* $\boldsymbol{R}$, *and hence* $\boldsymbol{R}$ *is undecidable. In fact, if we let*

$$\tau(a, f, g, h) \leftrightarrow f \neq 0 \wedge a + 1 \mid f \wedge \cdots \wedge a + l \mid f \wedge 1 + ag \mid h,$$

then

$$n \in \boldsymbol{N} \leftrightarrow \bigvee_{f,g,h} \left\{ \tau(0, f, g, h) \wedge \bigwedge_a \left[\tau(a, f, g, h) \rightarrow a = n \vee \tau(a+1, f, g, h) \right] \right\}.$$

PROOF. Suppose f, g, and h satisfy the right side of the equivalence for some n. By Lemma 1, there are only finitely many numbers a in $\boldsymbol{R}$ such that $\tau(a, f, g, h)$. The inductive form of the definition insures that $\tau(0, f, g, h), \tau(1, f, g, h), \cdots$, terminating only for $a=n$. Therefore, n must be a natural number.

Conversely, suppose that n is a natural number. We must show that f, g, and h exist which satisfy the condition on the right side of the definition. It will be sufficient to find f, g, and h so that

$$\tau(a, f, g, h) \leftrightarrow a = 0 \vee a = 1 \vee \cdots \vee a = n.$$

First put $f = (n + l)!$ and define the set $\boldsymbol{S}$ by

$$a \in \boldsymbol{S} \leftrightarrow a + 1 \mid f \wedge \cdots \wedge a + l \mid f.$$

By Lemma 1, $\boldsymbol{S}$ is finite. Now choose g a positive integer such that

$$a \in \boldsymbol{S} \wedge b \in \boldsymbol{S} \wedge a \neq b \rightarrow a - b \mid g$$

and

$$a \in \boldsymbol{S} \wedge a \neq 0 \rightarrow 1 + ag \nmid 1.$$

Clearly, we can choose g satisfying both conditions since the first condition holds if g is a multiple of all the rational integers $N(a-b)$ with a and b in $\boldsymbol{S}$ and $a \neq b$, and the second condition holds if g is so large a natural number that all the conjugates of $1+ag$ lie outside the unit circle. (Since $|N(1+ag)|>1$, then $1+ag$ is not a unit.)

For this particular choice of f and g, the numbers $1+ag$ and $1+bg$ for a and b in S and $a\neq b$ are relatively prime. Suppose this were not the case. If a prime ideal $\mathfrak{p}$ divided both $1+ag$ and $1+bg$, $\mathfrak{p}$ would divide the difference $(a-b)g$ and would also divide g. But $\mathfrak{p}$ cannot divide both g and $1+ag$.

Finally, put $h=(1+g)(1+2g)\cdots(1+ng)$. By our choice of f, g, and h, the relation $\tau(a, f, g, h)$ is satisfied for $a=0, \cdots, n$. If a is in S and not equal to $0, 1, \cdots, n$, then $1+ag$ is not a unit and is prime to h. Therefore, $1+ag\nmid h$ and $\tau(a, f, g, h)$ does not hold.

3. **The arithmetical definability of R in F.** In this section, we give a definition of the ring R within the arithmetic of F. Here, as before, R is the ring of the algebraic integers of a given field F of finite degree over the rationals.

Let $\mathfrak{p}$ be a valuation of F and $F_\mathfrak{p}$ be the completion of F with respect to $\mathfrak{p}$. Since non-Archimedean valuations of F are $\mathfrak{p}$-adic valuations with respect to some prime ideal $\mathfrak{p}$ of F, we will use the same letter $\mathfrak{p}$ for both the valuation and the prime ideal.

Two nonzero numbers a and b of F are said to be in the same $\mathfrak{p}$-adic class if a/b is the square of a number in $F_\mathfrak{p}$. There are only a finite number of $\mathfrak{p}$-adic classes for any valuation $\mathfrak{p}$. For an Archimedean valuation $\mathfrak{p}$, $F_\mathfrak{p}$ is either the field of complex numbers, which consists of just one $\mathfrak{p}$-adic class, or the field of real numbers, which consists of two $\mathfrak{p}$-adic classes. Corresponding to a prime ideal $\mathfrak{p}$ which does not divide 2, there are four $\mathfrak{p}$-adic classes. In the case of a prime ideal $\mathfrak{p}$ which divides 2, the number of classes is even—the exact number depending on the power to which $\mathfrak{p}$ divides 2.

We will use the Hilbert symbol $(a, b)_\mathfrak{p}$, which is defined for all nonzero numbers a and b in F and any valuation $\mathfrak{p}$ as follows:

$$(a, b)_\mathfrak{p} = \begin{cases} +1 & \text{if } ax^2 + by^2 = 1 \text{ is solvable in } F_\mathfrak{p}, \\ -1 & \text{otherwise.} \end{cases}$$

LEMMA 2. *Let a and b belong to R, $\mathfrak{p}$ be a prime ideal, and m be a positive integer such that $\mathfrak{p}^m\nmid 2$. If $a\not\equiv 0 \pmod{\mathfrak{p}^2}$ and $a\equiv b \pmod{\mathfrak{p}^{2m}}$, then a and b are in the same $\mathfrak{p}$ adic class.*

This lemma follows easily from Hensel's lemma as given by Artin [1, pp. 44–51].

It is clear that $(a, b)_\mathfrak{p}$ depends only on the $\mathfrak{p}$-adic classes to which a and b belong. Hasse [3; 4] gives formulas for evaluating the Hilbert symbol. Although we will not use these directly, we will need the following two lemmas which are immediate consequences of them.

LEMMA 3. *If a and b are nonzero numbers of $\boldsymbol{R}$, then $(a, b)_{\mathfrak{p}} = -1$ implies that either $\mathfrak{p}$ is an Archimedean valuation or the prime ideal $\mathfrak{p}$ divides $2ab$; hence, there are only a finite number of valuations $\mathfrak{p}$ for which $(a, b)_{\mathfrak{p}} = -1$.*

LEMMA 4. *If a belongs to $\boldsymbol{R}$ and contains $\mathfrak{p}$ to exactly the first power, then there exists a number b in $\boldsymbol{R}$ with $\mathfrak{p} \nmid b$ such that $(a, b)_{\mathfrak{p}} = -1$.*

Our next lemma was first proved by Furtwängler [**2**, p. 427].

LEMMA 5. *If a and b are nonzero numbers of $\boldsymbol{F}$, then $(a, b)_{\mathfrak{p}} = -1$ for an even number of valuations.*

Both Lemma 6 and Lemma 7 are special cases of a theorem given by Hasse [**5**, p. 32].

LEMMA 6. *There are infinitely many prime ideals in every ideal class.*

LEMMA 7. *If a is a number of $\boldsymbol{R}$ which is prime to an ideal $\mathfrak{m}$, there are infinitely many totally positive prime numbers p in $\boldsymbol{R}$ such that $p \equiv a \pmod{\mathfrak{m}}$.*

The next lemma is due to Hasse [**4**, p. 127]. Another proof is given by Witt [**10**, pp. 41–42].

LEMMA 8. *A nonzero number h of $\boldsymbol{F}$ can be represented by the ternary quadratic form $x^2 - ay^2 - bz^2$ in $\boldsymbol{F}$ if and only if h does not belong to the same $\mathfrak{p}$-adic class as $-ab$ for any valuation $\mathfrak{p}$ with $(a, b)_{\mathfrak{p}} = -1$.*

LEMMA 9. *Given any prime ideal $\mathfrak{p}_1$, there exist relatively prime numbers a and b in $\boldsymbol{R}$ such that $(a) = \mathfrak{p}_1 \cdots \mathfrak{p}_{2k}$, where $\mathfrak{p}_1, \cdots, \mathfrak{p}_{2k}$ are distinct prime ideals which include every prime ideal which divides 2, and b is a totally positive prime number such that $(a, b)_{\mathfrak{p}} = -1$ if and only if $\mathfrak{p} \mid a$.*

PROOF. Let $\mathfrak{p}_1, \cdots, \mathfrak{p}_{2k-1}$ be a set of distinct prime ideals which includes every prime ideal dividing 2. Let A be the ideal class which contains the product $\mathfrak{p}_1 \cdots \mathfrak{p}_{2k-1}$. By Lemma 6, we can choose a prime ideal $\mathfrak{p}_{2k}$ in the ideal class A^{-1} with $\mathfrak{p}_{2k} \neq \mathfrak{p}_i$ for $i = 1, \cdots, 2k-1$. Then $\mathfrak{p}_1 \cdots \mathfrak{p}_{2k}$ will be a principal ideal and we can take a so that $(a) = \mathfrak{p}_1 \cdots \mathfrak{p}_{2k}$.

For $i = 1, \cdots, 2k$, by Lemma 4 we can choose b_i in $\boldsymbol{R}$ and prime to $\mathfrak{p}_i$ so that $(a, b_i)_{\mathfrak{p}_i} = -1$. Let m be a positive integer so that $\mathfrak{p}^m \nmid 2$ for every prime ideal $\mathfrak{p}$. By Lemma 2, if

$$x \equiv b_i \pmod{\mathfrak{p}_i^{2m}} \qquad \text{for } i = 1, \cdots, 2k,$$

then $(a, x)_{\mathfrak{p}_i} = -1$ for $i = 1, \cdots, 2k$. By the analog of the Chinese Remainder Theorem, we can replace these $2k$ congruences by a single congruence:

$$x \equiv c \pmod{\mathfrak{p}_1^{2m} \cdots \mathfrak{p}_{2k}^{2m}}$$

where c is prime to the modulus. By Lemma 7, we can find a totally positive prime number b satisfying this congruence. Since c is prime to a, b will also be prime to a.

By our construction $(a, b)_{\mathfrak{p}_i} = -1$ for $i = 1, \cdots, 2k$. Since b is totally positive b is a square in $F_{\mathfrak{p}}$ for all Archimedean valuations $\mathfrak{p}$, and hence $(a, b)_{\mathfrak{p}} = +1$ for all Archimedean valuations. By Lemma 3, the only other valuation for which $(a, b)_{\mathfrak{p}} = -1$ could hold would be $\mathfrak{p} = (b)$; but, by Lemma 5, there are an even number of valuations $\mathfrak{p}$ such that $(a, b)_{\mathfrak{p}} = -1$. Therefore, in our case $(a, b)_{\mathfrak{p}} = -1$ if and only if $\mathfrak{p} = \mathfrak{p}_i$ for $i = 1, \cdots,$ or $2k$.

LEMMA 10. *If a and b satisfy Lemma 9 and m is a positive integer such that $\mathfrak{p}^m \nmid 2$ for all prime ideals $\mathfrak{p}$, then there exist x, y, and z in F such that $1 - abc^{2m} = x^2 - ay^2 - bz^2$ if and only if c is a $\mathfrak{p}$-adic integer for every prime ideal $\mathfrak{p}$ dividing a.*

PROOF. We can express $c = u/v$ with u and v in $\boldsymbol{R}$ and with no $\mathfrak{p}_i$ dividing both u and v. We then need to show that if u and v $(v \neq 0)$ are in $\boldsymbol{R}$ and no $\mathfrak{p}_i$ divides both u and v, $v^{2m} - abu^{2m} = x^2 - ay^2 - bz^2$ is solvable for x, y, and z in F if and only if v is prime to a.

Let $h = v^{2m} - abu^{2m}$. By Lemma 8, h can be represented by $x^2 - ay^2 - bz^2$ if and only if h does not lie in the same $\mathfrak{p}_i$-adic class as $-ab$ for $i = 1, \cdots, 2k$.

CASE I. Suppose $\mathfrak{p}_i \mid v$. Since $\mathfrak{p}_i$ does not divide u or b, $h \not\equiv 0 \pmod{\mathfrak{p}_i^2}$ and $h \equiv -abu^{2m} \pmod{\mathfrak{p}_i^{2m}}$. By Lemma 2, h is in the same $\mathfrak{p}_i$-adic class as $-abu^{2m}$. But this class is the same as the class of $-ab$; therefore, h cannot be represented by $x^2 - ay^2 - bz^2$.

CASE II. Suppose v is prime to a. Then h is prime to a and cannot be in the same $\mathfrak{p}_i$-adic class as $-ab$ for any $\mathfrak{p}_i$, since $\mathfrak{p}_i$ divides ab to exactly the first power. Therefore h can be represented by $x^2 - ay^2 - bz^2$ and the lemma follows.

THEOREM. *Let m be a positive integer such that $\mathfrak{p}^m \nmid 2$ for every prime ideal $\mathfrak{p}$ and let $\phi(a, b, c)$ be defined by the equivalence,*

$$\phi(a, b, c) \leftrightarrow \bigvee_{x,y,z} 1 - abc^{2m} = x^2 - ay^2 - bz^2.$$

If $\psi(t)$ is defined by

$$\psi(t) \leftrightarrow \bigwedge_{a,b} \left\{ \bigwedge_{c} \left[\phi(a, b, c) \rightarrow \phi(a, b, c + 1)\right] \rightarrow \phi(a, b, t)\right\},$$

then $t \in I \rightarrow \psi(t)$ and $\psi(t) \rightarrow t \in \boldsymbol{R}$.

PROOF. Since $\phi(a, b, 0)$ holds for every a and b, the inductive form of ψ insures that every natural number satisfies ψ. Since $\phi(a, b, c) \leftrightarrow \phi(a, b, -c)$, we see that every rational integer also satisfies ψ.

Suppose that t does not belong to R. Then there is some prime ideal $\mathfrak{p}_1$ such that t is not a $\mathfrak{p}_1$-adic integer. We choose a and b by Lemma 9. By Lemma 10, $\phi(a, b, c)$ holds if and only if c is a $\mathfrak{p}$-adic integer for every prime ideal $\mathfrak{p}$ which divides a. Clearly, for this a and b, $\phi(a, b, c) \rightarrow \phi(a, b, c+1)$; but $\phi(a, b, t)$ does not hold. Thus, if t is not in R, $\psi(t)$ does not hold.

It is not known whether $\psi(t) \rightarrow t \in I$; however, we can give an arithmetical definition of R with the help of ψ. Let $a_1, \cdots, a_l$ be an integral basis for R. (Here l is the degree of F over the rationals.) Then,

$$t \in R \leftrightarrow \bigvee_{x_1, \cdots, x_l} [t = a_1x_1 + \cdots + a_lx_l \wedge \psi(x_1) \wedge \cdots \wedge \psi(x_l)].$$

It would be sufficient for the x_i to range over I; but if the x_i range over R, no additional values of t are introduced. However, in general, $a_1, \cdots, a_l$ will not themselves be arithmetically definable.[3]

Let P_i be a polynomial with integer coefficients and leading coefficient equal to 1 such that $P_i(a_i) = 0$. Since every root of $P_i = 0$ which is in F is also in R, we obtain

$$t \in R \leftrightarrow \bigvee_{x_1, \cdots, x_l; y_1, \cdots, y_l} [t = x_1y_1 + \cdots + x_ly_l \wedge P_1(y_1) = 0$$
$$\wedge \cdots \wedge P_l(y_l) = 0 \wedge \psi(x_1) \wedge \cdots \wedge \psi(x_l)].$$

Since this formula gives a suitable arithmetical definition of R within the arithmetic of F, we have proved the following:

THEOREM. *If R is the ring of the algebraic integers of a field F of finite degree over the rationals, then R is arithmetically definable in F.*

This theorem with the final theorem of §2 gives us:

THEOREM. *If F is an algebraic field of finite degree over the rationals, the natural numbers are arithmetically definable in F and hence F is undecidable.*

REFERENCES

1. Emil Artin, *Lectures on modern higher algebra*, Part III, mimeographed notes, New York University, 1948.

2. Ph. Furtwängler, *Die Reziprozitätsgesetze für Potenzreste mit Primzahlexponenten in algebraischen Zahlkörpern*, III, Math. Ann. vol. 74 (1913) pp. 413–429.

3. Helmut Hasse, *Zur Theorie des quadratischen Hilbertschen Normenrestsymbols in algebraischen Körpern*, J. Reine Angew. Math. vol. 153 (1923) pp. 76–93.

4. ———, *Darstellbarkeit von Zahlen durch quadratische Formen in einem beliebigen algebraischen Zahlkörper*, J. Reine Angew. Math. vol. 153 (1923) pp. 113–130.

[3] For a description of numbers arithmetically definable in an algebraic field, see R. M. Robinson [8].

5. ———, *Bericht über neuere Untersuchungen und Probleme aus der Theorie der algebraischen Zahlkörper*, Teil I, Jahresbericht der Deutschen Mathematiker-Vereinigung vol. 35 (1926) pp. 1–55.

6. Julia Robinson, *Definability and decision problems in arithmetic*, J. Symb. Logic vol. 14 (1949) pp. 98–114.

7. R. M. Robinson, *Undecidable rings*, Trans. Amer. Math. Soc. vol. 70 (1951) pp. 137–159.

8. ———, *Arithmetical definability of field elements*, J. Symb. Logic vol. 16 (1951) pp. 125–126.

9. A. Tarski, A. Mostowski, and R. M. Robinson, *Undecidable theories*, Amsterdam, North-Holland Publishing Company, 1953.

10. Ernst Witt, *Theorie der quadratischen Formen in beliebigen Körpern*, J. Reine Angew. Math. vol. 176 (1936) pp. 31–44.

Contra Costa County, California

PROBLEMS OF NUMBER THEORY ARISING IN METAMATHEMATICS

by

Julia Robinson

We say a set A of natural numbers is <u>existential</u> or <u>Diophantine</u> if there is a polynomial $F(x_0, x_1, \ldots, x_n)$ with integer coefficients such that $a \in A$ if and only if the Diophantine equation $F(a, x_1, \ldots, x_n) = 0$ is solvable in natural numbers.* Here are the defining equations for some simple existential sets:

a is composite	$a - (x + 2)(y + 2) = 0$
a is square	$a - x^2 = 0$
a is not a square	$(x + 2)^2 - ay^2 - 1 = 0$
a is not a power of 2	$a - (2x + 3)y = 0$

The set of non-powers of six is also existential but the corresponding equation is quite complicated.

<u>Problem 1</u>. Is the set of primes existential? That is, is there a polynomial $P(x_0, x_1, \ldots, x_n)$ with integer coefficients such that p is a prime if and only if $P(p, x_1, \ldots x_n) = 0$ is solvable in natural numbers?

---

*Since an integer is a natural number if and only if it is the sum of four squares of integers, we could give our definition in terms of integers instead of natural numbers and we would obtain precisely the same sets of natural numbers.

Problem 2. Is there a polynomial $F(u, v, x_1, \ldots, x_n)$ with integer coefficients such that $v < u^u$ whenever $F(u, v, x_1, \ldots, x_n) = 0$ is solvable in natural numbers, and such that for every n there is some solution with $v > u^n$?

These two problems appear completely unrelated. However, I have shown (Tran. AMS 72, (1952) 437 - 449) that the existence of a polynomial satisfying Problem 2 implies that the set of primes is existential.

Problem 3. Find an existential set whose complement is not existential. It can be shown using metamathematical concepts that such sets exist but no example is known.

One of the classical problems of metamathematics is the decision problem for various mathematical systems. For example, the decision problem for the arithmetic of natural numbers is to answer the question: Is there a general method for determining the validity of all the statements of the arithmetic of natural numbers? By the arithmetic of natural numbers, I mean the mathematical theory whose statements are constructed from the logical symbols, + and • , and variables whose range is the natural numbers. Gödel, Church, and Rosser showed that no general method exists for the arithmetic of natural numbers.

Now in a ring R containing the natural numbers N, suppose we can give an arithmetical definition of the natural numbers, i.e. an equivalence with $n \in N$ on the left and an expression of the arithmetic of R on the right. In the case of the ring I of rational integers, we have for example

$$n \in N \leftrightarrow \bigvee_{x,y,z,w,} \quad n = x^2 + y^2 + z^2 + w^2 .$$

Then we can transform mechanically any statement of the arithmetic of N into an equivalent statement in the arithmetic of R. Hence R is also undecidable. In the case of the rational integers, suppose we are given a statement about natural numbers, we can obtain an equivalent statement about integers by replacing natural number variables by integer variables and adding the condition in each case that the integer is the sum of four integer squares. Using this method, I recently showed in a paper to be published in the Proceedings of the A. M. S. that the ring of algebraic integers of any field of finite degree over the rationals is undecidable.

In connection with the decision problem for rings, the following problems have come up:

Problem 4. In a given field of finite degree over the rationals, show that 1 cannot be represented as the difference of two units in infinitely many ways. More generally, show that an equation

$$a\,\varepsilon - b\,\varepsilon' = 1 \qquad a,\ b \text{ given}$$

is satisfied by only a finite number of units ε, ε' in the given field.

Kronecker showed that there are infinitely many sets of conjugate algebraic integers in the interval $0 < x < 4$ and only a finite number in any subinterval. In the smallest field containing the square root of each natural number, there are only a finite number of sets of conjugate integers of the field in any finite interval.

Problem 5. Show that in any totally real algebraic field, there is an interval $0 < x < t$ (t may be $+\infty$) containing infinitely many sets of conjugate algebraic integers of the field and such that there is only a finite number of sets of conjugate integers for any smaller t.

ANNALS OF MATHEMATICS
Vol. 74, No. 3, November, 1961
Printed in Japan

THE DECISION PROBLEM FOR EXPONENTIAL DIOPHANTINE EQUATIONS

BY MARTIN DAVIS[1], HILARY PUTNAM[1], AND JULIA ROBINSON

(Received July 26, 1960)

1. Introduction

We prove that every recursively enumerable set can be existentially defined in terms of exponentiation. Hence, there is no general algorithm for deciding whether or not an exponential diophantine equation has a solution in positive integers. We also obtain a general theorem about bounds for solutions of diophantine equations with a finite number of solutions[2].

The tenth problem on David Hilbert's famous list (cf., Hilbert [4]) is given as follows:

10. Entscheidung der Lösbarkeit einer diophantischen Gleichung.

Eine diophantische Gleichung mit irgendwelchen Unbekannten und mit ganzen rationalen Zahlkoeffizienten sei vorgelegt; man soll ein Verfahren angeben, nach welchem sich mittels einer endlichen Anzahl von Operationen entscheiden lässt, ob die Gleichung in ganzen rationalen Zahlen lösbar ist.

Our result provides a negative answer to a closely related problem.

A diophantine equation is an equation of the form $P(x_1, \cdots, x_n) = 0$ where P is a polynomial with integer coefficients and a solution in integers (or sometimes positive integers) is required.

An exponential diophantine equation is an equation which can be put in the form $E(x_1, \cdots, x_n) = F(x_1, \cdots, x_n)$ where E and F are functions built from $x_1, \cdots, x_n$ and particular positive integers by addition, multiplication and exponentiation. Such an equation is clearly equivalent to

[1] The contribution of Martin Davis and Hilary Putnam to this research was supported by the United States Air Force under Contract No. AF 49(638)-527.

[2] The result that every recursively enumerable set can be existentially defined in terms of exponentiation was obtained by Davis and Putnam, basing themselves on their [3], using the unproved number-theoretical hypothesis that there are arbitrarily long arithmetic progressions containing only prime numbers. They wish to acknowledge useful suggestions, made in conversation, by H. S. Shapiro. Their results are contained in the third part of an Airforce report (AFOSRTR 59-124, October 1959) and were presented to the American Mathematical Society, October 31, 1959. Abstract 560-10. Julia Robinson showed that the use of this hypothesis can be avoided. This result was presented to the American Mathematical Society, January 27, 1960. Abstract 564-275. Subsequently, she greatly simplified the entire proof. The theorem concerning bounds for solutions of diophantine equations is also due to Julia Robinson, and was presented to the American Mathematical Society, April 23, 1960. Abstract 569-23.

a system of equations, each of the form $\alpha + \beta = \lambda$, $\alpha\beta = \lambda$, or $\alpha^\beta = \lambda$, where α, β, and λ are variables or particular positive integers. Now a system of equations $A_i = B_i$ for $i = 1, \cdots, m$ is equivalent to the single equation $\sum (A_i - B_i)^2 = 0$. Hence every exponential diophantine equation is equivalent to an equation $E'(x_1, \cdots, x_n) = 0$, where E' is a linear combination (with integer coefficients) of products of terms of the form α^β and where α and β are either variables or particular positive integers. Without loss of generality, we may restrict ourselves to equations which have been reduced to this form. Examples of exponential diophantine equations from the current literature are $x^x y^y = z^z$, $2^y - 7 = x^2$, and $2^x + 11^y = 5^z$. Solutions of exponential diophantine equations are usually required to be positive integers.

An algorithm for determining the solvability of diophantine equations in integers would yield an algorithm for determining the solvability in positive integers and conversely. (See Davis [2, pp. 102-103] or Davis [1, pp. 35–36]). Hence we will limit our discussion to questions of solvability in positive integers. Lower-case Latin letters will always represent positive integers or variables whose range is the positive integers.

A relation $\rho(x_1, \cdots, x_n)$ *is existentially definable in terms of* $\varphi_1, \cdots, \varphi_k$ if there is an equivalence

$$\rho(x_1, \cdots, x_n) \longleftrightarrow \bigvee_{u_1, \cdots, u_m} \psi(x_1, \cdots, x_n, u_1, \cdots, u_m)$$

where ψ is an expression composed of the free variables $x_1, \cdots, x_n$; the bound variables $u_1, \cdots, u_m$; the logical connectives $\wedge, \vee, =$; $\varphi_1, \cdots, \varphi_k$; and symbols for particular positive integers. Here $\varphi_1, \cdots, \varphi_k$ can be either relations or operations (i.e., positive integer valued functions defined on the positive integers), on any number of variables. Clearly, if ρ can be existentially defined in terms of $\varphi_1, \cdots, \varphi_k$ and φ_k can be existentially defined in terms of $\varphi_1, \cdots, \varphi_{k-1}$, then ρ can be existentially defined in terms of $\varphi_1, \cdots, \varphi_{k-1}$.

It is easy to see that a set S of positive integers (i.e., a relation of one argument) is existentially definable in terms of addition and multiplication if and only if there is a polynomial P with integer coefficients (positive or negative) such that

$$s \in S \longleftrightarrow \bigvee_{u_1, \cdots, u_m} P(s, u_1, \cdots, u_m) = 0 \ .$$

(See Davis [2, pp. 104–106] or Davis [1, pp. 34–35]). Such a set is also called *diophantine* since it is the set of values of a parameter s for which the diophantine equation $P(s, u_1, \cdots, u_m) = 0$ is solvable in positive integers.

Similarly, a set S of positive integers is existentially definable in terms of exponentiation if and only if there is an exponential diophantine equation with one parameter s which is solvable if and only if s belongs to $\mathcal{S}$, and hence such a set is also called *exponential diophantine*. To show that the values of a parameter for which an exponential diophantine equation is solvable do form a set which is existentially definable in terms of exponentiation, we need only check that the relations given by $xy = z$ and $x + y = z$ are existentially definable in terms of exponentiation. But this follows from the equivalences

$$xy = z \longleftrightarrow (2^x)^y = 2^z \longleftrightarrow \bigvee_{u,v} (u = 2^x \wedge v = 2^z \wedge v = u^y)$$

and

$$x + y = z \longleftrightarrow 2^x 2^y = 2^z \,.$$

On the other hand, to show that every set which is existentially definable in terms of exponentiation is the set of values of a parameter for which some exponential diophantine equation is solvable, we can use the standard formulas:

$$A = 0 \wedge B = 0 \longleftrightarrow A^2 + B^2 = 0$$

and

$$A = 0 \vee B = 0 \longleftrightarrow AB = 0$$

to combine several equations into a single equation. Thus we can reduce an existential definition in terms of exponentiation to the form

$$s \in \mathcal{S} \longleftrightarrow \bigvee_{u_1, \cdots, u_m} E(s, u_1, \cdots, u_m) = 0$$

where $E = 0$ is an exponential diophantine equation.

Quite similarly, we can speak of *diophantine relations* and *exponential diophantine relations*. For example, the relations given by $x < y$ and $x \mid y$ (x divides y) are easily seen to be diophantine.

We now state the main theorem, but we will postpone the proof until § 3.

THEOREM. *Every recursively enumerable set (or relation) is exponential diophantine.*

In other words, every recursively enumerable set $\mathcal{S}$ can be expressed in the form

$$s \in \mathcal{S} \longleftrightarrow \bigvee_{u_1, \cdots, u_m} E(s, u_1, \cdots, u_m) = 0$$

where E is a linear combination (with integer coefficients) of products of terms of the form α^β where α and β are variables or particular positive integers. Since there are recursively enumerable sets which are not recursive, there is a non-recursive set $\mathcal{N}$ of positive integers which can be expressed in the form

$$n \in \mathcal{N} \longleftrightarrow \bigvee_{u_1,\cdots,u_m} N(n, u_1, \cdots, u_m) = 0$$

where $N=0$ is an exponential diophantine equation. That is, n belongs to $\mathcal{N}$ if and only if the exponential diophantine equation $N(n, u_1, \cdots, u_m)=0$ is solvable. Hence, if there were a general algorithm for determining the solvability of exponential diophantine equations, we could determine effectively whether or not $N(n, u_1, \cdots, u_m) = 0$ is solvable for a given n and hence whether or not $n \in \mathcal{N}$. But this is impossible, in general, since $\mathcal{N}$ is a non-recursive set. That is:

COROLLARY 1. *There is no general algorithm for determining whether or not a given exponential diophantine equation is solvable.*

Suppose all exponential diophantine equations are numbered effectively according to some definite scheme, and let $\mathcal{E}_i$ be the ith equation in this ordering. Also let x be some fixed variable and let $\mathcal{E}_i(u)$ be the equation obtained from $\mathcal{E}_i$ by substituting u for x throughout $\mathcal{E}_i$. We can easily give an effective procedure for listing all exponential diophantine equations which are solvable. Hence the set of positive integers i such that $\mathcal{E}_i$ has a solution is recursively enumerable. Hence, by Corollary 1, the set of positive integers i such that $\mathcal{E}_i$ has no solution is not recursively enumerable. We will now show that there is an effective procedure which, given a proposed effective listing of "unsolvable" exponential diophantine equations, will yield a particular equation which is either on the list (even though it has a solution) or is omitted from the list (even though it has no solution). The argument given is essentially the one used by Post [6] to prove the existence of creative sets.

Let the proposed list of "unsolvable" equations be $\mathcal{F}_1, \mathcal{F}_2, \cdots$. The set $\mathcal{U}$ of positive integers t such that $\mathcal{E}_t(t)$ is "unsolvable" according to the given list is clearly recursively enumerable, since for every s and t we can compare $\mathcal{E}_t(t)$ with $\mathcal{F}_s$ and put t in $\mathcal{U}$ whenever $\mathcal{E}_t(t)$ is $\mathcal{F}_s$. Hence[3], by our main theorem, there is a positive integer n such that $t \in \mathcal{U}$ if and only if $\mathcal{E}_n(t)$ has a solution. Then $\mathcal{E}_n(n)$ is the required counter-example since $\mathcal{E}_n(n)$ is solvable if and only if it is "unsolvable" according to the proposed list.

COROLLARY 2a. *There is a mechanical procedure which, given any*

[3] This argument has a lacuna at this point in that it is tacitly assumed that, from the possession of an algorithm which effectively enumerates $\mathcal{U}$, we can *effectively* obtain the required integer n. This difficulty can be overcome by verifying that the *proof* of our main theorem as well as the proof of Theorem B of § 3 is constructive.

axiomatization[4] *of number theory, will yield a particular exponential diophantine equation which has no solution, but which cannot be proved to be unsolvable from the given axioms.*

PROOF. We take as the set of "unsolvable" equations those which can be proved unsolvable in the given axiomatization of number theory and list them in order of proof in some systematic way. Then the counter-example corresponding to this list of unsolvable equations must be an equation which is unsolvable but which is not on our list.

COROLLARY 2b. *There is a mechanical procedure which, given any proposed algorithm for testing exponential diophantine equations for solvability, will yield a particular exponential diophantine equation as a counter-example.*

PROOF. Here we take for the list of "unsolvable" equations those equations which are given as "unsolvable" by the proposed algorithm ordered as they appear in the list $\mathcal{E}_1, \mathcal{E}_2, \cdots$.

2. Applications to ordinary diophantine equations

We now combine our main theorem with an earlier result of Julia Robinson (Theorem A below) to investigate the relationship of exponential diophantine equations and their decision problem to ordinary diophantine equations and Hilbert's tenth problem. We also obtain a curious theorem about possible bounds for solutions of ordinary diophantine equations.

Let $x * n$ be the nth super power of x. This can be defined recursively by $x * 1 = x$ and $x * (n + 1) = x^{(x*n)}$.

THEOREM A. *The relation given by $z = x^y$ can be defined existentially in terms of addition, multiplication and any relation $\varphi(u, v)$ satisfying the following two conditions:*

(1) *For some n, $\varphi(u, v)$ implies $v < u * n$.*

(2) *There is no n, such that $\varphi(u, v)$ implies $v < u^n$.*

For the proof of this theorem, see Julia Robinson [8]. Now Theorem A implies that if any relation $\varphi(u, v)$ which satisfies (1) and (2) is diophantine, then the relation given by $z = x^y$ is itself diophantine. This would mean that every exponential diophantine equation could be transformed mechanically into an ordinary diophantine equation with more variables. Hence there could be no algorithm for determining whether or not an arbitrary diophantine equation is solvable. On the other hand, if we could show that some particular recursively enumerable set is not diophantine,

[4] We use here only the two properties: (1) the axiomatization yields an algorithm for proving number-theoretic statements and (2) every exponential diophantine equation proved unsolvable has in fact no solution.

then we would obtain remarkable bounds on every diophantine relation. Corollaries 3 and 4 summarize these remarks and are immediate consequences of our main theorem and Theorem A.

COROLLARY 3. *If any diophantine relation* $\varphi(u, v)$ *satisfies* (1) *and* (2), *then every recursively enumerable set is diophantine and Hilbert's tenth problem is unsolvable* (i.e., *no algorithm exists for determining the solvability of an arbitrary diophantine equation*).

COROLLARY 4. *If there is any recursively enumerable set which is not diophantine, then given any diophantine equation* $P(x, y, u_1, \cdots, u_m) = 0$, *either for every* n *there is a solution with* $y > x * n$ *or there is an* n *such that every solution satisfies* $y < x^n$.

Clearly, the relation given by $v = 2^u \wedge u > 1$ satisfies (1) and (2). Hence the relation $z = x^y$ can be existentially defined in terms of addition, multiplication, and this relation. Since a conjunction or a disjunction of equations can be combined into a single equation using only addition and multiplication, we obtain a polynomial $Q(x, y, z, u_1, \cdots, u_k, v_1, \cdots, v_k)$ with integer coefficients such that

$$z = x^y \longleftrightarrow \bigvee_{u_1,\cdots,u_k} Q(x, y, z, u_1, \cdots, u_k, 2^{u_1}, \cdots, 2^{u_k}) = 0 \ .$$

Then Q can be used to eliminate the general exponential function x^y in terms of addition, multiplication, and the special exponential function 2^x. Hence we obtain the following reduction of exponential diophantine equations.

COROLLARY 5. *Every recursively enumerable set* $\mathcal{S}$ *can be given as the set of values of a parameter* s *for which an equation of the form* $P(s, x_1, \cdots, x_k, 2^{x_1}, \cdots, 2^{x_k}) = 0$, *where* P *is a polynomial with integer coefficients, is solvable in positive integers.*

By a straightforward use of the methods of Putnam [7], we have furthermore:

COROLLARY 6. *For every recursively enumerable set* $\mathcal{S}$, *there is a polynomial* $R(x_1, \cdots, x_n, y_1, \cdots, y_n)$ *with integer coefficients such that the range of the function* $R(x_1, \cdots, x_n, 2^{x_1}, \cdots, 2^{x_n})$ *for positive integral values of* $x_1, \cdots, x_n$ *consists of the members of* $\mathcal{S}$ *and all non-positive integers.*

PROOF. If $\mathcal{S}$ is the empty set, we can take $R = 1 - x_1$. Otherwise let $P(s, x_1, \cdots, x_k, 2^{x_1}, \cdots, 2^{x_k})$ be chosen by Corollary 5 so that

$$s \in \mathcal{S} \longleftrightarrow \bigvee_{x_1,\cdots,x_k} P(s, x_1, \cdots, x_k, 2^{x_1}, \cdots, 2^{x_k}) = 0 \ .$$

Let $Q(s, t, u, x_1, \cdots, x_k) = s(1 - P^2) - (t - 1)(s + u - 1)$. Since Q is a function of the desired form (with $n = k + 3$), it is sufficient to check that the range of Q is the union of $\mathcal{S}$ and the non-positive integers. For $Q > 0$, we must have $P = 0$ and $t = 1$; in which case, $Q = s$ where $s \in \mathcal{S}$. Conversely, if $s \in \mathcal{S}$ and we take $t = 1$, u arbitrary, and $x_1, \cdots, x_k$ so that $P = 0$, then $Q = s$. Finally, choose $s, x_1, \cdots, x_k$ so that $P = 0$ and take $t = 2$, then $Q = 1 - u$ which assumes all non-positive integers.

A rather amusing variation of Corollary 6 is obtained by considering the natural numbers (the non-negative integers) as the domain of the variables:

COROLLARY 7. *Every non-empty recursively enumerable set of natural numbers is the range of a function obtained from particular natural numbers and variables (whose domain is the natural numbers) by addition, subtraction, multiplication and exponentiation.*

PROOF. Again let $P(s, x_1, \cdots, x_k, 2^{x_1}, \cdots, 2^{x_k})$ be the function of Corollary 5 (modified so that the sets and variables range over the natural numbers) corresponding to a given non-empty set $\mathcal{S}$. Let a be any element of $\mathcal{S}$. Then the range of the function

$$(s - a)0^{P^2} + a$$

is exactly $\mathcal{S}$. (Here, it is understood that $0^0 = 1$.)

Finally, we obtain a theorem about possible bounds for solutions of a diophantine equation with a finite number of solutions. As far as we know, it is the first general theorem about bounds for solutions of arbitrary diophantine equations. Since our method of proof is quite roundabout, it seems likely that number theorists can come up with a direct proof and perhaps an improvement.

COROLLARY 8. *Consider diophantine equations $P(x, y, u_1, \cdots, u_m) = 0$ such that to every x there are but a finite number of values of y for which $P = 0$ is solvable. If to every such equation $P = 0$ there are k and n so that every solution of $P = 0$ satisfies $y < k(x * n)$, then to every such equation there are k' and n' so that each solution satisfies $y < k'x^{n'}$.*

PROOF. Suppose some relation φ satisfying (1) and (2) is diophantine. Then every recursively enumerable relation would be diophantine. In particular, there would be a diophantine equation $P(x, y, u_1, \cdots, u_m) = 0$ which is solvable if and only if $y = x * x$. But for such an equation, there could not be super-power bounds on y in terms of x. Hence the hypothesis of the corollary implies that no diophantine relation satisfies (1) and (2). But if $\varphi(u, v)$ is diophantine and $\varphi(u, v)$ implies $v < k(u * n)$, then the relation $\varphi(u, v) \wedge u > 1$ is also diophantine and satisfies (1) for some value of n. Hence it must fail to satisfy (2). That is, for some n', $\varphi(u, v) \wedge u > 1$ implies $v < u^{n'}$. Therefore for some k' and n', $\varphi(u, v)$ implies $v < k'u^{n'}$.

3. Proof of the main theorem

We show first that the binomial coefficient

$$\binom{\alpha}{k} = \alpha(\alpha - 1) \cdots (\alpha - k + 1)/k!$$

is existentially definable in terms of exponentiation for rational values of α greater than k. That is:

LEMMA 1. *The relation given by*

$$x/y = \binom{p/q}{k} \wedge p > qk$$

is exponential diophantine.

PROOF. Let $\alpha = p/q$, $\alpha > k$. Then, by Taylor's theorem with Lagrange's form for the remainder

$$a^{2k+1}(1 + a^{-2})^{\alpha} = \sum_{j=0}^{k} \binom{\alpha}{j} a^{2k-2j+1} + \binom{\alpha}{k+1} a^{-1}(1 + \theta a^{-2})^{\alpha-k-1}$$
$$= \sum_{j=0}^{k} \binom{\alpha}{j} a^{2k-2j+1} + \theta' \alpha^{k+1} a^{-1} 2^{\alpha-1}$$

where θ and θ' are some real numbers between 0 and 1. Let $\alpha = p/q$ and $S_k(a) = \sum_{j=0}^{k} \binom{\alpha}{j} a^{2k-2j+1}$. We wish to choose a so that

$$S_k(a) = [a^{2k+1}(1 + a^{-2})^{p/q}]$$

and

$$S_{k-1}(a) = [a^{2k-1}(1 + a^{-2})^{p/q}] .$$

Here $[\lambda]$ is the greatest integer in λ.

Suppose $q^k k! \mid a$ (i.e., $q^k k!$ divides a), then both $S_k(a)$ and $S_{k-1}(a)$ are integers, since the denominator of $\binom{p/q}{j}$ divides $q^k k!$ for all $j \leqq k$ and a comes in each term of the sums to at least the first power. Also if $a > 2^{p-1} p^{k+1}$, then the remainders in the expressions obtained by Taylor's theorem (for both k and $k - 1$) are less than 1. Hence if we put $a = 2^p p^{k+1} q^k k!$, then

$$u = S_k(a) \longleftrightarrow u \leqq a^{2k+1}(1 + a^{-2})^{p/q} > u + 1$$
$$\longleftrightarrow a^{2p} u^q \leqq a^{(2k+1)q}(a^2 + 1)^p < a^{2p}(u + 1)^q$$

and a similar formula holds for $S_{k-1}(a)$. Therefore both $S_k(a)$ and $S_{k-1}(a)$ are existentially definable in terms of exponentiation and factorial for this particular choice of a. Since the relation $y = x!$ is exponential diophantine, (see Julia Robinson [8]) $u = S_k(a)$ and $v = S_{k-1}(a)$ are also exponential diophantine relations of u, v, p, q, and k. The lemma follows from

$$\binom{p/q}{k} = a^{-1}S_k(a) - aS_{k-1}(a) .$$

LEMMA 2. *The relation* $y = \prod_{k \leq K}(a + bk)$ *is exponential diophantine*[5].

PROOF. $\prod_{k \leq K}(a + bk) = \binom{a/b + K}{K} b^K K!$.

We next recall the following basic theorem[6].

THEOREM B. *Every recursively enumerable set* $\mathcal{S}$ *can be expressed in the form*

$$x \in \mathcal{S} \longleftrightarrow \bigvee_y \bigwedge_{k \leq y} \bigvee_{z_1 \leq y} \cdots \bigvee_{z_m \leq y} P(x, y, k, z_1, \cdots, z_m) = 0$$

where P *is a polynomial with integer coefficients.*

From this it follows that we need only show that every relation of the form

$$\bigwedge_{k \leq y} \bigvee_{z_1 \leq y} \cdots \bigvee_{z_m \leq y} P(x, y, k, z_1, \cdots, z_m) = 0$$

where P is a polynomial, is exponential diophantine, in order to obtain our main theorem. But this last will follow at once from Lemma 2 and the following lemma:

LEMMA 3. *Let* $F(x, y, k, z_1, \cdots, z_m)$ *be any polynomial with integer coefficients. Let* $G(x, y)$ *be any polynomial*[7] *such that* $G(x, y) \geqq y$ *and*

$$\bigwedge_{k \leq y} \bigwedge_{z_1 \leq y} \cdots \bigwedge_{z_m \leq y} | F(x, y, k, z_1, \cdots, z_m) | \leqq G(x, y) .$$

Then

$$\begin{aligned} &\bigwedge_{k \leq y} \bigvee_{z_1 \leq y} \cdots \bigvee_{z_m \leq y} F(x, y, k, z_1, \cdots, z_m) = 0 \\ (3) \qquad &\longleftrightarrow \bigvee_c \bigvee_t \bigvee_{a_1} \cdots \bigvee_{a_m} \{t = G(x, y)! \wedge 1 + ct = \textstyle\prod_{k \leq y}(1 + kt) \\ &\qquad \wedge 1 + ct \mid F(x, y, c, a_1, \cdots, a_m) \wedge \textstyle\bigwedge_{i \leq m} 1 + ct \mid \prod_{j \leq y}(a_i - j)\} . \end{aligned}$$

PROOF. We shall first assume the right side of the equivalence and show that it implies the left side. Let k be any number not exceeding y and p_k be a prime dividing $1 + kt$. Then put

$$\operatorname{Rem}(a_i, p_k) = z_{ki} \qquad \text{for } i = 1, \cdots, m .$$

---

[5] That $\binom{a}{k}$ and $\prod_{k \leq K}(a + bk)$ are exponential diophantine was first proved by Davis and Putnam, extending the methods of Julia Robinson [8]. Note that the lemma remains true if a or b is zero.

[6] Cf., Davis [2], pp. 113-114. R.M. Robinson [9] gave a new proof of this result in which he showed that we can actually take $m = 4$. However, we shall not make use of this fact here.

[7] If $n > 0$ is the degree of F, then we can, for example, take $G(x, y) = cx^n y^n$ where c is the sum of the absolute values of the coefficients of F.

It will be sufficient to show that

$$F(x, y, k, z_{k1}, \cdots, z_{km}) = 0$$

and

$$0 < z_{ki} \leqq y \qquad \text{for all } i \leqq m .$$

Since $p_k \mid \prod_{j \leqq y}(a_i - j)$ we have $p_k \mid a_i - j$ for some j with $1 \leqq j \leqq y$. Thus $1 \leqq z_{ki} = \text{Rem}(a_i, p_k) \leqq y$. Furthermore $(p_k, t) = 1$ and $t = G(x, y)!$, so $p_k > G(x, y) \geqq y$. Hence, since $1 \leqq z_{ki} \leqq y$,

$$| F(x, y, k, z_{k1}, \cdots, z_{km}) | \leqq G(x, y) < p_k .$$

From the equation which determines c, we have $1 + ct \equiv 0 \pmod{1 + kt}$. Hence $c \equiv k \pmod{1 + kt}$ and finally, $c \equiv k \pmod{p_k}$. Therefore

$$\begin{aligned} F(x, y, k, z_{k1}, \cdots, z_{km}) &\equiv F(x, y, c, a_1, \cdots, a_m) && \pmod{p_k} \\ &\equiv 0 && \pmod{p_k} . \end{aligned}$$

Hence $F(x, y, k, z_{k1}, \cdots, z_{km}) = 0$ which was to be proved.

On the other hand, suppose the left side of the equivalence holds. That is, for every $k \leqq y$, there are $z_{k1}, \cdots, z_{km} \leqq y$ such that

$$F(x, y, k, z_{k1}, \cdots, z_{km}) = 0 .$$

We need to show that the right side can be satisfied. Let t and c be determined by the two equations on the right side. Then for every $k, j \leqq y$ and $k \neq j$, $(1 + kt, 1 + jt) = 1$. We see this by noticing that if a prime p divides both $1 + kt$ and $1 + jt$, then p divides $(k - j)t$. But any prime dividing $k - j$ also divides t since $t = G(x, y)!$ and $G(x, y) \geqq y$. Hence $p \mid t$ but then it cannot divide $1 + kt$. So $1 + kt$ and $1 + jt$ are relatively prime. Hence by the Chinese Remainder Theorem, there are $a_1, \cdots, a_m$ so that

$$(4) \qquad a_i \equiv z_{ki} \pmod{1 + kt} \qquad \text{for } k \leqq y \text{ and } i \leqq m .$$

We will show that these $a_1, \cdots, a_m$ satisfy the right side of the equivalence. As before $c \equiv k \pmod{1 + kt}$, so that

$$\begin{aligned} F(x, y, c, a_1, \cdots, a_m) &\equiv F(x, y, k, z_{k1}, \cdots, z_{km}) && \pmod{1 + kt} \\ &\equiv 0 && \pmod{1 + kt} \end{aligned}$$

for $k \leqq y$. Since all the moduli are relatively prime, it follows that

$$\prod_{k \leqq y}(1 + kt) \mid F(x, y, c, a_i, \cdots, a_m) .$$

But from (4) we have

$$1 + kt \mid a_i - z_{ki} \qquad \text{where } z_{ki} \leqq y .$$

Hence

$$1 + kt \mid \prod_{j \leq y}(a_i - j) \qquad \text{for every } k \leqq y \,.$$

But since these divisors are relatively prime, we obtain

$$\prod_{k \leq y}(1 + kt) \mid \prod_{j \leq y}(a_i - j) \qquad \text{for every } i \leqq m \,.$$

Hence the right side of the equivalence holds and the lemma follows.

In order to complete the proof of our theorem, we need only check that the relations appearing on the right side of the equivalence in Lemma 3 are exponential diophantine.

We have already seen that the relations $u = v!$ and $u = \prod_{k \leqq y}(1 + kt)$ are exponential diophantine. The relation $u \mid v$ is of course equivalent to $\bigvee_w uw = v$. Finally,

$$1 + ct \mid \prod_{j \leqq y}(a_i - j) \longleftrightarrow a_i \leqq y \vee \bigvee_u\{a_i > y \wedge (1 + ct)u = \prod_{j \leqq y}(a_i - y - 1 + j)\} \,.$$

Hence the relations appearing on the right side of (3) are exponential diophantine. Thus, every relation of the form

$$\bigwedge_{k \leq y} \bigvee_{z_1 \leq y} \cdots \bigvee_{z_m \leq y} F(x, y, k, z_1, \cdots, z_m) = 0 \,,$$

where F is a polynomial with integer coefficients, is exponential diophantine. Finally, by Theorem B, every recursively enumerable set is exponential diophantine.

The fact that every recursively enumerable relation $\rho(x_1, \cdots, x_n)$ is exponential diophantine follows from the result for sets. For, let the set $\mathcal{R}$ be defined by the equivalence:

$$r \in \mathcal{R} \longleftrightarrow \bigvee_{x_1, \cdots, x_n}(r = 2^{x_1}3^{x_2} \cdots p_n^{x_n} \wedge \rho(x_1, \cdots, x_n))$$

where p_n is the nth prime. If ρ is a recursively enumerable relation, then $\mathcal{R}$ is a recursively enumerable set. Hence $\mathcal{R}$ is exponential diophantine. But

$$\rho(x_1, \cdots, x_n) \longleftrightarrow 2^{x_1}3^{x_2} \cdots p_n^{x_n} \in \mathcal{R} \,.$$

Hence ρ is also exponential diophantine.

NEW YORK UNIVERSITY
PRINCETON UNIVERSITY
UNIVERSITY OF CALIFORNIA, BERKELEY

BIBLIOGRAPHY

1. MARTIN DAVIS, *Arithmetical problems and recursively enumerable predicates*, J. Symb. Logic, 18 (1953), 33-41.
2. ———, Computability and Unsolvability, McGraw-Hill, 1958.
3. ——— and HILARY PUTNAM, *Reduction of Hilbert's tenth problem*, J. Symb. Logic, 23 (1958), 183-187.

4. DAVID HILBERT, *Mathematische Probleme. Vortrag, gehalten auf dem internationalen Mathematiker-Kongress zu Paris 1900.* Nachr. K. Ges. Wiss. Göttingen, Math.-Phys. Kl. 1900, 253-297. Reprinted, Arch. Math. Phys., 3s, 1 (1901), 44-63, 213-237. English translation, Bull. Amer. Math. Soc., 8 (1901-1902), 437-479.
5. S. C. KLEENE, Introduction to Metamathematics, D. Van Nostrand Co., Inc., 1952.
6. E. L. POST, *Recursively enumerable sets of positive integers and their decision problems,* Bull. Amer. Math. Soc., 50 (1944), 284-316.
7. HILARY PUTNAM, *An unsolvable problem in number theory,* to appear in the J. Symb. Logic.
8. JULIA ROBINSON, *Existential definability in arithmetic,* Trans. Amer. Math. Soc., 72 (1952), 437-449.
9. R. M. ROBINSON, *Arithmetical representation of recursively enumerable sets,* J. Symb. Logic, 21 (1956), 162-186.

THE UNDECIDABILITY OF EXPONENTIAL DIOPHANTINE EQUATIONS

JULIA ROBINSON

University of California, Berkeley, California, U.S.A.

Hilbert's tenth problem asks for an algorithm to determine whether or not an arbitrary Diophantine equation has a solution in integers. The corresponding problem for exponential Diophantine equations (i.e., equations in which exponentiation as well as addition and multiplication is permitted) can now be answered negatively: *There is no algorithm to determine whether or not an arbitrary exponential Diophantine equation is solvable in positive integers.* This result is an immediate consequence of the following theorem:[1]

Every recursively enumerable set (or relation) is exponential Diophantine.

That is, to every recursively enumerable set S of positive integers, there is an exponential Diophantine equation $E(s, u_1, \ldots, u_m)$ such that $s \in S$ if and only if $E(s, u_1, \ldots, u_m)$ is solvable in positive integers.

The proof of this result is constructive. Given a recursive enumeration of S in any of the standard forms, a corresponding exponential Diophantine equation $E(s, u_1, \ldots, u_m)$ can be constructed. Hence we can use the argument of Post [2] to obtain two corollaries:

There is a mechanical procedure which, given any axiomatization of number theory, will yield a particular exponential Diophantine equation which has no solution but which cannot be proved to be unsolvable from the given axioms.

There is a mechanical procedure which, given any proposed algorithm for testing exponential Diophantine equations for solvability, will yield a particular exponential Diophantine equation for which the algorithm gives the wrong answer.

It is not known whether exponential Diophantine sets are necessarily Diophantine. However it was shown in [3] that every exponential Diophantine equation could be transformed mechanically into an equivalent ordinary Diophantine equation in more unknowns provided that there is a Diophantine equation $D(x, y, u_1, \ldots, u_m)$ such that (1) all solutions of D satisfy $y < x^x$ and (2) for every n, there is some solution with $y > x^n$.

Hence if such a Diophantine equation exists, then every recursively enumerable set would be Diophantine and Hilbert's tenth problem would be unsolvable. On the other hand, if some particular recursively enumerable set is not Diophantine, we would obtain surprising bounds on solutions of Diophantine equations. In fact, a sharpening of the above result yields:

---

[1] Martin Davis and Hilary Putnam [1] obtained this theorem under the assumption that there exist arbitrarily long arithmetic progressions containing only prime numbers. In [4], I later extended and modified their argument to obtain the theorem without using the hypothesis. The proofs of the results in this summary will appear in a joint paper by Davis, Putnam, and myself in the *Annals of Mathematics*.

If there is any recursively enumerable set which is not Diophantine, then each Diophantine equation $F(x, y, u_1, \ldots, u_m)$ either has a solution with y greater than the nth super power of x for every n or there is an n such that every solution satisfies $y < x^n$.

References

[1] DAVIS, M., and H. PUTNAM. On Hilbert's tenth problem. *Notices of the American Mathematical Society*, Vol. 6 (1959), p. 544.

[2] POST, E. L. Recursively enumerable sets of positive integers and their decision problems. *Bulletin of the American Mathematical Society*, Vol. 50 (1944), pp. 284–316.

[3] ROBINSON, JULIA. Existential definability in arithmetic. *Transactions of the American Mathematical Society*, Vol. 72 (1952), pp. 437–449.

[4] ROBINSON, JULIA. The undecidability of exponential Diophantine equations. *Notices of the American Mathematical Society*, Vol. 7 (1960), p. 75.

42

On the Decision Problem for Algebraic Rings

Julia Robinson

Our main result is that the ring of totally real algebraic integers is undecidable.* That is, there is no general method of mechanically determining the truth of statements of the arithmetic of totally real algebraic integers (i.e., algebraic integers that together with all their conjugates are real).

The classical work of Gödel, Church, and Rosser established that the arithmetic of natural numbers (and hence the ring $\boldsymbol{I}$ of rational integers) is undecidable. Hence any ring $\boldsymbol{R}$ in which the natural numbers (or a model of the arithmetic of natural numbers) can be defined arithmetically is itself undecidable. For in either case there is a mechanical method of translating an arbitrary sentence of the arithmetic of natural numbers into an equivalent sentence of $\boldsymbol{R}$. Thus the arithmetic of $\boldsymbol{R}$ contains all the mystery of number theory. This has been the method used to prove the undecidability of rings† except for $\boldsymbol{I}$ itself.

An arithmetical definition of the natural numbers in the field of rational numbers was given in [4] to show that the arithmetic of rationals is undecidable. R. M. Robinson [6] showed that the ring of integers of any quadratic field or of any cubic or quartic field with one fundamental unit is undecidable. These results were extended in [5] to prove that any finite extension field over the rationals, as well as the corresponding ring of integers, is undecidable.

In addition, R. M. Robinson [6] showed that various polynomial rings and rings of formal power series are undecidable. Malcev [2] has recently proved the undecidability of rational functions of one variable over a real closed field.

On the other hand, the fundamental work of Tarski [9] includes a decision

* The metamathematical terms and results used here are explained in [10]. For a discussion applied to rings, see [6].

† It is of almost no interest that undecidable rings can be artificially constructed, for example, by adjoining the square roots of a non-recursive set of prime numbers to the ring of rational integers.

method for the arithmetic of real closed fields showing that the field of all real algebraic numbers, as well as the field of all algebraic numbers, is decidable.

In this paper, we shall study the decision problem for the ring of integers of a field of algebraic numbers (often of infinite degree over the rationals). These rings will be called *algebraic integer rings*. The degree of an algebraic integer ring is the degree of its quotient field.

A family $\mathscr{F}$ of sets of numbers of a ring $\boldsymbol{R}$ is *defined arithmetically* by an arithmetical formula $\Phi(r, r_1, \cdots, r_k)$ of $\boldsymbol{R}$ if and only if

$$\varphi \in \mathscr{F} \leftrightarrow (\bigvee \mathfrak{r})(\bigwedge r)[r \in \varphi \leftrightarrow \Phi(r, \mathfrak{r})] .$$

Here $\mathfrak{r}$ stands for $r_1, \cdots, r_k$. Throughout, we use the corresponding German letter to stand for a finite sequence of Latin letters, relying on the context to determine the appropriate length. By a family of sets, we shall always mean a family of sets of numbers of the ring under discussion.

We shall show that the undecidability of an algebraic integer ring follows from the arithmetical definability in the ring of a family of sets with certain properties.

THEOREM 1. *If a family $\mathscr{F}$ of finite sets including all finite initial segments of the natural numbers is arithmetically definable in an algebraic integer ring $\boldsymbol{R}$, then the set of natural numbers $\boldsymbol{N}$ is arithmetically definable in $\boldsymbol{R}$, and $\boldsymbol{R}$ is undecidable.*

PROOF. Clearly,

$$n \in \boldsymbol{N} \leftrightarrow (\bigvee \varphi \in \mathscr{F})[0 \in \varphi \wedge (\bigwedge r \in \varphi)(r \in \varphi \rightarrow r + 1 \in \varphi \vee r = n)] . \tag{1}$$

If $\mathscr{F}$ is arithmetically defined by $\Phi(r, \mathfrak{r})$ in $\boldsymbol{R}$, then we can obtain an arithmetical definition of $\boldsymbol{N}$ by making the usual translation:

$$n \in \boldsymbol{N} \leftrightarrow (\bigvee \mathfrak{r})\{\Phi(0, \mathfrak{r}) \wedge (\bigwedge r)[\Phi(r, \mathfrak{r}) \rightarrow \Phi(r + 1, \mathfrak{r}) \vee r = n]\} . \tag{2}$$

Before applying Theorem 1, we need two lemmas.

LEMMA 1. *Let α be a finite set of non-zero algebraic integers and t be a positive integer. Then there is a positive integer g that can be taken arbitrarily large so that t and all numbers $1 + ag$ for $a \in \alpha$ are relatively prime and the numbers $1 + ag$ are not units or zero.*

PROOF. To ensure the condition of relative primeness, it is sufficient to choose g to be a multiple of t and of all the differences $a - a'$ for distinct a and $a' \in \alpha$. Then t and $1 + ag$ are certainly relatively prime. Also, if an algebraic integer x divides both $1 + ag$ and $1 + a'g$, then $x \mid (a - a')g$, so $\sqrt{x} \mid g$. But then $x \mid 1$, so x is a unit.

Now since the numbers of α together with their conjugates form a finite set of non-zero algebraic integers, if we take g to be a positive integer sufficiently large, then all the numbers $1 + ag$ for $a \in \alpha$ and their conjugates will lie outside the unit circle and hence are not units or zero. Thus, if we take for g a sufficiently large positive integer divisible by t and all the differences $a - a'$ for a and $a' \in \alpha$, and $a \neq a'$, then the conditions of the lemma will be satisfied.

LEMMA 2. *If a family $\mathscr{F}$ of sets of an algebraic integer ring is arithmetically definable, so also is a family $\mathscr{G}$ of subsets of the sets of $\mathscr{F}$ that contains all the subsets of the finite sets of $\mathscr{F}$.*

PROOF. Consider the family $\mathscr{G}$ of sets defined by

$$\Phi(r, \mathfrak{r}) \wedge 1 + rr_{k+1} \mid r_{k+2} \wedge r \neq r_{k+3} .$$

Certainly, $\mathscr{G}$ is a family of subsets of sets of $\mathscr{F}$. Now suppose $\rho \in \mathscr{F}$ and ρ is finite. By Lemma 1, choose r_{k+1} so that the numbers $1 + rr_{k+1}$ are relatively prime non-units except for $r = 0$, and are not zero. Let $\sigma \subset \rho$ and set $r_{k+2} = \prod_{r \in \sigma}(1 + rr_{k+1})$. Then for this choice of r_{k+1} and r_{k+2},

$$\Phi(r, \mathfrak{r}) \wedge 1 + rr_{k+1} \mid r_{k+2} \leftrightarrow r \in \sigma \vee (r = 0 \wedge 0 \in \rho - \sigma) .$$

Finally, by taking $r_{k+3} = 0$, we obtain the subsets of ρ without 0, and by taking $r_{k+3} \neq 0$ we obtain those subsets (if any) containing 0.

COROLLARY. *The natural numbers can be defined arithmetically in any algebraic integer ring of finite degree by a particular arithmetical formula that does not depend on the ring. Hence the theory of algebraic integer rings of finite degree is undecidable.*

PROOF. A theorem of Pólya [3] states that if $f(x)$ is an irreducible quadratic polynomial with integer coefficients, then the greatest prime divisor of $f(n)$ tends to infinity with n. Siegel [7, pp. 204–5] generalized Pólya's theorem to algebraic integer rings of finite degree arbitrary polynomials. *Let $\boldsymbol{R}$ be an algebraic integer ring of finite degree and t be a positive integer. Given a polynomial $f(x)$ with coefficients in $\boldsymbol{R}$ such that $f(x) = 0$ has at least two distinct roots (not necessarily in $\boldsymbol{R}$), there are only finitely many $x \in \boldsymbol{R}$ such that $N\mathfrak{p} < t$ for all prime ideals $\mathfrak{p}$ that divide $f(x)$.*

Let $\mathscr{S}$ be the family of sets defined by $\Sigma(r, s)$ where

$$\Sigma(r, s) \leftrightarrow \{[r(r + 1) \mid s \wedge s \neq 0] \vee r = 0 \vee r = -1\} , \tag{3}$$

in an algebraic integer ring $\boldsymbol{R}$ of finite degree. By Siegel's theorem applied to the polynomial $f(x) = x(x + 1)$, there are at most a finite number of $r \in \boldsymbol{R}$ satisfying $\Sigma(r, s)$ for a fixed s, so that $\mathscr{S}$ is a family of finite sets. Let ρ be any finite set not containing 0 or -1 and set $s = \prod_{r \in \rho} r(r + 1)$. Then

$$(\bigwedge r \in \rho)\Sigma(r, s) .$$

Since every set of $\mathscr{S}$ contains 0 and -1, we see that every finite set of $\boldsymbol{R}$ is a subset of a set in $\mathscr{S}$. Hence, by Lemma 2, the family $\mathscr{F}$ of all finite sets of $\boldsymbol{R}$ can be defined. In fact, the arithmetical formula $\Phi(r, r_1, r_2, r_3, r_4)$ given by

$$\Sigma(r, r_1) \wedge 1 + rr_2 \mid r_3 \wedge r \neq r_4 , \tag{4}$$

where Σ is defined by (3), defines the family of all finite sets of $\boldsymbol{R}$.

Hence by Theorem 1, the natural numbers are arithmetically definable. Indeed, we may use $\Phi(r, \mathfrak{r})$ defined by (4) and (3) in (2) to obtain an arithmetical formula that defines the natural numbers whenever the range of the variables is an algebraic integer ring of finite degree.

Thus any algebraic integer ring of finite degree is undecidable. Also, the theory of such rings is undecidable; i.e., there is no general method of determining whether or not a given arithmetical statement holds in all algebraic integer rings of finite degree.

We now turn to totally real rings and obtain a less stringent condition on the family of sets needed to show the ring undecidable.

THEOREM 2. *If a family $\mathscr{F}$ of sets of a totally real algebraic integer ring $\boldsymbol{R}$ containing arbitrarily large finite sets can be arithmetically defined in $\boldsymbol{R}$, then the natural numbers can be defined arithmetically in $\boldsymbol{R}$. Hence $\boldsymbol{R}$ is undecidable.*

A totally real number x is *totally non-negative* ($x \gg 0$) if and only if x and all its conjugates are non-negative. We write $x \ll y$ to indicate that $y - x$ is totally non-negative.

Siegel [8] proved that a number in a totally real algebraic field is the sum of four squares of numbers of the field if and only if it is totally non-negative. Hence in any totally real algebraic integer ring,

$$x \ll y \leftrightarrow (\mathsf{V}t, u, v, w, z)[t^2(y - x) = u^2 + v^2 + w^2 + z^2 \wedge t \neq 0] .$$

In the ring of all totally real integers, we have the simpler formula

$$x \ll y \leftrightarrow (\mathsf{V}z)(y - x = z^2) .$$

Thus the relation $x \ll y$ can be arithmetically defined in any totally real algebraic integer ring.

We next show that in a totally real ring we can define arithmetically a unique representative of certain residue classes.

LEMMA 3. *Given two totally real integers m and t, there is at most one totally real integer a such that $t \equiv a \bmod m$ and $0 \ll a \ll m - 1$.*

PROOF. Suppose

$$a \equiv b \bmod m \wedge 0 \ll a \ll m - 1 \wedge 0 \ll b \ll m - 1 ,$$

where a, b, and m are totally real integers. Let l be the degree of the smallest totally real algebraic integer ring $\boldsymbol{R}$ containing a, b, and m. Also, let $x^{(1)}, \cdots, x^{(l)}$ be the l conjugates of x for $x \in \boldsymbol{R}$ in some fixed order. Then

$$\begin{aligned} &0 \leqq a^{(k)} < m^{(k)} \\ &0 \leqq b^{(k)} < m^{(k)} \qquad (k = 1, \cdots, l) . \end{aligned}$$

Since $a \equiv b \bmod m$, there exists $v \in \boldsymbol{R}$ such that $a = b + mv$. Hence

$$| m^{(k)} v^{(k)} | = | a^{(k)} - b^{(k)} | < m^{(k)} .$$

Thus $| v^{(k)} | < 1$ $(k = 1, \cdots, l)$, so that $v = 0$ and $a = b$.

PROOF OF THEOREM 2. Without loss of generality, we take the sets of $\mathscr{F}$

to contain only totally non-negative numbers. Otherwise, we could define the family of sets whose elements are the squares of those of sets of $\mathscr{F}$. Let

$$\Sigma(a, \varphi, t, g) \leftrightarrow (\bigvee f \in \varphi)(0 \ll a \ll fg \wedge t = a \bmod 1 + fg) ;$$

then

$$(5) \qquad n \in \boldsymbol{N} \leftrightarrow (\bigvee \varphi \in \mathscr{F})(\bigvee t, g)\{\Sigma(n, \varphi, t, g) \wedge (\bigwedge a)[\Sigma(a, \varphi, t, g) \rightarrow a = 0 \vee \Sigma(a-1, \varphi, t, g)]\}$$

is a definition of the natural numbers that can be transformed into an arithmetical definition in the usual way, i.e., as we obtained (2) from (1).

For any n, there are only a finite number of totally non-negative numbers in the sequence $n, n-1, n-2, \cdots$. Thus, if n satisfies the right side of (5), the sequence $n, n-1, n-2, \cdots$ must terminate in 0. Hence n is a natural number.

On the other hand, suppose n is a natural number. Let φ be a finite set in $\mathscr{F}$ with more than n elements. We shall find t and g such that

$$(6) \qquad \sum (a, \varphi, t, g) \leftrightarrow a = 0 \vee a = 1 \vee \cdots \vee a = n .$$

By Lemma 1, we can choose a positive integer g such that the numbers $1+fg$ for $f \in \varphi$ are relatively prime non-units and g so large that

$$n \ll fg \qquad (f \in \varphi, f \neq 0) .$$

(This last condition can be satisfied because all $f \in \varphi$ are totally non-negative.)

Let $f_1, \cdots, f_n$ be distinct positive elements of φ. By the Chinese remainder theorem, we can choose t such that

$$t \equiv j \bmod 1 + f_j g \qquad \text{for} \quad j = 1, \cdots, n ,$$
$$t \equiv 0 \bmod 1 + fg \qquad \text{for} \quad \text{all other } f \text{ in } \varphi .$$

Now for this choice of φ, t, and g, we see that $\Sigma(0, \varphi, t, g)$, $\Sigma(1, \varphi, t, g), \cdots$, $\Sigma(n, \varphi, t, g)$ all hold. By Lemma 3, these are the only values of a such that $\Sigma(a, \varphi, t, g)$ holds. Hence (6) holds, so that n satisfies the right side of (5), which was to be proved.

COROLLARY. *The natural numbers can be defined arithmetically in any totally real algebraic integer ring* $\boldsymbol{R}$ *such that there is a smallest interval* $(0, s)$, s *real or* ∞, *which contains infinitely many sets of conjugates of numbers of* $\boldsymbol{R}$, *i.e., infinitely many* $x \in \boldsymbol{R}$ *with* $0 \ll x \ll s$.

Remark. This condition may in fact hold for all totally real algebraic integer rings.

PROOF. The corollary follows immediately from Theorem 2, since the family of sets defined by

$$\Phi(r, r_1, r_2) \leftrightarrow 0 \ll qf \ll p$$

satisfies the hypothesis of the theorem because we can choose positive integers p and q so that p/q is as close as we like to but less than s. We need not worry about defining s itself, since the hypothesis of Theorem 2 does not demand that all the sets of $\mathscr{F}$ are finite.

COROLLARY. *The natural numbers can be defined in the ring of all totally real algebraic integers and hence it is undecidable.*

PROOF. It follows from a theorem of Kronecker [1] that the interval $(0, 4)$ contains infinitely many sets of conjugates of totally real algebraic integers and that no sub-interval does.

Next we turn to the ring $\boldsymbol{R}$ of integers of the field obtained from the rational field $\boldsymbol{Q}$ by adjoining the square roots of positive integers. Let $\boldsymbol{Q}_n$ be the field $\boldsymbol{Q}(\sqrt{2}, \sqrt{3}, \sqrt{5}, \cdots, \sqrt{p_n})$ and $\boldsymbol{R}_n$ the ring of integers of $\boldsymbol{Q}_n$.

Since the square roots of square-free positive integers are linearly independent with respect to rational coefficients, each number of $\boldsymbol{R}_n$ can be written uniquely as $A + B\sqrt{p_n}$, where A and B are in $\boldsymbol{Q}_{n-1}$. It can be shown by induction that $2A$ and $2B$ must be in $\boldsymbol{R}_{n-1}$.

Let $\lceil x \rceil$ be the maximum absolute value of the conjugates of x. We shall now show that for $t > 0$ the inequality $\lceil x \rceil < t$ has only a finite number of solutions with $x \in \boldsymbol{R}$. Suppose $x \in \boldsymbol{R}_n$ such that $x = \frac{1}{2}(a + b\sqrt{p_n})$ with $a, b \in \boldsymbol{R}_{n-1}$. The conjugates of x in $\boldsymbol{R}_n$ are given by

$$\tfrac{1}{2}(a^{(k)} + b^{(k)}\sqrt{p_n})$$

and

$$\tfrac{1}{2}(a^{(k)} - b^{(k)}\sqrt{p_n}) \qquad (k = 1, 2, \cdots, 2^{n-1}),$$

where $a^{(k)}$ and $b^{(k)}$ are corresponding conjugates of a and b in $\boldsymbol{R}_{n-1}$.

Thus $\lceil x \rceil \geqq \frac{1}{2}\lceil b \rceil \sqrt{p_n}$. Hence, if $\frac{1}{2}\sqrt{p_n} \geqq t$, then $\lceil x \rceil \geqq \lceil b \rceil t \geqq t \vee b = 0$. Therefore all solutions of $\lceil x \rceil < t$ must lie in $\boldsymbol{R}_k$, where k is the largest number such that the kth prime p_k satisfies $\frac{1}{2}\sqrt{p_k} < t$. But $\boldsymbol{R}_k$ has finite degree, so there are only a finite number of solutions of $\lceil x \rceil < t$ in $\boldsymbol{R}_k$ and hence in $\boldsymbol{R}$ itself.

Hence the smallest interval $(0, s)$ that contains infinitely many sets of conjugates of integers of $\boldsymbol{R}$ is $(0, \infty)$. This is also true of any algebraic integer ring included in $\boldsymbol{R}$. Thus we have proved the following corollary:

COROLLARY. *The natural numbers can be defined in the ring of integers of any field contained in $\boldsymbol{Q}(\sqrt{2}, \sqrt{3}, \sqrt{5}, \cdots)$, and hence all such rings are undecidable.*

Returning to rings that are not necessarily totally real, we prove a final theorem in which both hypothesis and conclusion are weaker than in Theorem 1.

THEOREM 3. *A model of the arithmetic of natural numbers can be defined arithmetically in any algebraic integer ring $\boldsymbol{R}$ such that a family $\mathscr{F}$ of finite sets containing arbitrarily large finite sets is arithmetically definable in $\boldsymbol{R}$.*

PROOF. Let $\rho, \sigma, \tau, \cdots$ be finite sets of non-zero numbers of $\boldsymbol{R}$, and $N\rho$ be the number of elements of ρ.

A. $N\rho = N\sigma$ if there exist g and m such that $1 + sg \mid m - r$ establishes a one-to-one correspondence between the elements r of ρ and s of σ.

PROOF. Since any one-to-one correspondence yields $N\rho = N\sigma$, it is sufficient to assume $N\rho = N\sigma$ and to find g and m such that $1 + sg \mid m - r$ gives such a correspondence. By Lemma 1, choose g so that the numbers $1 + sg$ for $s \in \sigma$ are relatively prime non-units prime to some positive integer that is a multiple of all differences $r - r'$ of distinct pairs of elements of ρ. Let $f(s)$ be some one-to-one function from σ to ρ. By the Chinese remainder theorem, we can choose m satisfying

$$m \equiv f(s) \mod 1 + sg \quad \text{for} \quad s \in \sigma ,$$

since the moduli are relatively prime. Also, for each $s \in \sigma$, there is exactly one $r \in \rho$ such that $1 + sg \mid m - r$, since $1 + sg \nmid r - r'$ for distinct numbers $r, r' \in \rho$. Therefore, $1 + sg \mid m - r$ does establish the desired one-to-one correspondence.

B. $$N\rho + N\sigma = N\tau \leftrightarrow (\bigvee a)(\bigwedge s \in \sigma)(\bigwedge r \in \rho)(r \neq s + a \wedge a \neq -s \wedge N\tau = N\{r : r \in \rho \wedge r - a \in \sigma\}) .$$

PROOF. Clear since ρ and σ are finite sets.

C. $N\rho \cdot N\sigma = N\tau$ if there exists an a such that the numbers $r + sa$ for $r \in \rho$ and $s \in \sigma$ are all distinct and not zero, and $N\tau = N\{r + sa : r \in \rho \wedge s \in \sigma\}$.

PROOF. We need only show that we can choose a such that the numbers $r + sa$ for $r \in \rho$ and $s \in \sigma$ will be distinct and not zero. Let a be a natural number such that

$$\max_{\substack{r, r' \in \rho \\ r \neq r'}} | r - r' | < \min_{\substack{s, s' \in \sigma \\ s \neq s'}} a \, | s - s' | .$$

Now $r + sa = r' + s'a$ implies $| r - r' | = a \, | s - s' |$, so that $r = r'$ and $s = s'$.

Returning to the proof of Theorem 3, we can by Lemma 2 define a family $\mathscr{G}$ of all subsets that do not contain 0 of sets of $\mathscr{F}$. $\mathscr{G}$ will, of course, contain sets with exactly n elements for every natural number n.

Our model will be constructed by identifying the natural number n with the sets of $\mathscr{G}$ that have exactly n elements. Every statement of the arithmetic of natural numbers can mechanically be put in the form

$$(W_1 n_1) \cdots (W_m n_m) \Theta(n_1, \cdots, n_m) , \tag{7}$$

where $W_1, \cdots, W_m$ is some sequence of quantifiers binding the variables $n_1, \cdots, n_m$ whose range is the natural numbers, and Θ is a conjunction of equations of the three types,

$$n_i = n_j , \qquad n_i + n_j = n_k , \qquad n_i \cdot n_j = n_k . \tag{8}$$

Proposition (7) is clearly equivalent to

$$(W_1 \rho_1 \in \mathscr{G}) \cdots (W_m \rho_m \in \mathscr{G}) \Theta(N\rho_1, \cdots, N\rho_m) . \tag{9}$$

By using A, B, and C, we can write formulas $\Phi_1(\rho, \sigma)$, $\Phi_2(\rho, \sigma, \tau)$, and

$\Phi_3(\rho, \sigma, \tau)$, which differ from arithmetical formulas of $\boldsymbol{R}$ only in that some of the variables may be restricted to ρ, σ, or τ, and such that

$$\Phi_1(\rho, \sigma) \leftrightarrow N\rho = N\sigma ,$$
$$\Phi_2(\rho, \sigma, \tau) \leftrightarrow N\rho + N\sigma = N\tau ,$$
$$\Phi_3(\rho, \sigma, \tau) \leftrightarrow N\rho \cdot N\sigma = N\tau .$$

Then if we substitute Φ_1, Φ_2, and Φ_3 for the corresponding equations in Θ in (9), it becomes

$$(W_1\rho_1 \in \mathscr{G}) \cdots (W_m\rho_m \in \mathscr{G})\Gamma , \tag{10}$$

where Γ differs from an arithmetical formula of $\boldsymbol{R}$ only in that some of its bound variables are restricted to $\rho_1, \cdots, \rho_m$. Finally, by introducing the arithmetical definition of the family $\mathscr{G}$ and eliminating $\rho_1, \cdots, \rho_m$ from (10), we obtain an arithmetical formula of $\boldsymbol{R}$ equivalent to the arbitrary proposition of number theory (7). Hence we have succeeded in defining a model of number theory in $\boldsymbol{R}$, so $\boldsymbol{R}$ is undecidable.

Kensington, California

References

[1] KRONECKER, L., Zwei Sätze über Gleichungen mit ganzzahligen Coefficienten, *J. Reine Angew. Math.*, **53** (1857), 173-75.

[2] MALCEV, A. I., On the Undecidability of the Elementary Theory of Certain Fields (Russian), *Sibirskiĭ Mat. Z.*, **1** (1960), 71-77.

[3] PÓLYA, GEORGE, Zur arithmetischen Untersuchung der Polynome, *Math. Z.*, **1** (1918), 143-48.

[4] ROBINSON, JULIA, Definability and Decision Problems in Arithmetic, *J. Symbol. Logic*, **14** (1949), 98-114.

[5] ROBINSON, JULIA, The Undecidability of Algebraic Rings and Fields, *Proc. Amer. Math. Soc.*, **10** (1959), 950-57.

[6] ROBINSON, R. M., Undecidable Rings, *Trans. Amer. Math. Soc.*, **70** (1951), 137-59.

[7] SIEGEL, CARL, Approximation algebraischer Zahlen, *Math. Z.*, **10** (1921), 173-213.

[8] SIEGEL, CARL, Darstellung total positiver Zahlen durch Quadrate, *Math. Z.*, **11** (1921) 246-75.

[9] TARSKI, ALFRED, *A Decision Method for Elementary Algebra and Geometry*, 2d ed. Berkeley: Univ. California Press, 1951.

[10] TARSKI, ALFRED, A. MOSTOWSKI, and R. M. ROBINSON, *Undecidable Theories*. Amsterdam: North-Holland Publishing Company, 1953.

Symposium on the Theory of Models; North-Holland Publ. Co., Amsterdam (1965)

THE DECISION PROBLEM FOR FIELDS

JULIA ROBINSON
University of California, Berkeley, California, U.S.A.

1. Introduction. One aspect of the theory of models which has been developing in the last 30 years is the decision problem for various mathematical theories and the related problems of definability. The decision problem for fields furnishes an excellent illustration of the methods and results.

The decision problem for a given field $\mathfrak{F}$ is the problem of determining whether or not the set Σ of true (first-order) sentences of $\mathfrak{F}$ is recursive. If Σ is recursive, we say $\mathfrak{F}$ is *decidable*; if not, we say $\mathfrak{F}$ is *undecidable*. More generally, the decision problem for a class $\mathscr{C}$ of fields is the problem of determining whether or not the set Γ of sentences true in all fields of $\mathscr{C}$ is recursive. The known decidable fields are:

D1. any finite field,
D2. any real closed field,
D3. any algebraically closed field.

The known undecidable fields are:

U1. the rational field,
U2. any finite extension of the rational field,
U3. any pure transcendental extension of a formally real field [1],
U4. certain fields of formal power series over an undecidable field (in particular the field of formal power series over the rational field),
U5. certain fields whose definitions depend on non-recursive sets,
U6. fields elementarily equivalent to any of the above fields.

We will discuss these results and single out some of the open problems which are either of particular interest or appear likely of solution. We consider formulas in the first-order language L with equality whose mathematical symbols are $+$, $\cdot$, 0, and 1. Thus a field $\mathfrak{F}$ is a structure

[1] In general, we will omit the words 'pure' and 'formally'. By a *transcendental extension*, we will always mean a *pure* transcendental extension and by a *real field*, we will always mean a *formally* real field.

$<F, +, \cdot, 0, 1>$ which satisfies the usual finite set of axioms. We say a set S is *definable* in $\mathfrak{F}$ if there is a formula φ with one free variable such that a satisfies φ in $\mathfrak{F}$ if and only if $a \in S$. Any formula ψ with free variables $x_1, \ldots, x_n$ is equivalent to a formula of the form

$$\underset{u_1}{X_1} \ldots \underset{u_k}{X_k} \quad P(x_1, \ldots, x_n, u_1, \ldots, u_k) = Q(x_1, \ldots, x_n, u_1, \ldots, u_k)$$

where $X_1, \ldots, X_k$ are quantifiers, P and Q are polynomials with natural number coefficients (i.e., sums of 1's). We can reduce ψ to this form by first putting ψ in prenex form, then applying the following equivalences as needed, and finally transposing the terms so that they have positive coefficients:

$$u = 0 \vee w = 0 \leftrightarrow u \cdot w = 0,$$

$$u = 0 \wedge w = 0 \leftrightarrow \bigwedge_{x,y} u \cdot x = w \cdot y,$$

$$u \neq 0 \leftrightarrow \bigvee_{x} u \cdot x = 1.$$

Hence any set definable in $\mathfrak{F}$ can be obtained from an algebraic surface in an appropriate space over $\mathfrak{F}$ by taking projections and complements.

2. Decidable fields. A finite field $\mathfrak{F}$ is clearly decidable since any sentence can be systematically reduced to a Boolean combination of sentences of the diagram of $\mathfrak{F}$, which is finite.

A field is *formally real* if and only if -1 is not the sum of squares of elements of the field. A formally real field is *real closed* if and only if it has no proper algebraic extension which is formally real. Examples are the field of real numbers and the field of real algebraic numbers. This definition of a real closed field does not appear to be a first-order property of fields. However Artin and Schreier [27], who introduced the concept, also showed that a real closed field $\mathfrak{F}$ could be characterized by the conditions:

1. $\mathfrak{F}$ is a formally real field;
2. every element of $\mathfrak{F}$ is either a square or the negative of a square;
3. every polynomial equation in one variable over $\mathfrak{F}$ of odd degree has a solution in $\mathfrak{F}$.

These conditions can be written without difficulty as an infinite set of sentences of L.

A real closed field can be ordered in one and only one way. For, in a real closed ordered field, we must have true

$$\bigwedge_{x,y} (x<y \leftrightarrow \bigvee_{z} (y=x+z^2 \wedge z \neq 0)). \tag{1}$$

Let L' be the language obtained from L by adding the relation symbol "$<$". Let P be the set of axioms for a real closed field together with (1), i.e., P is the set of axioms for a real closed ordered field. In 1930 [2], Tarski found a decision method for determining whether or not an arbitrary sentence of L' is derivable from P. His method of systematically eliminating the quantifiers of a given formula culminates in the theorem:

Every formula φ of L' is P*-equivalent to a formula θ of L' with no free variables other than those of φ and no bound variables.*

Thus θ is a Boolean combination of formulas of the form $\alpha>\beta$ and $\alpha=\beta$ where each α and β are obtained from the free variables of φ, 0, and 1 by additions and multiplications. Tarski gave explicit rules for obtaining θ from a formula φ by using a generalization of Sturm's theorem. In case φ is a sentence, θ is a combination of equalities and inequalities among natural numbers. Hence θ can easily be verified or disproved as the case may be on the basis of P. Thus P is complete.

We also obtain as a corollary to Tarski's theorem the following characterization of sets definable in the field of real numbers:

A set S is definable in the field of real numbers if and only if S is a finite union of (open, closed, half-open) intervals with algebraic or infinite endpoints.

This decision method, or algorithm, of Tarski is on the edge of practicality. With continued improvement in machines and programming techniques, we can expect someday to have a machine carry out this algorithm (or variants of it[3]) to solve, for example, certain packing and covering problems of elementary geometry which are now beyond the reach of mathematicians.

Since the axioms for real closed fields are also complete, as we see by using (1) to eliminate inequalities, any pair of real closed fields are

---

[2] This work was not published in full until 1948, when the monograph (see Tarski [51]) was prepared for publication by J. C. C. McKinsey.

[3] Seidenberg [54] has given another algorithm based on a single application of Sturm's theorem which should shorten machine time.

elementarily equivalent. For example, any sentence true in the field of real numbers is also true in the field of real algebraic numbers or any other real closed field. This is not the case for two simple transcendental extensions of real closed fields, as we shall see below.

A simpler proof of the completeness of the axioms for real closed fields has been given by A. Robinson [56; 63] using model-theoretic methods. However his results do not give as much information about definable sets nor such a direct algorithm for testing the truth of sentences of L in a real closed field.

A. Robinson's method of *model completeness* [4] consists in first showing that every formula φ of L' is P-equivalent to an existential formula ψ, i.e., a formula in prenex form without universal quantifiers, and then using the persistence under extension of such existential formulas to show completeness. It is well known and easy to see that a complete axiomatizable theory is also decidable.

Kochen [61] gave another proof of the completeness of the theory of real closed fields using ultraproducts.

A decision method in the easier case of algebraically closed fields was also obtained first by Tarski by the same method of elimination of quantifiers (unpublished, see Seidenberg [54]). Additional proofs of the completeness of the theory of algebraically closed fields of a given characteristic have been given by A. Robinson [56], Kochen [61], and Vaught [54].

A possible strengthening of these theorems may lie in the open decision problems for the ring of all real algebraic integers and the ring of all algebraic integers. In general, if an integral domain is decidable, so also is its quotient field. For it is clear that every statement in the quotient field can be transformed mechanically into an equivalent statement in the integral domain. We do not know whether or not the converse holds. On the other hand, to show the undecidability of these rings, it would be sufficient to define a family of arbitrarily large finite sets, i.e., to find a formula $\varphi(x, y)$ such that for each x, there are finitely many y's satisfying $\varphi(x, y)$ and for each natural number n, there is some x for which there are more than n numbers y satisfying $\varphi(x, y)$, see J. Robinson [62]. However it seems unlikely that this can be done.

A number of years ago, Tarski conjectured that the fields D1–D3

[4] See A. Robinson [63] for an account of this method which is applicable in some cases where the elimination of quantifiers is not and often is easier to carry out.

were the only decidable fields. In each of these cases, the fields are decidable because in some sense "so many" equations are solvable. This suggests that the p-adic fields are promising candidates for counter-examples to Tarski's conjecture. For like the real field, the p-adic fields are completions of the rational field; i.e., Cauchy sequences in a p-adic field under the appropriate definition of absolute value have a limit.

Hensel introduced p-adic numbers as formal power series

$$\sum_{n=k}^{\infty} c_n p^n \qquad \text{for } k \text{ an integer and } c_n \in \{0, 1, \ldots, p-1\}.$$

The p-adic integers are those p-adic numbers whose expansion does not contain negative powers. Hensel showed that for every prime p, these formal power series form a field. He also found that while their number theory is not trivial, it is simpler and more regular than that of the rational field or of the rational integers. For example, a number is a p-adic square for an odd prime p if and only if the first non-zero coefficient in its expansion corresponds to an even power of p and is a quadratic residue (mod p); a number is a 2-adic square if and only if it is a power of 4 times a number of the form $1+c_3 2^3+c_4 2^4+\ldots$.

We can easily define the p-adic integers in the field of p-adic numbers by the formulas:

$$\text{Int } \mathsf{x} \leftrightarrow \bigvee_{\mathsf{y}} \mathsf{y}^2 = 1 + \mathsf{p}\mathsf{x}^2 \qquad \text{for odd } p,$$

$$\text{Int } \mathsf{x} \leftrightarrow \bigvee_{\mathsf{y}} \mathsf{y}^2 = 1 + 8\mathsf{x}^2 \qquad \text{for } p = 2.$$

Hence the field of p-adic numbers is decidable if and only if the ring of p-adic integers is decidable. Nerode [63] has recently shown that the existential theory of a p-adic field is decidable.

3. Undecidable fields. The result of Gödel [31], Church [36] and Rosser [36] that the arithmetic of natural numbers is undecidable is the *sine qua non* of known proofs of the undecidability of fields. To show that a given field $\mathfrak{F}$ is undecidable, we merely find a formula φ of L with one free variable such that an element a of F satisfies φ in $\mathfrak{F}$ if and only if a is a natural number. Then any sentence σ of the arithmetic of natural numbers is equivalent to the sentence σ' obtained from σ by restricting the range of the variables to elements satisfying φ. Of course it would also be enough to define a model of the arithmetic of natural numbers in $\mathfrak{F}$ in such a way that any sentence of number theory could be translated mechanically into an equivalent sentence of $\mathfrak{F}$.

The field first shown to be undecidable was the field of rational numbers. Let $\varphi(\mathsf{a}, \mathsf{b}, \mathsf{k})$ be the formula

$$\bigvee_{\mathsf{x}, \mathsf{y}, \mathsf{z}} \mathsf{x}^2 + \mathsf{a}\mathsf{y}^2 = \mathsf{b}\mathsf{z}^2 + \mathsf{a}\mathsf{b}\mathsf{k}^2 + 2.$$

Then the integers can be defined in the field of rationals by the formula

$$\operatorname{Int} \mathsf{n} \leftrightarrow \bigwedge_{\mathsf{a}, \mathsf{b}} \{\varphi(\mathsf{a}, \mathsf{b}, 0) \wedge \bigwedge_{\mathsf{m}} [\varphi(\mathsf{a}, \mathsf{b}, \mathsf{m}) \rightarrow \varphi(\mathsf{a}, \mathsf{b}, \mathsf{m}+1)] \rightarrow \varphi(\mathsf{a}, \mathsf{b}, \mathsf{n})\}.$$

Since $\varphi(a, b, -n)$ whenever $\varphi(a, b, n)$, it is easy to see that every integer n satisfies the right side of the equivalence. In the other direction, it is necessary to use some known results about the representation of numbers by quadratic forms in the rational field. Here we will just note that for every prime p, there are numbers a and b so that $\varphi(a, b, 0)$, the set of t satisfying $\varphi(\mathsf{a}, \mathsf{b}, \mathsf{t})$ is inductive, and $\varphi(a, b, t)$ implies that t is a rational p-adic integer, i.e., the denominator of t in lowest terms is not divisible by p. Thus the denominator of n is not divisible by any prime p so n is an integer. We can use Lagrange's theorem to define the natural numbers as those integers which are the sum of four squares. Hence the field of rationals is undecidable [5].

From this definition of the set of natural numbers in the rational field, we can find an algebraic surface in a many-dimensional space over the rationals such that by starting with this surface and successively taking projections and complements, we will obtain the set of natural numbers. In contrast, we have already seen from Tarski's theorem on real closed fields that no matter what algebraic surface we start with in a space over the real numbers, the only sets which will be obtained by taking successive projections and complements are unions of a finite number of intervals with algebraic end points.

Let θ be a formula of L with one free variable such that a rational number a satisfies θ in the field of rationals if and only if a is a natural number. We can then find an essentially undecidable sentence σ of L which is true in the field of rational numbers [6]. Namely, we obtain σ by taking the conjunction of the axioms of R. M. Robinson's theory Q and then relativizing the quantifiers to θ in the resulting sentence. Hence by Tarski's argument [6], any class $\mathscr{C}$ of fields which contains the

[5] See J. Robinson [49].

[6] See Tarski-Mostowski-Robinson [53].

rational field is undecidable since the set of sentences true in all fields of $\mathscr{C}$ is compatible with the essentially undecidable theory of σ.

In particular, the theory of fields is undecidable, i.e., the set of sentences derivable from the axioms for a field is not recursive. This result can be obtained from the definability of the set of natural numbers in any field. We shall give a proof below of the definability of natural numbers in a pure transcendental extension of a real closed field which does not depend on such deep results of number theory as needed in the rational field. Other classes of fields which we see are undecidable are the real fields, the fields of characteristic 0, the infinite fields, etc.

The definition of the set of natural numbers given for the rational field can be modified so as to hold in any given field of finite degree over the rational field. In this case, we go through an intermediate step of first defining the ring of all algebraic integers of the field and then showing that the subring of rational integers can be defined in this ring [7].

The only infinite algebraic extensions of the rational field known to be undecidable are fields whose descriptions depend on non-recursive sets. For example, if we adjoin to the rational field the square-roots of a non-recursive set of prime numbers, then the set of true sentences in the resulting field of the special form

$$\bigvee_{x} x^2 = n \qquad \text{for } n \text{ a natural number,}$$

is non-recursive. Certainly, then, the field is undecidable.

Hence there are many interesting open cases of the decision problem for algebraic fields of infinite degree over the rational field. In particular, Tarski has mentioned the case of the field of all numbers expressible in terms of radicals as well as that of the field of constructible numbers, i.e., real algebraic numbers which can be expressed in terms of square-roots. Perhaps the most promising case to work on is the field of all totally real algebraic numbers [8]. It is known that the ring of all totally real algebraic integers is undecidable [9].

Now it is not true that the set of natural numbers can be defined in every undecidable field or even that a model of the arithmetic of natural numbers can be defined. In fact, we shall construct an undecidable field $\mathfrak{F}$ such that each sentence true in $\mathfrak{F}$ is also true in some finite field.

---

[7] See J. Robinson [59].

[8] An algebraic number is *totally* real if it and all of its conjugates are real.

[9] See J. Robinson [62].

Hence no essentially undecidable sentence σ can hold in $\mathfrak{F}$, whereas a translation of Q can be made whenever a model of the arithmetic of natural numbers can be defined.

Let S be any set of natural numbers and Σ be the set of sentences true in all finite fields. Consider the set Γ_S of the following sentences:

$$(2)\qquad \begin{cases} \bigvee_{\mathrm{x}} \mathrm{x}^2 = \mathrm{p}_k & \text{for } k \in S, \\ \neg \bigvee_{\mathrm{x}} \mathrm{x}^2 = \mathrm{p}_k & \text{for } k \notin S, \end{cases}$$

where p_k is an abbreviation for the kth prime number written as a sum of 1's. We can assign the quadratic character of a finite number of primes arbitrarily and find a prime q for which the Legendre symbol (p/q) will have the desired value for each of the given primes p. Thus the field with q elements will satisfy the corresponding sentences of (2). Hence $\Sigma \cup \Gamma_S$ is consistent, since every finite subset is consistent. Now Σ contains the negation of any sentence which has only infinite models. Therefore any sentence true in a model of $\Sigma \cup \Gamma_S$ must have a finite model. Furthermore if S is non-recursive, then every model of $\Sigma \cup \Gamma_S$ is undecidable. In fact, we have shown that for any set S of natural numbers, there is a field $\mathfrak{F}_S$ such that S is recursive in the set of true sentences of $\mathfrak{F}_S$ and every true sentence of $\mathfrak{F}_S$ is true in some finite field. The decision problem for the theory of finite fields, i.e., the question of whether or not Σ is recursive, is an interesting open problem [10].

Before turning to the case of transcendental extensions of real fields, we will discuss an earlier result. R. M. Robinson [51a, section 2] showed that a polynomial ring over a field is undecidable. We shall be concerned here only with the case in which the characteristic of the field is 0. If we define divisibility by

$$\mathrm{x} | \mathrm{y} \leftrightarrow \bigvee_{\mathrm{z}} \mathrm{y} = \mathrm{xz},$$

then the constants (i.e., the elements of the base field) are defined by

$$\mathrm{Con}\, \mathrm{x} \leftrightarrow \mathrm{x} = 0 \vee \mathrm{x} | 1$$

[10] This problem may be closely connected with unpublished results of Cobham and with extensions of Cobham's work by Vaught [62]. In particular, Cobham strengthens results of Tarski-Mostowski-Robinson [53] concerning the theory R and uses this to show the undecidability of the theory of finite groups. See also a related paper by Malcev [61].

It is not difficult to see that the set of natural numbers is defined by

$$\mathrm{Nat}\, x \leftrightarrow \bigvee_{u,w} \{\neg\, \mathrm{Con}\, u \wedge w \neq 0 \wedge u|w$$

$$\wedge \bigwedge_{y} (\mathrm{Con}\, y \wedge u+y|w \rightarrow u+y+1|w \vee y=x)\}.$$

This definition depends essentially on the concept of divisibility in the ring of polynomials. But in a field of rational functions the relation of divisibility is trivial and some other concept will have to be used.

The first special case of the undecidability of a transcendental extension of a real field is also due to R. M. Robinson [11]. He showed in 1950 that any pure transcendental extension of the field of rational numbers is undecidable.

Malcev [60, 61a] first showed the undecidability of a simple transcendental extension of a real closed Archimedean [12] field by defining the set of natural numbers.

He defined the subring of functions bounded by a constant. In this ring, the relation of divisibility is not trivial and can play roughly the same role as divisibility in rings of polynomials.

In this same paper, Malcev also showed that certain fields of formal power series over undecidable fields are undecidable. In particular, the field of formal power series over the rational field is undecidable.

---

[11] An account of this work is to be included in R. M. Robinson [64].

[12] Let $N(x)$ be the formula constructed by Malcev which defines the set of natural numbers in a simple transcendental extension of an Archimedean real closed field. Tarski has shown that $N(x)$ always defines a wider class in case the ground field is a non-Archimedean real closed field. He constructed the following formula which does define the set of natural numbers in a simple transcendental extension of an arbitrary real closed field:

$$N(x) \wedge \bigvee_{u} \bigvee_{v} \bigwedge_{w} (\mathrm{Pos}\, w \rightarrow (u+w||v \leftrightarrow N(w) \wedge w \leqq x)),$$

where $\mathrm{Pos}\, u$ and $u \leqq v$ are defined by Malcev and $u||v$ is given by

$$u||v \leftrightarrow \bigvee_{w} \bigvee_{z} (uw=v \wedge w \leqq z \wedge \mathrm{Pos}\, z).$$

Tarski has also shown that the following formula containing an indeterminate α defines the set of natural numbers in a pure transcendental extension of any real closed field:

$$N(x) \wedge \bigvee_{v} \bigwedge_{w} (\mathrm{Pos}\, w \rightarrow (\alpha+w||v \leftrightarrow N(w) \wedge w \leqq x)).$$

The indeterminate α can be eliminated from this formula (cf. formula (5) in this section).

Let $\mathfrak{R}$ and $\mathfrak{R}'$ be real closed fields and let $\mathfrak{R}_1$ and $\mathfrak{R}_1'$ be simple transcendental extensions of $\mathfrak{R}$ and $\mathfrak{R}'$ respectively. Tarski [63] pointed out that there is a sentence σ of L which is true in $\mathfrak{R}_1$ if and only if $\mathfrak{R}$ is Archimedean, since both the set of natural numbers and the set of elements of $\mathfrak{R}$ are definable in $\mathfrak{R}_1$. Hence if $\mathfrak{R}$ is Archimedean and $\mathfrak{R}'$ is non-Archimedean, then σ will be true in $\mathfrak{R}_1$ but false in $\mathfrak{R}_1'$. Thus $\mathfrak{R}_1$ and $\mathfrak{R}_1'$ are not elementarily equivalent, even though $\mathfrak{R}$ and $\mathfrak{R}'$ are elementarily equivalent, as we have seen.

Since we can define the natural numbers in $\mathfrak{R}_1$, we can also define the set of rational numbers which are less than the real number e and the set of rational numbers which are greater than e. (Both of these sets are recursive and hence arithmetically definable.) So there is a sentence θ of L such that θ is true in $\mathfrak{R}_1$ if and only if $\mathfrak{R}$ contains an element in the Dedekind cut of the rationals corresponding to e. Let $\mathfrak{R}$ be the field of real numbers and $\mathfrak{R}'$ be the field of all real algebraic numbers. Then θ is true in $\mathfrak{R}_1$ but false in $\mathfrak{R}_1'$, so $\mathfrak{R}_1$ and $\mathfrak{R}_1'$ are not elementarily equivalent.

Tarski and Vaught [57] have shown a related, contrasting result that any two transcendental extensions of infinite degree over the same field are elementarily equivalent.

We will now turn to a new proof of the results of Malcev and Tarski. Let $\mathfrak{F}$ be any real closed field and α a transcendental over $\mathfrak{F}$. ($\mathfrak{F}$ may be Archimedean or non-Archimedean.) We will show that the set of natural numbers can be defined in $\mathfrak{F}(\alpha)$. At first we will allow the symbol α in the definition but later show that it can be eliminated.

In the first place, we can easily define the constants by the formula,

$$\text{Con}\, \mathsf{x} \leftrightarrow \bigvee_{\mathsf{y}} 1+\mathsf{x}^4=\mathsf{y}^4. \tag{3}$$

Clearly, if x is in $\mathfrak{F}$, then there is a y satisfying the right side, since $\mathfrak{F}$ is real closed. Conversely, it is known that the curve $1+\mathsf{x}^4=\mathsf{y}^4$ does not have a rational parametrization so that any x satisfying the right side must be in the ground field.

Let $\mathsf{x}\leqq_n \mathsf{y}$ be defined by the formula,

$$\mathsf{x}\leqq_n \mathsf{y} \leftrightarrow \bigvee_{\mathsf{u}_1,\ldots,\mathsf{u}_n} \mathsf{x}+\mathsf{u}_1^2+\ldots+\mathsf{u}_n^2=\mathsf{y}.$$

Now it is well known and easily proved that any positive definite rational function in $\mathfrak{F}(\alpha)$ is the sum of two squares. Thus an inequality $\mathsf{u}\leqq\mathsf{v}$, assumed to hold between the two rational functions u and v for all values of α for which they are defined, is equivalent to $\mathsf{u} \leqq_2 \mathsf{v}$ and hence is definable in $\mathfrak{F}(\alpha)$. These facts are used by Malcev in his proof.

If a is a constant, the inequality $\mathsf{u}^2 \leqq (\alpha - \mathsf{a})^2$ guarantees that the rational function u has a zero at a. On the other hand, we can satisfy the inequalities

$$\mathsf{u}^2 \leqq (\alpha - \mathsf{a}_1)^2, \quad \mathsf{u}^2 \leqq (\alpha - \mathsf{a}_2)^2, \quad \ldots, \quad \mathsf{u}^2 \leqq (\alpha - \mathsf{a}_k)^2,$$

simultaneously for any constants $a_1, a_2, \ldots, a_k$ by a rational function u which has zeros only at these points. In fact, it is sufficient to take

$$u = \frac{c(\alpha - a_1)(\alpha - a_2) \ldots (\alpha - a_k)}{1 + (\alpha - a_1)^2 (\alpha - a_2)^2 \ldots (\alpha - a_k)^2},$$

where c is a sufficiently small constant.

We can now define the natural numbers by the formula

$$(4) \qquad \mathrm{Nat}\ \mathsf{n} \leftrightarrow \bigvee_{\mathsf{u}} [\mathsf{u} \neq 0 \wedge \mathsf{u}^2 \leqq_2 \alpha^2 \wedge \bigwedge_{\mathsf{k}} (\mathrm{Con}\ \mathsf{k} \wedge \mathsf{k} \neq \mathsf{n} \wedge \mathsf{u}^2 \leqq_2 (\alpha - \mathsf{k})^2 \rightarrow \mathsf{u}^2 \leqq_2 (\alpha - \mathsf{k} - 1)^2)].$$

Indeed, if n is a natural number, then we can find u so that $u^2 \leqq_2 (\alpha - k)^2$ for $k = 0, 1, \ldots, n$ and u has no other zeros. Hence the condition on the right is satisfied. On the other hand, if n is not a natural number, then u would have to have zeros at all the natural numbers in order to satisfy the right side. This is impossible.

We still must eliminate α from the definition of natural numbers. Let $\varphi(\alpha, \mathsf{n})$ be the formula on the right side of the equivalence sign in (4). Then

$$(5) \qquad \mathrm{Nat}\ \mathsf{n} \leftrightarrow \bigwedge_{\mathsf{t}} \{\varphi(\mathsf{t}, 0) \wedge \bigwedge_{\mathsf{k}} [\varphi(\mathsf{t}, \mathsf{k}) \rightarrow \varphi(\mathsf{t}, \mathsf{k}+1)] \rightarrow \varphi(\mathsf{t}, \mathsf{n})\}$$

is a suitable definition of the natural numbers. The condition on the right is clearly necessary and its sufficiency is seen by taking $t = \alpha$.

This completes the proof that the natural numbers can be defined in a simple transcendental extension of a real closed field $\mathfrak{F}$ and hence this field is undecidable. As mentioned before, this example is also sufficient to prove the undecidability of the elementary theory of fields.

The proof can be extended at once to the case of a transcendental extension $\mathfrak{F}(\alpha)$ of a real field $\mathfrak{F}$ which may not be real closed but which has a real closed subfield $\mathfrak{F}_0$. It will still be true that the equation $1+x^4=y^4$ can be satisfied for all x in $\mathfrak{F}_0$. The formula $u^2 \leqq_2 (\alpha - a)^2$ for a in $\mathfrak{F}$ will still guarantee that u has a zero at a. A rational function can have only a finite number of zeros, and we can still satisfy any finite number of conditions of the form

$$u^2 \leqq_2 (\alpha - a_j)^2,$$

at least for a_j in $\mathfrak{F}_0$, which is sufficient. Hence the same definition of Nat is valid. As a special case of this result, we see that if any number of indeterminates are adjoined to a real closed field $\mathfrak{F}_0$, we can still define the natural numbers in the resulting field.

Only a slight modification of the above definition of natural numbers is needed in a transcendental extension $\mathfrak{F}(\alpha)$ of a completely arbitrary real field $\mathfrak{F}$ if the set of elements of $\mathfrak{F}$ can be defined (or is considered given). In fact, it is sufficient to define some subset of the constants which includes the natural numbers. Suppose such a set of constants is defined by a formula $\gamma(x)$ so that

$$(\text{Nat } x \rightarrow \gamma(x)) \wedge (\gamma(x) \rightarrow \text{Con } x).$$

Then we can define the set of natural numbers by

$$\text{Nat } n \leftrightarrow \bigvee_u [u \neq 0 \wedge u^2 \leqq_8 \alpha^2 \wedge \bigwedge_k (\gamma(k) \wedge k \neq n \wedge u^2 \leqq_8 (\alpha - k)^2 \rightarrow u^2 \leqq_8 (\alpha - k - 1)^2)].$$

The argument is the same as before, except that now we use the theorem of Landau [06] that any positive definite rational function of one variable with rational coefficients can be expressed as the sum of eight squares of rational functions with rational coefficients. Hence we can make up a function u with rational coefficients and prescribed rational zeros $a_1, \ldots, a_k$ so that $u^2 \leqq_8 (\alpha - a_j)^2$ for $j=1, \ldots, k$. Thus the whole problem for a transcendental extension $\mathfrak{F}(\alpha)$ would be solved if we could define a suitable set of constants.

Recently, R. M. Robinson [64] has carried on these methods to show that *every* pure transcendental extension of a real field is undecidable.

The difficult case for transcendental extensions of real fields arises because we do not have a suitable definition of a sufficiently wide class of constants. In the case of a transcendental extension $\mathfrak{C}(\alpha)$ of the field of all complex numbers $\mathfrak{C}$, we can immediately define the complex numbers by the formula

$$\text{Con}\ x \leftrightarrow \bigvee_{y} 1+x^4=y^4.$$

However at present we know of no way of defining the natural numbers in $\mathfrak{C}(\alpha)$.

The case of a transcendental extension of a finite field is an interesting open problem. Here, of course, there is no question of defining the natural numbers but there may be a definable model of arithmetic nevertheless. In the case of the ring of polynomials over a finite field, R. M. Robinson [51a] defined a model consisting of powers of a transcendental element. Quite possibly something similar can be done for the field of rational functions over a finite field. As yet there are no known undecidable fields of characteristic different from 0 except those defined using non-recursive sets.

THE JOURNAL OF SYMBOLIC LOGIC
Volume 32, Number 3, Sept. 1967

AN INTRODUCTION TO HYPERARITHMETICAL FUNCTIONS

JULIA ROBINSON

§1. Introduction. By a *functional equation* we mean an equation of the form

$$A_1 \cdots A_k = B_1 \cdots B_l. \tag{1}$$

Here the A's and B's are functions of one variable from and to the natural numbers and FG is the function obtained from F and G by composition, i.e. $FG(x) = F(G(x))$ for all natural numbers x. We wish to investigate finite systems of functional equations. Now if all the A's and B's of (1) are equal to the identity function I (or all equal to the zero function O), then the equation (1) is satisfied trivially. Thus, in order to make the problem of solvability of systems of equations interesting, we must have some function given which will be held fixed throughout. We take the successor function S to be this given function.

Let $\Sigma(S, F, U_1, \cdots)$ be a finite system of functional equations in $S, F, U_1, \cdots$. We say that Σ defines F_0 if F_0 is the unique function F for which there are functions $U_1, \cdots$ such that $\Sigma(S, F, U_1, \cdots)$ holds. The functions $U_1, \cdots$ are called *auxiliary functions* and they need not be uniquely determined by Σ. A function is called *hyperarithmetical* if there is some system of functional equations which defines it. A set of natural numbers is *hyperarithmetical* if its characteristic function is hyperarithmetical.

We say $\Sigma(S, G_1, \cdots, F, U_1, \cdots)$ *defines* F_0 *in terms of* $G_1, \cdots$ if F_0 is the unique function F for which there are functions $U_1, \cdots$ such that $\Sigma(S, G_1, \cdots, F, U_1, \cdots)$ holds. Here $G_1, \cdots$ are either particular functions considered given or are parameters. In the latter case, for every assignment of $G_1, \cdots$ there is exactly one function F for which Σ can be satisfied. If F is definable in terms of $G_1, \cdots$, we say F is *hyperarithmetical in* $G_1, \cdots$.

EXAMPLES. The single equation

$$SI = S \tag{2}$$

defines the identity function I. That is, I is the unique function F such that $SFx = Sx$ for all x. Note that $FS = S$ does not by itself define any function since $F0$ is not determined.

Both the double function D and the zero function O are defined by the system

$$DO = O, \qquad DS = SSD. \tag{3}$$

By induction, we see that $Dx = 2x + D0$ from the second equation. Hence the only possible fixed point for D is at $x = 0$. Therefore by the first equation, the range of O (written $\mathscr{R}(O)$) is $\{0\}$. Hence O is identically 0. Thus $D0 = 0$ and $Dx = 2x$. Conversely, if D is the double function and O is the zero function, (3) is clearly satisfied.

Received December 15, 1966.

Also Z the function O^x is defined by

$$ZO = SO, \qquad ZS = O \tag{4}$$

in terms of O. Note that sgn $(x) = ZZx$.

The predecessor function P, i.e. the function with $P0 = 0$ and $P(x + 1) = x$, is defined by the system

$$PO = O, \qquad PS = I \tag{5}$$

in terms of O and I. Hence (2), (3), and (4) together define P. So P is hyperarithmetical.

In general, if Σ defines F in terms of $G_1, \cdots$ and Σ_i defines G_i then Σ together with $\Sigma_1, \cdots$ define F provided the symbols for the auxiliary functions in each system are different. Hence if F is hyperarithmetical in $G_1, \cdots$ and $G_1, \cdots$ are hyperarithmetical functions then F is hyperarithmetical.

In §4, there is a system of equations that defines the satisfaction function W which gives the truth value of every arithmetical statement. It follows (§5) that every arithmetical function is hyperarithmetical but W itself is not arithmetical (Tarski). Thus the hyperarithmetical functions form a wider class of functions than the arithmetical functions which are known to include all recursive or computable functions. In §7, we show that F_0 is hyperarithmetical if and only if there is a formula $\mathscr{A}(F, G)$ arithmetical in F and G such that F_0 is the unique function for which there is a function G such that $\mathscr{A}(F, G)$ holds. As a corollary, we see that a set is hyperarithmetical if and only if it can be expressed in both one-function-quantifier forms (Kleene). In §§8–9, we use methods of Kleene, Markwald, Spector [8], [9] to show that if F is defined by a system of functional equations, we can construct another system which also defines F and which has the further property that its auxiliary functions are also uniquely defined. Finally, in §10 we show that there is a system of functional equations which has solutions but has no hyperarithmetical solution (Kleene). The paper is almost entirely self-contained.

Earlier work on hyperarithmetical functions and definability by functional equations has been for functions of an arbitrary number of variables. Herbrand first proposed in a letter to Gödel the class of functions definable by functional equations in the unknown function F and known, i.e. previously defined and initial functions. (See [1, p. 70].) This suggestion was modified by Gödel to become the Herbrand-Gödel definition of recursive functions.

Kleene [5], [6] introduced hyperarithmetical functions and predicates as the classes of functions and predicates which can be built up from arithmetical predicates by means of constructions based on a system of recursive ordinal notations. Davis, Mostowski, and Post independently considered such extensions of arithmetical predicates (see historical references in [6]). Kleene showed that the hyperarithmetical predicates were precisely those expressible in both one-function-quantifier forms in the analytic hierarchy. Grzegorczyk, Mostowski, and Ryll-Nardzewski [4] defined hyperarithmetical functions (of any number of variables) to be those functions whose graphs can be expressed in both one-function-quantifier form and proved that every such function can be defined by a system of functional equations. The proof given is due to Grzegorczyk and uses the theory of the Kleene hierarchy. Addison and Kuznecov obtained this result independently.

Kleene [7] showed that there are recursive predicates $\mathscr{R}(x, F)$ such that there is an F with $(\bigwedge x)\mathscr{R}(x, F)$ but no hyperarithmetical F.

It would be relatively simple to derive the results of this paper from these known results together with a system of equations which defines suitable pairing functions. However functions definable by systems of functional equations is a mathematically natural class of functions to study so it seems worthwhile to derive basic facts about this class directly. The important role of hyperarithmetical functions in current research in foundations is not discussed.

§2. **Pairing functions.** If $J(x, y)$ maps the set of ordered pairs of natural numbers onto the set of natural numbers in a one-to-one way, then there exist unique functions K and L such that

(1) $$KJ(x, y) = x, \qquad LJ(x, y) = y.$$

Alternatively, K and L may be determined from

(2) $$J(Kz, Lz) = z.$$

Such functions K and L will be called *pairing functions*. They may be characterized by the condition that for every x and y there is a unique z such that $Kz = x$ and $Lz = y$. Indeed, given such K and L, either (1) or (2) determines the corresponding J.

A suitable mapping function J is

(3) $$J(x, y) = (2x + 1)2^y - 1.$$

Notice that

$$J(x, 0) = 2x, \qquad J(x, y + 1) = 2J(x, y) + 1.$$

Hence

$$K(2x) = x, \qquad L(2x) = 0, \qquad K(2z + 1) = Kz, \qquad L(2z + 1) = L(z) + 1.$$

These equations determine K and L. Thus K and L are defined in terms of I, D, and O by

(4) $$KD = I, \qquad LD = O, \qquad KSD = K, \qquad LSD = SL.$$

The pairing functions defined by (4) will be used throughout this paper. However any recursive pairing functions J, K, and L which satisfy the following conditions could equally well be used:

(i) if $u \leqq v$ then $J(x, u) \leqq J(x, v)$.

(ii) $J(x, y) \geqq x$ with equality holding only for $x = y = 0$.

Clearly, if $J(x, y) = (2x + 1)2^y - 1$, then J is an increasing function of y for each x so (i) holds. Furthermore $(2x + 1)2^y - 1 \geqq 2x \geqq x$ with the first equality holding only if $y = 0$ and the second only if $x = 0$. Hence the pairing functions given by (3) and (4) satisfy (i) and (ii). In this case, J, K, and L are easily recognized as primitive recursive functions. The pairing functions corresponding to $J(x, y) = \frac{1}{2}\{(x + y)^2 + 3x + y\}$ could also be used but the system of equations defining them would be more complicated than (4).

If F is a function of two variables, and A and B are functions of one variable, then we define $F(A, B)$ to be the function of one variable such that $F(A, B)x = F(Ax, Bx)$ for all x. For example, we have $J(K, L) = I$ from (2). Also $J(I, O) = D$ and $J(K, SL) = SD$ by (4).

For any functions F and G, there is a unique function H such that

$$KH = F, \qquad LH = G. \tag{5}$$

Indeed, $H = J(F, G)$ is equivalent to (5) for all F and G. In order to save letters and space, we will write $J(F, G)$ when convenient and omit the defining equations (5).

Furthermore, if F is a function of two variables then $F(K, L)J(x, y) = F(x, y)$. In this way the information carried by a function of two variables is given by a function of one variable. For example, let E correspond to the characteristic function of equality. Thus E is given by

$$\begin{aligned} EJ(x, y) &= 1 \text{ if } x = y, \\ &= 0 \text{ otherwise.} \end{aligned}$$

Then E is defined by

$$EO = SO, \qquad EJ(O, S) = EJ(S, O) = O, \qquad EJ(SK, SL) = E, \tag{6}$$

in terms of $O, K, L, J(O, S), J(S, O)$ and $J(SK, SL)$. These last three functions can be defined as in (5), e.g. $J(O, S)$ is defined by

$$KJ(O, S) = O, \qquad LJ(O, S) = S.$$

Next we wish to define $F + G$, $F \dot{-} G$ (where $x \dot{-} y = \max\{x - y, 0\}$), and $F \cdot G$ in terms of F and G. It is sufficient to define $K + L$, $K \dot{-} L$, and $K \cdot L$ because

$$\begin{gathered} F + G = (K + L)J(F, G), \qquad F \dot{-} G = (K \dot{-} L)J(F, G), \\ F \cdot G = (K \cdot L)J(F, G). \end{gathered} \tag{7}$$

Addition is characterized by the identities $x + 0 = x$ and $x + Sy = S(x + y)$ which are equivalent to the functional equations

$$(K + L)J(I, O) = I, \qquad (K + L)J(K, SL) = S(K + L). \tag{8}$$

Indeed,

$$\begin{gathered} (K + L)J(I, O)x = (K + L)J(x, 0) = x, \qquad Ix = x, \\ (K + L)J(K, SL)J(x, y) = (K + L)J(x, Sy) = x + Sy, \\ S(K + L)J(x, y) = S(x + y). \end{gathered}$$

Notice that $\mathscr{R}(J(x, y))$ is the set of all natural numbers. So $AJ(x, y) = BJ(x, y)$ for all x and y implies that $A = B$.

Since $J(I, O) = D$ and $J(K, SL) = SD$ for our pairing functions, $K + L$ is defined by

$$(K + L)D = I, \qquad (K + L)SD = S(K + L) \tag{9}$$

in terms of D and I.

Similarly, $x \dot{-} 0 = x$ and $x \dot{-} Sy = P(x \dot{-} y)$ so $K \dot{-} L$ is defined by

$$(K \dot{-} L)D = I, \qquad (K \dot{-} L)SD = P(K \dot{-} L).$$

Multiplication is characterized by the identities $x \cdot 0 = 0$ and $x \cdot Sy = x \cdot y + x$ which are equivalent to the functional equations

$$(K \cdot L)D = O, \qquad (K \cdot L)SD = (K \cdot L) + K.$$

Hence $K \cdot L$ is the unique function for which there is a function A such that

$$(K \cdot L)D = O, \qquad (K \cdot L)SD = (K + L)A, \qquad KA = K \cdot L, \qquad LA = K.$$

Here $D, O, K + L, K, L$ are all considered given since they have been defined previously. We could adjoin (4), (9), and the defining equations for D, O, and I and obtain a system of equations in which $D, O, I, K, L, K + L$ would be uniquely determined auxiliary functions.

All the functions so far have been defined by systems of equations with unique solutions (or in terms of functions so defined). We will continue to make definitions with this property in order to show that every hyperarithmetical function can be defined by a system of equations with a unique solution in §9.

The letters I, D, Z, O, P, E, K, and L will be reserved for these particular functions.

§3. Sequences. We wish to represent infinite sequences of natural numbers which are eventually 0 by a single number. Let x represent $\langle x_0, x_1, \cdots\rangle$ where

$$x_n = LK^n x.$$

For example, the sequence $\langle 1, 2, 3, 0, 0, \cdots\rangle$ is represented by the number $J(J(J(0, 3), 2), 1)$. Clearly, every sequence which is eventually 0 is represented by some number. Furthermore, the number is unique. Indeed, if x represents $\langle x_0, x_1, \cdots\rangle$ then Kx represents $\langle x_1, x_2, \cdots\rangle$ and $Lx = x_0$. Hence it is sufficient to check that 0 is the only number which represents the zero sequence $\langle 0, 0, \cdots\rangle$. Let y represent $\langle 0, 0, \cdots\rangle$. Now $y \geqq Ky \geqq KKy \geqq \cdots$. We have seen in the last section that $z > Kz$ unless $z = 0$. Hence if $y \neq 0$, there is some t with $K^t y \neq 0$ and $K^{t+1}y = 0$. But this is impossible since $K^t y$ represents $\langle 0, 0, \cdots\rangle$ so $L(K^t y) = 0$ and we would have $K^t y = J(0, 0) \neq 0$. Hence 0 is the only number which represents the zero sequence. (Note this is not true if we interchange K and L and take $x_n = KL^n x$.)

The term function T given by

$$TJ(n, x) = x_n,$$

where x represents $\langle x_0, x_1, \cdots\rangle$, is hyperarithmetical. In fact, T satisfies the recursive conditions

$$(1) \qquad TJ(0, x) = x_0 = Lx,$$

$$(2) \qquad TJ(Sn, x) = x_{n+1} = TJ(n, Kx).$$

Hence T is uniquely defined by

$$(3) \qquad TJ(O, I) = L, \qquad TJ(SK, L) = TJ(K, KL).$$

To check the second equation of (3), apply both sides to $J(n, x)$ which yields (2). We shall sometimes write $T_n x$ for $TJ(n, x)$. Thus x represents the sequence $\langle T_0 x, T_1 x, \cdots\rangle$.

Let N be the function given by

$$NJ(t, J(x, y)) = 1 \text{ if } T_k x = T_k y \text{ for all } k \neq t, \quad = 0 \text{ otherwise.}$$

Then

$$(4) \qquad NJ(0, J(x, y)) = 1 \text{ if } Kx = Ky, \quad = 0 \text{ otherwise.}$$

and

$$(5) \qquad NJ(St, J(x, y)) = 1 \text{ if } NJ(t, J(Kx, Ky)) = 1 \text{ and } Lx = Ly, \quad = 0 \text{ otherwise.}$$

Hence N is uniquely defined by

(6) $$NJ(O, I) = EJ(KK, KL),$$

(7) $$NJ(SK, L) = NJ(K, J(KKL, KLL)) \cdot EJ(LKL, LLL).$$

These equations follow from (4) and (5) by applying (6) to $J(x, y)$ and (7) to $J(t, J(x, y))$.

§4. Arithmetical formulas and satisfaction. A formula is arithmetical if it can be built up from equations of the form $\alpha = \beta$, $\alpha + \beta = \gamma$, and $\alpha \cdot \beta = \gamma$ where α, β, and γ are variables or symbols for particular natural numbers by means of the logical symbols $\neg$ (not), $\vee$ (or), $\wedge$ (and), $\rightarrow$ (if, then), $\leftrightarrow$ (if and only if), $\bigvee$ (there exists), and $\bigwedge$ (for all). Here the variables range over the natural numbers.

A formula is arithmetical in $F, G, \cdots$ and $\mathscr{S}, \mathscr{T}, \cdots$ if, in addition to the equations of the form $\alpha = \beta$, $\alpha + \beta = \gamma$, and $\alpha \cdot \beta = \gamma$, formulas of the form $F\alpha = \beta$, $G\alpha = \beta, \cdots$ and $\alpha \in \mathscr{S}$, $\alpha \in \mathscr{T}, \cdots$ are included among the atomic formulas. Note that $F, G, \cdots$ and $\mathscr{S}, \mathscr{T}, \cdots$ occur as parameters and are not subject to quantification.

A set $\mathscr{S}$ is arithmetical if there is an arithmetical formula $\phi(x)$ with one free number variable x and no function parameters such that

$$x \in \mathscr{S} \leftrightarrow \phi(x).$$

Similarly, a set $\mathscr{S}$ is arithmetical in F_0 if there is an arithmetical formula $\phi(x, F)$ with one free number variable x and one function parameter F so that

$$x \in \mathscr{S} \leftrightarrow \phi(x, F_0).$$

Also a function F is arithmetical if there is an arithmetical formula $\phi(x, y)$ with two free variables and no function parameters such that

$$Fx = y \leftrightarrow \phi(x, y, G_0).$$

Finally, a function F is arithmetical in G_0 if there is an arithmetical formula $\phi(x, y, G)$ such that

$$Fx = y \leftrightarrow \phi(x, y).$$

It is easily seen that if F is arithmetical in G_0 and G_0 is itself an arithmetical function then F is arithmetical. Similarly, if F is arithmetical in an arithmetical set $\mathscr{S}$ then F is arithmetical, etc. In particular, a set $\mathscr{S}$ is arithmetical if and only if its characteristic function H is arithmetical. Clearly,

$$x \in \mathscr{S} \leftrightarrow Hx = 1$$

and

$$Hx = y \leftrightarrow (x \in \mathscr{S} \wedge y = 1) \vee (\neg x \in \mathscr{S} \wedge y = 0).$$

We now wish to define the concept of a sequence $\langle x_0, x_1, \cdots \rangle$ satisfying an arithmetical formula without function or set parameters. It is easily seen that any such formula is equivalent to a formula built up from equations of the form $\alpha = \beta$, $\alpha + \beta = \gamma$ and $\alpha \cdot \beta = \gamma$ by means of $\neg$, $\vee$, $\bigvee$. Hence it is sufficient to consider these formulas and take $v_0, v_1, \cdots$ as the variables. We define satisfaction first for all equations and then for the remaining formulas in terms of earlier formulas.

DEFINITION. 1. $\langle x_0, x_1, \cdots \rangle$ satisfies $v_i = v_j$ iff $x_i = x_j$;

2. $\langle x_0, x_1, \cdots \rangle$ satisfies $v_i + v_j = v_k$ iff $x_i + x_j = x_k$;

3. $\langle x_0, x_1, \cdots \rangle$ satisfies $v_i \cdot v_j = v_k$ iff $x_i \cdot x_j = x_k$;
4. $\langle x_0, x_1, \cdots \rangle$ satisfies $\neg\phi$ iff $\langle x_0, x_1, \cdots \rangle$ does not satisfy ϕ;
5. $\langle x_0, x_1, \cdots \rangle$ satisfies $\phi \vee \psi$ iff $\langle x_0, x_1, \cdots \rangle$ satisfies ϕ or $\langle x_0, x_1, \cdots \rangle$ satisfies ψ;
6. $\langle x_0, x_1, \cdots \rangle$ satisfies $(\bigvee v_k)\phi$ iff there is some y such that $\langle x_0, \cdots, x_{k-1}, y, x_{k+1}, \cdots \rangle$ satisfies ϕ.

Clearly, whether or not a sequence satisfies a particular formula depends only on a finite number of terms of the sequence. Hence we shall only consider sequences which are ultimately 0. Each such sequence is represented by a single natural number.

We number these formulas as follows

$$\begin{aligned}
&\phi_{6t}: v_{Kt} = v_{Lt},\\
&\phi_{6t+1}: v_{KKt} + v_{LKt} = v_{Lt},\\
&\phi_{6t+2}: v_{KKt} \cdot v_{LKt} = v_{Lt},\\
&\phi_{6t+3}: \neg\phi_t,\\
&\phi_{6t+4}: \phi_{Kt} \vee \phi_{Lt},\\
&\phi_{6t+5}: (\bigvee v_{Kt})\phi_{Lt}.
\end{aligned}$$

We call the function W given by

$$\begin{aligned}
WJ(n, x) &= 1 \text{ if } \langle T_0x, T_1x, \cdots \rangle \text{ satisfies } \phi_n,\\
&= 0 \text{ otherwise}
\end{aligned}$$

the *satisfaction function.* An arithmetical formula ϕ_n with no free variables is true iff $\langle 0, 0, \cdots \rangle$ satisfies ϕ_n, i.e. $WJ(n, 0) = 1$. (Since ϕ_n contains no free variables, if any sequences of natural numbers satisfy ϕ_n then all sequences will satisfy ϕ_n and ϕ_n is true.)

Hence if we could calculate the values of W we could answer any question of number theory. But this is impossible. In the next section, we show that W is not computable or even arithmetical.

In order to show that W is definable by a system of functional equations, we show that each of the six conditions in the definition of satisfaction can be replaced by identities involving W and previously defined functions except for the last condition which is expressed in terms of the existence of an auxiliary function which satisfies certain identities.

1. $WJ(6t, x) = EJ(TJ(Kt, x), TJ(Lt, x))$.
2. $WJ(6t + 1, x) = EJ(TJ(KKt, x) + TJ(LKt, x), TJ(Lt, x))$.
3. $WJ(6t + 2, x) = EJ(TJ(KKt, x) \cdot TJ(LKt, x), TJ(Lt, x))$.
4. $WJ(6t + 3, x) = ZWJ(t, x)$.
5. $WJ(6t + 4, x) = ZZJ(WJ(Kt, x), WJ(Lt, x))$.
6. $WJ(6t + 5, x) \cdot NJ(Kt, J(x, y)) \cdot WJ(Lt, y) = NJ(Kt, J(x, y)) \cdot WJ(Lt, y)$,

$$NJ(Kt, J(x, CJ(t, x))) = 1, \; WJ(6t + 5, x) = WJ(Lt, CJ(t, x)).$$

Here C is an auxiliary function. The first equation of 6 insures that the 'if' part of condition 6 in the definition of satisfaction holds. The last two equations insure that for every t, x there is a z (given here by the auxiliary function $CJ(t, x)$) such that the sequences $\langle T_0x, T_1x, \cdots\rangle$ and $\langle T_0z, T_1z, \cdots\rangle$ differ at most at the Ktth term and $\langle T_0x, T_1x, \cdots\rangle$ satisfies $(\bigvee v_{Kt})\phi_{Lt}$ iff $\langle T_0z, T_1z, \cdots\rangle$ satisfies ϕ_{Lt}.

Thus W is the unique function such that for some function C, 1–6 holds for all t, x, and y. However C is not uniquely determined. Hence we add conditions that insure that for every t and x, $CJ(t, x)$ has the least possible value consistent with the last two equations of 6. Namely, if C satisfies the additional condition

$$\begin{aligned} CJ(t, x) \dot{-} y = 0 \lor\ & NJ(Kt, J(x, y)) = 0 \\ \lor\ & WJ(6t + 5, x) \neq WJ(Lt, y) \end{aligned}$$

for all t, x, and y then C will have the least possible value for each t and x. This condition is equivalent to

7. $(CJ(t, x) \dot{-} y) \cdot NJ(Kt, J(x, y)) \cdot EJ(WJ(6t + 5, x), WJ(Lt, y)) = 0.$

In order to translate the identities 1–7 into functional equations, we need to define $6t$. Let $At = 6t$ then A is uniquely determined by the equations

$$AO = O, \qquad AS = S^6A.$$

For the record, here is the system of functional equations which defines W in terms of previously defined functions and has a unique solution:

1. $WJ(AK, L) = EJ(TJ(KK, L), TJ(LK, L)),$
2. $WJ(SAK, L) = EJ(TJ(K^3, L) + TJ(LK^2, L), TJ(LK, L)),$
3. $WJ(S^2AK, L) = EJ(TJ(K^3, L) \cdot TJ(LK^2, L), TJ(LK, L)),$
4. $WJ(S^3AK, L) = ZW,$
5. $WJ(S^4AK, L) - ZZJ(WJ(KK, L), WJ(LK, L)),$
6. $WJ(S^5AK, KL) \cdot NJ(KK, L) \cdot WJ(LK, LL) = NJ(KK, L) \cdot WJ(LK, LL),$
   $NJ(KK, J(L, C) = SO,$
   $WJ(S^5AK, L) = WJ(LK, C),$
7. $(CJ(K, KL) \dot{-} LL) \cdot NJ(KK, L) \cdot EJ(WJ(S^5AK, KL), WJ(LK, LL)) = 0.$

Hence W is hyperarithmetical.

§5. Arithmetical functions and sets. I. *Every arithmetical function is hyperarithmetical.* If F is arithmetical there is some k so that

$$Fm = n \leftrightarrow \langle m, n, 0, \cdots\rangle \text{ satisfies } \phi_k.$$

Hence F is the unique function such that

$$WJ(S^kO, J(J(O, F), I)) = SO.$$

II. *Every arithmetical set is hyperarithmetical.* Suppose $\mathscr{S}$ is arithmetical then its characteristic function H is arithmetical and hence hyperarithmetical.

III. *The satisfaction function W is not arithmetical* (Tarski [10]). If W were arithmetical, then the set $\mathscr{U}$ given by

$$n \in \mathscr{U} \leftrightarrow WJ(n, J(0, n)) = 0 \tag{1}$$

would be arithmetical. Indeed,

$$x \in \mathscr{U} \leftrightarrow \bigvee_{y,z} (Wy = 0 \land Ky = x \land Ly = z \land Kz = 0 \land Lz = x)$$

and K and L are arithmetical functions by Gödel's Lemma [2] since they are primitive recursive. But if $\mathscr{U}$ is arithmetical, there is some k such that

$$n \in \mathscr{U} \leftrightarrow \langle n, 0, \cdots\rangle \text{ satisfies } \phi_k.$$

Hence

(2) $$n \in \mathcal{U} \leftrightarrow WJ(k, J(0, n)) = 1$$

for all n. But this is impossible since $n = k$ in (1) and (2) gives a contradiction.

Thus every arithmetical function is hyperarithmetical but not conversely.

§6. **Satisfaction for arithmetical formulas with one function parameter.** We wish to define satisfaction for arithmetical formulas with one function parameter F. We simply adjoin one more clause in the definition in §4. Namely,

0. $\langle x_0, x_1, \cdots \rangle$ satisfies $Fv_i = v_j$ iff $Fx_i = x_j$.

We number the formulas arithmetical in F similarly to the numbering of arithmetical formulas. Namely,

$$\begin{aligned}
&\theta_{7t}(F)\colon Fv_{Kt} = v_{Lt},\\
&\theta_{7t+1}(F)\colon v_{Kt} = v_{Lt},\\
&\theta_{7t+2}(F)\colon v_{KKt} + v_{LKt} = v_{Lt},\\
&\theta_{7t+3}(F)\colon v_{KKt}\cdot v_{LKt} = v_{Lt},\\
&\theta_{7t+4}(F)\colon \neg\theta_t(F),\\
&\theta_{7t+5}(F)\colon \theta_{Kt}(F) \vee \theta_{Lt}(F),\\
&\theta_{7t+6}(F)\colon (\bigvee v_{Kt})\theta_{Lt}(F).
\end{aligned}$$

The satisfaction function W_F for these formulas is given by

$$\begin{aligned}
W_FJ(n, x) &= 1 \text{ if } \langle T_0x, T_1x, \cdots \rangle \text{ satisfies } \theta_n(F),\\
&= 0 \text{ otherwise.}
\end{aligned}$$

We want to show that W_F is hyperarithmetical in F. We introduce a new function A with $Au = 7u + 1$, i.e.

(1) $$AO = SO, \qquad AS = S^7A.$$

Now

$$W_FJ(7t, x) = EJ(FT_{Kt}x, T_{Lt}x).$$

The corresponding functional equation is

0. $$W_FJ(PAK, L) = EJ(FTJ(KK, L), TJ(LK, L)).$$

The remaining functional equations are exactly the equations 1–7 of §4 where W is replaced with W_F. (A of course is now the function $7t + 1$ defined by (1) above.)

Hence there is a system Ω of equations in S, F, W, and auxiliary functions $U_1, \cdots$ such that for each function F there is a unique solution for $W, U_1, \cdots$ and in this solution $W = W_F$.

§7. **Hyperarithmetical functions and sets.** In this section, we will show that other characterizations of hyperarithmetical functions and sets are equivalent to our definition.

THEOREM. *H_0 is hyperarithmetical if and only if there is a formula $\mathcal{A}(H, U_1, \cdots)$ arithmetical in $H, U_1, \cdots$ such that H_0 is the unique function H for which there are functions $U_1, \cdots$ so that $\mathcal{A}(H, U_1, \cdots)$ holds.*

PROOF. An equation $A_1 \cdots A_k = B_1 \cdots B_l$ holds iff

$$\begin{aligned}
(\bigwedge x)(\bigvee y_1, \cdots, y_k, z_1, \cdots, z_l)(y_1 = A_kx \wedge y_2 = A_{k-1}y_1\\
\wedge \cdots \wedge y_k = A_1y_{k-1} \wedge z_1 = B_lx \wedge z_2 = B_{l-1}z_1\\
\wedge \cdots \wedge z_l = B_1z_{l-1} \wedge z_l = y_k).
\end{aligned}$$

Hence a functional equation is equivalent to a formula arithmetical in the functions occurring in it. Therefore if H_0 is defined by $\Sigma(S, H, U_1, \cdots)$ then there is an arithmetical formula $\mathscr{A}$ in $H, U_1, \cdots$ such that H_0 is the unique function for which there are $U_1, \cdots$ so that $\mathscr{A}(H, U_1, \cdots)$ holds. (Here we use the fact that S is arithmetical.)

On the other hand, suppose H_0 is the unique H for which there are $U_1, \cdots$ so that $\mathscr{A}(H, U_1, \cdots)$. Let $F = J(L, J(H, J(U_1, \cdots)))$, so $L = KF$, $H = KLF = KKFF$, $U_1 = KL^2F = K(KF)^2F, \cdots$. Hence H_0 is the unique function H for which there is a function F such that

$$H = KKFF \wedge \mathscr{A}(KKFF, K(KF)^2F, \cdots).$$

Since K is arithmetical, $\mathscr{A}(KKFF, K(KF)^2F, \cdots)$ is arithmetical in F. Hence from §6, it is equivalent to $\theta_k(F)$ for some k where $\theta_k(F)$ contains no free variables. Therefore H_0 is the unique function H for which there is an F such that

$$H = KKFF \wedge W_F J(S^k O, O) = SO.$$

(Since $\theta_k(F)$ has no free variables, $\langle 0, 0, \cdots \rangle$ satisfies $\theta_k(F)$ if and only if $\theta_k(F)$ holds.) Hence H_0 is hyperarithmetical, since W_F can be defined by the system Ω of the last section.

COROLLARY. *A function H_0 is hyperarithmetical if and only if there is a formula $\mathscr{B}(H, G)$ arithmetical in H and G such that H_0 is the unique function H for which there is a function G so that $B(H, G)$ holds.*

REMARK. The auxiliary function G in the corollary cannot be eliminated. Feferman [3] has shown that a generic set $\mathscr{S}$ is not uniquely determined by any arithmetical formula $\mathscr{A}(\mathscr{S})$. Hence the characteristic function of a generic set is not uniquely determined by any arithmetical formula although it is hyperarithmetical.

PROOF. Let $G = J(U_1, J(U_2, \cdots))$ so $U_1 = KG$, $U_2 = KLG, \cdots$. By the theorem, H_0 is determined uniquely by $\mathscr{A}(H, U_1, \cdots)$. Hence we can take for B an arithmetical formula equivalent to $\mathscr{A}(H, KG, KLG, \cdots)$.

THEOREM. *A set $\mathscr{S}$ is hyperarithmetical if and only if there are arithmetical formulas $\mathscr{A}(x, F)$ and $\mathscr{B}(x, F)$ with one free variable x and one function parameter F such that*

$$x \in \mathscr{S} \leftrightarrow (\bigvee F)\mathscr{A}(x, F), \tag{1}$$

$$x \notin \mathscr{S} \leftrightarrow (\bigvee F)\mathscr{B}(x, F). \tag{2}$$

PROOF. Suppose $\mathscr{S}$ satisfies (1) and (2). Let H_0 be the characteristic function of $\mathscr{S}$. Then H_0 is the unique function H such that

$$(\bigwedge x)(\bigvee F)((\mathscr{A}(x, F) \wedge Hx = 1) \vee (\mathscr{B}(x, F) \wedge Hx = 0)). \tag{3}$$

Let $FJ(x, y) = F_x y$. In this way, one function F represents a sequence of functions $F_0, F_1, \cdots$. Hence (3) is equivalent to

$$(\bigvee F)(\bigwedge x)((\mathscr{A}(x, F_x) \wedge Hx = 1) \vee (\mathscr{B}(x, F_x) \wedge Hx = 0)).$$

Now each equation of the form $F_x u = v$ which occurs in $\mathscr{A}(x, F_x)$ or $\mathscr{B}(x, F_x)$ can be replaced by a formula of the form

$$(\bigvee w)(Fw = v \wedge Kw = x \wedge Lw = u).$$

Hence there is an arithmetical formula $\mathscr{C}(H, F)$ such that H_0 is the unique function

H for which there is a function F so $\mathscr{C}(H, F)$ holds. Therefore H is hyperarithmetical.

Conversely, let H_0 be the characteristic function of $\mathscr{S}$. Suppose H_0 is defined by a system $\Sigma(S, H, U_1, \cdots)$ of equations. Then

$$x \in \mathscr{S} \leftrightarrow (\bigvee H, U_1, \cdots)(\Sigma(S, H, U_1, \cdots) \wedge Hx = 1)$$

and

$$x \notin \mathscr{S} \leftrightarrow (\bigvee H, U_1, \cdots)(\Sigma(S, H, U_1, \cdots) \wedge Hx = 0).$$

By pairing $H, U_1, \cdots$, we can easily see that $\mathscr{S}$ can be put in the forms (1) and (2).

THEOREM. *A function H is hyperarithmetical if and only if it can be put in the form*

$$Hx = y \leftrightarrow (\bigvee F)\mathscr{A}(x, y, F) \tag{4}$$

where $\mathscr{A}$ is arithmetical in x, y, and F.

PROOF. If (4) holds, then H is the unique function such that

$$(\bigwedge x)(\bigvee F)\mathscr{A}(x, Hx, F).$$

But this is equivalent to

$$(\bigvee F)(\bigwedge x)\mathscr{A}(x, Hx, F_x).$$

Hence there is an arithmetical formula $\mathscr{C}(H, F)$ such that H is the unique function for which there is an F with $\mathscr{C}(H, F)$. Therefore H is hyperarithmetical.

Conversely, if H is the unique function for which $\mathscr{C}(H, F)$ holds for some F then

$$Hx = y \leftrightarrow (\bigvee G)(\mathscr{C}(KG, LG) \wedge KGx = y)$$

where the formula in parentheses is arithmetical in G.

REMARK. If (4) holds, then

$$\neg Hx = y \leftrightarrow (\bigvee F)(\bigvee z)(\mathscr{A}(x, z, F) \wedge \neg z = y).$$

This corresponds to the fact that if the graph of a function is recursively enumerable, then it is recursive and the function is recursive.

THEOREM. *A nonempty set $\mathscr{S}$ is hyperarithmetical if and only if there is a system $\Sigma(S, F, U_1, \cdots)$ of functional equations which has a solution and is such that $\mathscr{R}(F) = \mathscr{S}$ for every F for which there are $U_1, \cdots$ with $\Sigma(S, F, U_1, \cdots)$.*

PROOF. Let H be the characteristic function of $\mathscr{S}$ and k be an element of $\mathscr{S}$. Then put

$$F = I \cdot H + S^k O \cdot ZH.$$

Then $\mathscr{R}(F) = \mathscr{S}$ and F is hyperarithmetical if H is.

On the other hand, suppose $\mathscr{S}$ is $\mathscr{R}(F)$ for every function F for which $\Sigma(S, F, U_1, \cdots)$ holds for some $U_1, \cdots$. Without loss of generality, we may assume $\mathscr{S}$ is not the entire set of natural numbers. Then the characteristic function H of $\mathscr{S}$ is defined by Σ together with

$$HF = SO, \qquad F = AD, \qquad HASD = O, \qquad AB = I. \tag{5}$$

Indeed, $AB = I$ insures that A assumes all values. Hence $HAD = SO$ and $HASD = O$ imply that $\mathscr{R}(AD)$ and $\mathscr{R}(ASD)$ are complementary sets. Thus $Hx = 1$ if $x \in \mathscr{R}(F)$ and $Hx = 0$ otherwise. Moreover for any F such that $\mathscr{R}(F)$ is not the entire set of natural numbers, there are functions A and B satisfying (5).

§8. Finite sequences. In general there is no way to compute the values of a function F from a system Σ of functional equations which defines it since if F is not recursive it is not computable. Nevertheless we can answer certain questions about systems of functional equations by considering approximations to solutions of Σ by means of finite sequences, i.e. number-theoretic functions whose domain is an initial segment of the natural numbers. Small capitals $\mathrm{F}, \mathrm{G}, \ldots$ will be used to represent finite sequences.

If a system $\Sigma(S, F, G, \cdots)$ of functional equations has a solution, then for every t there are finite sequences $\mathrm{F}, \mathrm{G}, \cdots$ so that

$$\Sigma(S, \mathrm{F}, \mathrm{G}, \cdots)n \text{ for all } n \leqq t.$$

(Here $\Sigma(S, \mathrm{F}, \mathrm{G}, \cdots)n$ indicates that each equation of Σ holds at n and $\mathrm{F}, \mathrm{G}, \cdots$ are defined for all the relevant arguments.) Indeed, we can choose for $\mathrm{F}, \mathrm{G}, \cdots$ suitable restrictions of some solution $F, G, \cdots$ of Σ.

What can be said in the converse direction? Let $\Sigma(S, F, G, \cdots)n$ indicate that each equation of Σ holds at n. Then

$$\Sigma(S, F, G, \cdots) \leftrightarrow (\bigwedge n)\Sigma(S, F, G, \cdots)n.$$

But it is not true that

$$(1) \qquad (\bigwedge t)(\bigvee \mathrm{F}, \mathrm{G}, \cdots)(\bigwedge n \leqq t)\Sigma(S, \mathrm{F}, \mathrm{G}, \cdots)n$$

implies that Σ has a solution. For example, let Σ be the system

$$(2) \qquad SI = S, \qquad FS = I, \qquad GF = I.$$

The first equation insures that I is the identity function. Since $FS = I$, F must assume all values for positive arguments. Hence F is not univalent so there is no G with $GF = I$. Hence Σ has no solution.

On the other hand, let t be given and put

$$\begin{aligned} \mathrm{F}x &= t \text{ if } x = 0, & \mathrm{G}x &= x + 1 \text{ if } x < t,\\ &= x - 1 \text{ if } 0 < x \leqq t + 1, & &= 0 \text{ if } x = t, \end{aligned}$$

then $\mathrm{F}Sn = n$ and $\mathrm{GF}n = n$ for all $n \leqq t$. Hence if I is the restriction of I to $\{0, \cdots, t\}$, then (1) holds for Σ but Σ has no solution.

We introduce the notation $\alpha \doteq \beta$ where α and β are partial or total functions for $\alpha t = \beta t$ whenever both sides of the equation are defined. Also if Σ is a system of equations then we define $\dot{\Sigma}$ to be the system obtained by replacing $=$ by $\doteq$ throughout Σ. Thus $\Sigma(\mathrm{F}_1, \cdots, \mathrm{G}_1, \cdots)t$ implies that each equation of Σ holds at t with all the relevant function values defined while $\dot{\Sigma}(\mathrm{F}_1, \cdots, \mathrm{G}_1, \cdots)t$ implies only that each equation either holds at t or at least one side of it is undefined.

We say F is a part of F, i.e. $\mathrm{F} \subset F$, if F agrees with F whenever F is defined. Similarly F is a part of G if G is an extension of F.

LEMMA. $\Sigma(F, G_1, \cdots) \leftrightarrow (\bigwedge \mathrm{F})\{\mathrm{F} \subset F \rightarrow \dot{\Sigma}(\mathrm{F}, G_1, \cdots)\}$.

PROOF. If the left side holds and $\mathrm{F} \subset F$, then F does not violate any equation of Σ so $\dot{\Sigma}(\mathrm{F}, G_1, \cdots)$. If the right side holds, then for every n there is an $\mathrm{F} \subset F$ such that $\Sigma(\mathrm{F}, G_1, \cdots)n$. Hence F does not violate any equation in Σ so $\Sigma(F, G_1, \cdots)t$ for all t. Thus $\Sigma(F, G_1, \cdots)$.

Consider the order relation $\prec$ defined over the set of all nonempty finite sequences as follows: $\mathrm{F} \prec \mathrm{G}$ if F is a proper extension of G or if at the first place t where F and G differ $\mathrm{F}t < \mathrm{G}t$. This is the lexicographical ordering if we first extend each sequence by a single extra term ∞. Under $\prec$ the set of nonempty finite sequences have the order type of the rationals η. This is the only ordering of finite sequences which we shall use. Hence if we say that a set of finite sequences is well-ordered or one finite sequence precedes another, we shall always refer to this ordering.

LEMMA. *Let $\mathscr{S}$ be a set of finite sequences such that if $G \in \mathscr{S}$ and F is a part of G then $F \in \mathscr{S}$. Then $\mathscr{S}$ contains every part of some function if and only if $\mathscr{S}$ is not well-ordered.*

PROOF. Suppose $\mathscr{S}$ contains every part of F. Let $F \restriction n$ be F restricted to $\{0, \cdots, n-1\}$. Then

$$\cdots \prec F \restriction 3 \prec F \restriction 2 \prec F \restriction 1$$

is a descending sequence contained in $\mathscr{S}$, so $\mathscr{S}$ is not well-ordered.

Conversely, if $\mathscr{S}$ is not well-ordered there is a descending sequence $\mathscr{C}$ consisting of $\cdots \prec G_2 \prec G_1 \prec G_0$ contained in $\mathscr{S}$. We shall prove by induction on x that $\lim_{k\to\infty} G_k x$ exists. If $G_k \preccurlyeq G_l$ then $G_k 0 \leqq G_l 0$. Hence $\lim_{k\to\infty} G_k 0$ exists since a decreasing sequence of natural numbers always converges. Suppose $\lim_{k\to\infty} G_k y$ exists for $y < x$. Then there is an n such that for all $k \geqq n$, $G_k \restriction x = G_n \restriction x$. Hence for $k, l \geqq n$, if $G_k \prec G_l$ then $G_k x \leqq G_l x$. Therefore $\lim_{k\to\infty} G_k x$ exists. Let

$$Fx = \lim_{k\to\infty} G_k x.$$

Now every part of F is a part of some G in $\mathscr{C}$. Hence every part of F belongs to $\mathscr{S}$ which was to be proved.

THEOREM. *Let $G_1, \cdots$ be any given functions. A system of equations $\Sigma(G_1, \cdots, F)$ has a solution for F if and only if $\{F: \dot{\Sigma}(G_1, \cdots, F)\}$ is not well-ordered.*

PROOF. Let $\mathscr{M} = \{F: \dot{\Sigma}(G_1, \cdots, F)\}$. By the first lemma, Σ has a solution F if and only if every part of F belongs to $\mathscr{M}$. Hence to obtain the theorem from the second lemma, we need only check that if $F \in \mathscr{M}$ and $F' \subset F$ then $F' \in \mathscr{M}$. But this is clear since if F does not violate any equation of Σ then no part of F can violate an equation of Σ.

COROLLARY. *To every hyperarithmetical function H_0 there is a system Θ of functional equations in S, K, L, H, and G such that H_0 is the unique function H for which $\Theta(S, K, L, H, G)$ has no solution for G. Hence H_0 is the unique function H for which $\{G: \dot{\Theta}(S, K, L, H, G)\}$ is well-ordered.*

PROOF. Consider a system of functional equations in $S, H, U_1, \cdots$ which defines H_0. We can make up a system with just one auxiliary function U if we allow K and L as well as S to be considered given functions. Thus if H_0 is hyperarithmetical, then there is a system of equations Σ in S, K, L, H, and U such that H_0 is the unique function for which $\Sigma(S, K, L, H, U)$ has a solution for U. Hence H_0 is the unique function H for which $\Sigma(S, K, L, F, U)$, $F \neq H$ cannot be satisfied for any F and U. Therefore H_0 is the unique function H for which there are no functions F, U, A, B such that

$$\Sigma(S, K, L, F, U), \qquad AFB = SAHB. \tag{3}$$

Indeed, if $Fn \neq Hn$, let $B = S^n O$ and

$$\begin{aligned} Ax &= 1 \text{ if } x = Fn, \\ &= 0 \text{ otherwise,} \end{aligned}$$

then $AFB = SAHB$. On the other hand, if $AFB = SAHB$, then $F \neq H$.

4—J.S.L.

By successive pairing, we can replace the auxiliary functions of (3), i.e. F, U, A, and B, by a single auxiliary function G. Hence the required Θ can be obtained by replacing F, U, A, B in (3) by KG, KLG, KL^2G, L^3G respectively.

§9. **Uniqueness of auxiliary functions.** In this section we will show that from a system $\Sigma(S, F, U_1, \cdots)$ of functional equations which determine F uniquely, we can construct a new system $\Sigma'(S, F, V_1, \cdots)$ which determines not only F but also $V_1, \cdots$ uniquely. We cannot always accomplish this by merely adjoining more equations to Σ. Indeed, we shall see in the next section that there is a system Ψ of equations which has a solution but no hyperarithmetical solution. Obviously no extension of $\Psi \wedge \Sigma$ can uniquely determine all its auxiliary functions since they are not all hyperarithmetical.

Let F_0 be hyperarithmetical then there are systems of functional equations $\Sigma(S, K, L, F, G)$, written $\Sigma(F, G)$ for short, and $\Theta(S, K, L. F, H)$, written $\Theta(F, H)$, such that (1) F_0 is the unique function F for which $\Sigma(F, G)$ has a solution for G and (2) F_0 is the unique function F for which $\Theta(F, H)$ has no solution for H. Then the combined system $\Sigma(F, G) \wedge \Theta(F, H)$ has no solution. Let $\Phi(M)$ be the system

$$\Psi(KKM, LKM) \wedge \Theta(KKM, LM).$$

Hence by the theorem of the last section, the set $\mathscr{M} = \{M: \dot{\Phi}(M)\}$ is well-ordered by $\prec$. Let $\mathscr{M}(F) = \{H: \dot{\Theta}(F, H)\}$. Now $\mathscr{M}(F)$ will be well-ordered by $\prec$ if and only if $F = F_0$. Furthermore it turns out that the order type (considered as an ordinal) of $\mathscr{M}(F_0)$ does not exceed that of $\mathscr{M}$. Hence F_0 is the unique function F such that $\mathscr{M}(F)$ can be mapped by a one-to-one order-preserving function onto an initial segment of $\mathscr{M}$. Note that the mapping function is also uniquely determined. It is this characterization of F_0 that we use to construct Σ'.

An *initial segment* of a set $\mathscr{S}$ of finite sequences is a set $\mathscr{T} \subseteq \mathscr{S}$ such that for all X if $X \in \mathscr{T}$ and $Y \prec X$ then $Y \in \mathscr{T}$. We will write $F: \mathscr{A} \to \mathscr{S}$ to stand for 'F maps $\mathscr{A}$ in a one-to-one order-preserving way onto an initial segment of $\mathscr{S}$'.

Let G_0 be any solution of $\Sigma(F_0, G)$. If $H \in \mathscr{M}(F_0)$ and the domain of H is $\mathscr{D}(H$ then let M in $\mathscr{M}$ correspond to H where

$$KKM = F_0 \restriction \mathscr{D}(H), \qquad LKM = G_0 \restriction \mathscr{D}(H), \qquad LM = H.$$

We wish to show that this mapping of $\mathscr{M}(F_0)$ into $\mathscr{M}$ is one-to-one and order-preserving. We will show first that $M \in \mathscr{M}$. Since KKM and LKM are restrictions of a solution of $\Sigma(F, G)$, it follows that $\dot{\Sigma}(KKM, LKM)$. Also $\dot{\Theta}(F_0, H)$, KKM is a restriction of F_0, and $H = LM$ so $\dot{\Theta}(KKM, LM)$. Hence $M \in \mathscr{M}$. Now suppose $H \prec H'$ for some H and H' in $\mathscr{M}(F_0)$. Case I. H is an extension of H'. By definition then M is an extension of M' so $M \prec M'$. Case II. $Hx < H'x$ and $Hy = H'y$ for all $y < x$. Then $My = M'y$ for all $y < x$ and $Mx < M'x$. (Here we use the fact that if $v < v'$ then $J(u, v) < J(u, v')$.) Hence $M \prec M'$. Therefore $\mathscr{M}(F_0)$ is mapped in a one-to-one order-preserving way into $\mathscr{M}$. So the ordinal of $\mathscr{M}(F_0)$ does not exceed that of $\mathscr{M}$.

Next we will number all (nonempty) finite sequences and express the characterization of F_0 in terms of this numbering. Let $\alpha_0, \alpha_1, \cdots$ be a list of all finite sequences without repetition. We call m the number of the sequence α_m. Now redefine $\mathcal{M}$, $\mathcal{M}(F)$ to be the corresponding sets of numbers and '$\prec$' to be a binary relation over the natural numbers such that $m \prec n$ if and only if $\alpha_m \prec \alpha_n$. Hence

(1) $$m \in \mathcal{M} \leftrightarrow \dot{\Phi}(\alpha_m),$$

(2) $$m \in \mathcal{M}(F) \leftrightarrow \dot{\Theta}(F, \alpha_m),$$

(3) $$m \preceq n \leftrightarrow (\bigwedge t \in \mathscr{D}(\alpha_n))\alpha_n t = \alpha_m t \vee (\bigvee x)(\alpha_m x < \alpha_n x \wedge (\bigwedge y < x)\alpha_m y = \alpha_n y)$$

Also an *initial segment* of a set $\mathscr{S}$ of natural numbers is the set of numbers of the elements of some initial segment of the corresponding set of finite sequences. Lastly, '$N: \mathscr{S} \to \mathscr{T}$' where $\mathscr{S}$ and $\mathscr{T}$ are sets of natural numbers stands for 'N maps $\mathscr{S}$ onto an initial segment of $\mathscr{T}$ in a one-to-one order-preserving way'. Hence

(4) $$N: \mathscr{S} \to \mathscr{T} \leftrightarrow (\bigwedge n, n' \in \mathscr{S})(n \prec n' \leftrightarrow Nn \prec Nn') \wedge (\bigwedge n \in \mathscr{S})(Nn \in \mathscr{T}) \wedge (\bigwedge n \in \mathscr{T})(\bigwedge m \in \mathscr{S})(n = Nm \to (\bigwedge t \in \mathscr{T})(t \prec n \to (\bigvee r \in \mathscr{S})(t = Nr))).$$

Now F_0 is the unique function F such that for some function N, $N: \mathcal{M}(F) \to \mathcal{M}$. Furthermore if we put the additional condition on N that $Nx = 0$ if $x \notin \mathcal{M}(F)$, then N will also be uniquely determined.

It is sufficient to show that there is a formula $\mathscr{A}(F, N)$ arithmetical in F and N which is equivalent to

(5) $$N: \mathcal{M}(F) \to \mathcal{M} \wedge (\bigwedge x \notin \mathcal{M}(F))Nx = 0.$$

Indeed, if this is the case, let $\mathscr{A}(KA, LA)$ be $\theta_k(A)$. Of course $\mathscr{A}$ will have no free number variables so $\theta_k(A)$ is equivalent to $\langle 0, 0, \cdots \rangle$ satisfies $\theta_k(A)$. By §6, there is a system $\Omega(S, A, W, U_1, \cdots)$ which defines W_A and has a unique solution. Thus, the system

$$\Omega(S, A, W, U_1, \cdots) \wedge KA = F \wedge WJ(S^k O, O) = SO$$

or more simply

(6) $$\Omega(S, A, W, U_1, \cdots) \wedge KA = F \wedge WDS^k O = SO$$

has a unique solution and in this solution $F = F_0$. Hence for Σ' we can take (6) together with defining equations for K, D, and O. The value of k can be determined for a given Σ by translating (5) into an arithmetical formula by the method outlined below.

Hence it remains to show that (5) is equivalent to a formula arithmetical in F and N. By (4), it is enough to show that $\mathcal{M}$ and $\prec$ are arithmetical and $\mathcal{M}(F)$ is arithmetical in F. We can build up a formula equivalent to $\dot{\Phi}(\alpha_m)$ from formulas of the forms: $Sx = y$, $Kx = y$, $Lx = y$, $x \in \mathscr{D}(\alpha_m)$, and $\alpha_m x = y$. Since S, K, and L are arithmetical and $x \in \mathscr{D}(\alpha_m)$ is equivalent to $(\bigvee y)\alpha_m x = y$, we see that $\dot{\Phi}(\alpha_m)$ i.e. $\mathcal{M}$, will be arithmetical if $\alpha_m x = y$ is a ternary arithmetical relation. Similarly, $\mathcal{M}(F)$ will be arithmetical. Finally, the binary relation $\prec$ will be arithmetical provided $\alpha_m x = y$ is arithmetical by means of (3).

We now number the nonempty finite sequences as follows: the finite sequence $\langle F0, F1, \cdots, Fn \rangle$ has number

$$J(n, J(F0, J(F1, \cdots J(F(n-1), Fn) \cdots))).$$

Thus, α_m is the sequence whose domain is $\{0, \cdots, Km\}$ and whose values are given by

$$\begin{aligned} \alpha_m x &= KL^{x+1}m \quad \text{if } x < Km, \\ &= L^{x+1}m \quad \text{if } x = Km. \end{aligned}$$

Hence $\alpha_m x$ is a primitive recursive function of two variables and is arithmetical by Gödel's Lemma. Thus (5) is equivalent to a formula arithmetical in F and N and (6) is the required system of equations. Hence we obtain the following results.

THEOREM. *F_0 is hyperarithmetical if and only if there is a system $\Sigma(S, F, U_1, \cdots)$ of functional equations which has a solution only if $F = F_0$ and the solution is unique.*

We can replace $\Sigma(S, F, U_1, \cdots)$ with a system of equations in S, K, and G where $G = J(L, J(F, J(U_1, \ldots)))$. Here $KG = L$, $K(KG)G = F$, so we obtain:

COROLLARY. *F is hyperarithmetical if and only if there is a system Σ of equations in S, K, and one unknown function G such that $\Sigma(S, K, G)$ has a unique solution G and $F = KKGG$.*

§10. Hyperarithmetical solvability of functional equations. In this section we shall prove the following theorem due to Kleene [7].

THEOREM. *There is a system $\Sigma(S, K, L, F)$ of functional equations which has a solution for F but has no hyperarithmetical solution.*

For any function G, let $G_n x = GJ(n, x)$. We say that G *lists* G_n. If G lists every hyperarithmetical function then G is not hyperarithmetical itself. This follows by the usual diagonal argument, since $SGJ(I, I)$ differs from G_n at n and hence is not listed by G.

From the last section, we know that to every hyperarithmetical function H, there is a function F which is the unique solution of $\theta_k(F)$ for some k and $H = KF$. Here $\theta_k(F)$ has no free variables hence we may as well say that F is the unique function such that $\langle 0, 0, \cdots \rangle$ satisfies $\theta_k(F)$. Clearly, no new functions F are defined if we consider all such functions. Let $\mathscr{U}$ be the class of all functions F such that for some n, F is the unique function for which $\langle 0, 0, \cdots \rangle$ satisfies $\theta_n(F)$. Then if G lists every function in $\mathscr{U}$, then KG lists every hyperarithmetical function. Therefore G is not hyperarithmetical. The proof of Kleene's theorem will exhibit a system of equations Σ such that if F satisfies Σ then KF lists all functions in $\mathscr{U}$.

PROOF OF THE THEOREM. Let $\Psi(F, W, U)$ be a system of functional equations in S, K, L, F, W, and U where S, K, L are given, F is a parameter, and W, U are uniquely determined in terms of F, and such that $\Psi(F, W, U)$ implies that W is the satisfaction function W_F. We can easily construct such a system by pairing the auxiliary functions in Ω. Then

(1) $\quad [\langle 0, 0, \cdots \rangle \text{ satisfies } \theta_n(F)] \leftrightarrow (\bigvee W, U)(\Psi(F, W, U) \wedge WDn = 1)$.

Let $\Psi_n(G)$ be the system of equations

(2) $$\Psi(KG, KLG, LLG) \wedge KLGDS^nO = SO.$$

Since S, K, L, D, and O are arithmetical functions, $\Psi_n(G)$ is arithmetical in n and G. Furthermore

$$[\langle 0, 0, \cdots \rangle \text{ satisfies } \theta_n(F)] \leftrightarrow (\bigvee G)[F = KG \wedge \Psi_n(G)].$$

Let

$$\mathscr{M}_n = \{m: \dot{\Psi}_n(\alpha_m)\}.$$

Then $\mathscr{M}_n$ is well-ordered by $\prec$ if and only if there is no G for which $\Psi_n(G)$, i.e. no F for which $\langle 0, 0, \cdots \rangle$ satisfies $\theta_n(F)$. Furthermore the binary relation $m \in \mathscr{M}_n$ is arithmetical. We can check this by the same method used in §9. Finally, let

$$\mathscr{N}_n(G) = \{m: m \in \mathscr{M}_n \wedge \neg\alpha_m \subset G\}.$$

Clearly, $\mathcal{N}_n(G)$ is arithmetical in n and G. Also $\mathcal{N}_n(G)$ is obtained by deleting one descending sequence from $\mathcal{M}_n$. Hence $\mathcal{N}_n(G)$ is well-ordered if and only if $\mathcal{M}_n$ itself is well-ordered or if G is the unique function such that all parts of G belong to $\mathcal{M}_n$. Hence the two conditions $\Psi_n(G)$ and $\mathcal{N}_n(G)$ is well-ordered imply that KG is the unique function such that $\langle 0, 0, \cdots \rangle$ satisfies $\theta_n(KG)$.

Therefore there is exactly one function H for which there are functions G and G' such that

$$(3) \qquad \begin{aligned} &(\bigwedge n)\{(\Psi_n(G_n) \wedge \Psi_n(G'_n) \wedge G_n \neq G'_n \wedge H_n = O) \\ &\quad \vee (\Psi_n(G_n) \wedge \mathcal{N}_n(G_n) \text{ is well-ordered} \wedge H_n = KG_n) \\ &\quad \vee (\mathcal{M}_n \text{ is well-ordered} \wedge H_n = O)\}, \end{aligned}$$

since for each n exactly one of the three alternatives holds. H is certainly not hyperarithmetical since H lists every function in $\mathcal{U}$.

Consider the formula

$$(4) \qquad \begin{aligned} &(\bigwedge n)\{(\Psi_n(G_n) \wedge \Psi_n(G'_n) \wedge G_n \neq G_n \wedge H'_n = O) \\ &\quad \vee (\Psi_n(G_n) \wedge G'_n : \mathcal{N}_n(G_n) \to \mathcal{T} \wedge H_n = KG_n) \\ &\quad \vee (G'_n : \mathcal{M}_n \to \mathcal{T} \wedge H_n = O)\}. \end{aligned}$$

It is arithmetical in G, G', H, and $\mathcal{T}$. If $\mathcal{T}$ is a well-ordered set whose ordinal is sufficiently large, i.e. at least as great as the ordinals of any set $\mathcal{N}_n(G_n)$ or $\mathcal{M}_n$ which is well-ordered, then (4) uniquely determines H as before. Hence $\mathcal{T}$ cannot be hyperarithmetical for otherwise H would be.

Now consider the class $\mathcal{C}$ of functions H for which there are G, G', and $\mathcal{T}$ such that (4) holds. In the first place, $\mathcal{C}$ is not empty. Indeed, if we take for $\mathcal{T}$ a well-ordered set with sufficiently large ordinal, then the function H defined by (3) satisfies (4) for suitable G and G'. On the other hand, if $H \in \mathcal{C}$, then we will show that H lists every function in $\mathcal{U}$. Suppose $U \in \mathcal{U}$, say U is the unique function such that $\langle 0, 0, \cdots \rangle$ satisfies $\theta_n(U)$. Then for every H, G, G', and $\mathcal{T}$ which satisfies (4), $U = H_n$. Otherwise we must have

$$G'_n : \mathcal{M}_n \to \mathcal{T}$$

since the first alternative of (4) does not hold. Let $\hat{G}$ be the unique function such that $\Psi_n(G)$ holds. Then $\mathcal{N}_n(\hat{G})$ is well-ordered and hyperarithmetical. By hypothesis $\mathcal{M}_n$ is not well-ordered, hence every well-ordered initial segment of $\mathcal{T}$ is the map of a well-ordered initial segment of $\mathcal{M}_n$, and this is also an initial segment of $\mathcal{N}_n(\hat{G})$. Hence the ordinal of $\mathcal{N}_n(\hat{G})$ could be used instead of $\mathcal{T}$. But this is impossible since $\mathcal{N}_n(\hat{G})$ is hyperarithmetical and well-ordered. Hence H lists every function in $\mathcal{U}$ and is not hyperarithmetical. Replace $\mathcal{T}$ in (4) by $\mathcal{R}(B)$, then (4) is arithmetical in H, G, G', and B.

Finally, let $\hat{F} = J(H, J(G, J(G', B)))$. Then (4) is equivalent to a formula $\mathcal{B}(\hat{F})$ arithmetical in $\hat{F}$, say $\theta_r(\hat{F})$ where $\theta_r(\hat{F})$ has no free number variables. Thus the system of equations

$$\Psi(\hat{F}, W, U) \wedge WDS^r O = SO$$

has solutions but none in which $\hat{F}$ is hyperarithmetical. We now obtain Kleene's theorem by pairing D, O, $\hat{F}$, W, and U.

REFERENCES

[1] MARTIN DAVIS, *The undecidable*, Raven Press, Hewlett, New York, 1965.

[2] K. GÖDEL, *Über formal unentscheidbare Sätze der Principia Mathematica und verwandter System*. I, ***Monatshefte für Mathematik und Physik***, vol. 38 (1931), pp. 173–198.

[3] SOLOMON FEFERMAN, *Some applications of the notions of forcing and generic sets*, ***Fundamenta mathematicae***, vol. 56 (1964/65), pp. 325–345.

[4] A. GRZEGORCZYK, A. MOSTOWSKI, and C. RYLL-NARDZEWSKI, *The classical and the ω-complete arithmetic*, this JOURNAL, vol. 23 (1958), pp. 188–206.

[5] S. C. KLEENE, *Hierarchies of number-theoretic predicates*, ***Bulletin of the American Mathematical Society***, vol. 61 (1955), pp. 193–213.

[6] S. C. KLEENE, *Arithmetical predicates and function quantifiers*, ***Transactions of the American Mathematical Society***, vol. 79 (1955), pp. 312–340.

[7] S. C. KLEENE, *Quantification of number-theoretic functions*, ***Composito mathematica***, vol. 14 (1959), pp. 23–40.

[8] CLIFFORD SPECTOR, *Recursive well-orderings*, this JOURNAL, vol. 20 (1955), pp. 151–163.

[9] CLIFFORD SPECTOR, *Hyperarithmetical quantifiers*, ***Fundamenta mathematicae***, vol. 48 (1960), pp. 313–320.

[10] A. TARSKI, A. MOSTOWSKI, and R. M. ROBINSON, ***Undecidable theories***, North-Holland Publishing Company, Amsterdam, 1953.

UNIVERSITY OF CALIFORNIA, BERKELEY

RECURSIVE FUNCTIONS OF ONE VARIABLE

JULIA ROBINSON

We say that the function F is obtained by *general recursion* from A, B, M, and N if

(i) $FA=M$, $FB=NF$ and

(ii) every natural number n belongs to $\mathcal{R}(B^kA)$ for some $k\geqq 0$.

If F satisfies (i), then $F(B^kAt)=N^kMt$. If A and B satisfy (ii), then every natural number n is B^kAt for some k and t. Hence F is uniquely determined. Furthermore if A, B, M, and N are computable functions, then given n, we can obtain k and t effectively so F is computable.

Condition (ii) is satisfied if A is the zero function O and B is the successor function S. There is a function obtained by general recursion from O, S, M, and N if and only if M is a constant function, say $M=S^mO$. Then $Ft=N^tm$. Thus, iteration is included in general recursion.

If G is a permutation, then $F=G^{-1}$ can be obtained by general recursion from GS, GO, S, and O since $FGS=S$, $FGO=OF$, and every n belongs to $\mathcal{R}((GO)^kGS)$ for $k=0$ or $k=1$. More generally, if F is determined by $FG=H$ where G assumes all values, then F is obtained by general recursion from GS, GO, HS, and HO since $FGS=HS$, $FGO=HOF$, and the side condition holds as before.

THEOREM. *Every (general) recursive function of one variable can be obtained from O and S by repeated compositions and general recursions from previously defined functions.*

We need to prove a lemma first. Let $J(x, y)$ be a one-to-one mapping of the set of ordered pairs of natural numbers onto the set of natural numbers and let K and L be the inverse functions so $KJ(x, y)=x$ and $LJ(x, y)=y$. Any such function J will be called a *pairing function.*

LEMMA. *Every primitive recursive function of one variable can be obtained from O, S, K, and L by repeatedly constructing a new function F from previously obtained functions A and B by composition ($F=AB$), pairing ($F=J(A, B)$), or general recursion of the form $FJ(I, O)=A$ and $FJ(K, SL)=BF$.*

Received by the editors April 1, 1967.

REMARK. The general recursion of the lemma always defines a function. In fact, for every n, there is a unique pair r, s with $n=J(r, s)$, so $Fn=B^sAr$.

PROOF. If J, K, and L are adjoined to the usual initial functions in the standard definition of primitive recursive functions, then recursion can be replaced by iterations of the form

$$G(x, 0) = Ax, \qquad G(x, Sy) = BG(x, y). \tag{1}$$

(See R. M. Robinson [3, p. 932].) Since J is an initial function, we can replace this iteration scheme by the general recursion of the lemma. Namely, if F is defined by $FJ(I, O)=A$ and $FJ(K, SL)=BF$, then the function G defined by (1) is $FJ(x, y)$. Hence every primitive recursive function is obtained from the usual initial functions together with J, K, and L by general recursion of the form in the lemma and substitution. To obtain functions of one variable by substitution, it is always sufficient to use substitution only to define functions of one variable. The only initial function of more than one variable which yields new functions of one variable by substitution is J. Hence besides compositions of functions of one variable, we must allow $F=J(A, B)$ to be constructed from previously obtained functions A and B. Hence all primitive recursive functions of one variable are obtained from O, I, S, K, and L by compositions, pairings, and general recursions of the form given in the lemma. Since $I=J(K, L)$, we can omit I from the initial functions. (This proof closely parallels the proof given by R. M. Robinson in §7 of [3]. A more detailed discussion will be found there.)

PROOF OF THE THEOREM. Let $\mathfrak{R}$ be the class of functions obtainable from O and S by composition and general recursion. We will show first that every primitive recursive function belongs to $\mathfrak{R}$.

Suppose that J is a pairing function such that $J(I, O)$ and $J(K, SL)$ belong to $\mathfrak{R}$. Then I, K, and L are in $\mathfrak{R}$ as we see by the general recursions:

$$IO = O, \qquad IS = SI;$$
$$KJ(I, O) = I, \qquad KJ(K, SL) = IK;$$
$$LJ(I, O) = O, \qquad LJ(K, SL) = SL.$$

In order to use the lemma, we need to show that $\mathfrak{R}$ is closed under pairing. Let $F^*=J(K, FL)$ for each F. We will show by induction that $\mathfrak{R}$ is closed under the *-operation.

For the initial functions O and S, we have

$$O^* = J(K, OL) = J(I, O)K, \qquad S^* = J(K, SL),$$

so O^* and S^* are both in $\mathfrak{R}$.

Furthermore,

$$A^*B^* = J(K, AL)J(K, BL) = J(K, ABL) = (AB)^*.$$

Hence if $F=AB$ and A^*, $B^* \in \mathcal{R}$, then F^* is in $\mathcal{R}$.

Suppose F is obtained from A, B, M, and N by general recursion $FA=M$, $FB=NF$, and A^*, B^*, M^*, and N^* belong to $\mathcal{R}$. Then

$$F^*A^* = (FA)^* = M^*, \qquad F^*B^* = (FB)^* = (NF)^* = N^*F^*.$$

We still must check that A^* and B^* satisfy (ii) given that A and B do. For every n, there are r and s so that $n=J(r, s)$. By hypothesis $s=B^kAt$ for some k and t, so $(B^*)^kA^*J(r, t)=J(r, B^kAt)=J(r, s)=n$. Hence n belongs to $\mathcal{R}((B^*)^kA^*)$ and (ii) holds. Therefore if F is in $\mathcal{R}$, so also is F^*.

Next we will show that if A and B are in $\mathcal{R}$, then $J(A, B)$ is in $\mathcal{R}$. Now

$$J(A, B) = J(L, K)A^*J(L, K)B^*J(I, I).$$

To check this, notice that

$$J(L, K)A^* = J(L, K)J(K, AL) = J(AL, K), \qquad J(L, K)B^* = J(BL, K).$$

Hence

$$\begin{aligned} J(L, K)A^*J(L, K)B^*J(I, I) &= J(AL, K)J(BL, K)J(I, I) \\ &= J(AK, BL)J(I, I) = J(A, B). \end{aligned}$$

Both $J(L, K)$ and $J(I, I)$ are in $\mathcal{R}$ as we see from the following series of definitions by general recursions:

$$\begin{aligned} J(O, I)O &= J(I, O)O, & J(O, I)S &= J(K, SL)J(O, I); \\ J(SK, L)J(I, O) &= J(I, O)S, & J(SK, L)J(K, SL) &= J(K, SL)J(SK, L); \\ J(L, K)J(I, O) &= J(O, I), & J(L, K)J(K, SL) &= J(SK, L)J(L, K); \\ J(I, I)O &= J(I, O)O, & J(I, I)S &= J(SK, L)J(K, SL)J(I, I). \end{aligned}$$

We have already seen that (ii) holds in each of these case, so the definitions are legitimate. Hence $J(A, B)$ is in $\mathcal{R}$.

Hence by the lemma, every primitive recursive function belongs to $\mathcal{R}$ provided there is a pairing function J such that $J(I, O)$ and $J(K, SL)$ are in $\mathcal{R}$. One such J is given by $J(x, y)=2^y(2x+1)-1$. For this J, $J(x, 0)=2x$ and $J(x, Sy)=2J(x, y)+1$. Hence $J(I, O)$ is the double function D and $J(K, SL)$ is SD. Of course D belongs to $\mathcal{R}$ since it is defined by the general recursion: $DO=O$ and $DS=SSD$. Therefore every primitive recursive function belongs to $\mathcal{R}$.

Finally, if F is recursive and J is the function given above (or any recursive pairing function) then $\{y: y=J(x, Fx)\}$ is a recursive set and hence the range of some primitive recursive function G. So F is defined by the equation $FKG=LG$, where KG assumes all values. We have already seen that F is obtained by the general recursion $FKGS=LGS$ and $FKGO=LGOF$. Hence every recursive function of one variable is in $\mathcal{R}$.

A *functional equation* is an equation of the form

$$A_1 \cdots A_k = B_1 \cdots B_l.$$

We say that the equation $\alpha=\beta$ is *derivable* from a system Σ of functional equations if there is a chain of equations $\alpha_1= \cdots =\alpha_n$ $(n\geqq 1)$ such that $\alpha_1=\alpha$, $\alpha_n=\beta$, and for each t, α_{t+1} is obtained from α_t by replacing a part γ of α_t by δ where $\gamma=\delta$ or $\delta=\gamma$ belongs to Σ.

A system $\Sigma(S, F, U_1, \cdots)$ of functional equations *defines* a particular function if Σ has a unique solution in which S is the successor function, and in this solution F is the required function. Even though Σ defines F, it may not provide any way to compute the values of F. Indeed, every hyperarithmetical function can be defined by some system of functional equations [**2**]. Hence in general there is no way at all to compute a function defined by a system of functional equations.

EXAMPLE. The equations

$$DO = O, \qquad DS = SSD \tag{2}$$

define both the zero function O and the double function D. The second equation gives $Dx=2x+D0$. Hence the only possible fixed point for D is at $x=0$. Therefore the first equation implies $\mathcal{R}(O)=\{0\}$. Hence O is identically 0 and $Dx=2x$. Furthermore, it is easily seen that every equation of the form $DS^kO=S^{2k}O$ can be derived from (2). (Here S^k is an abbreviation for $S \cdots S$ with kS's.) However, the equation $OO=O$ cannot be derived from (2), since any equation derived from (2) will have the same number of O's on each side.

On the other hand, all equations of the form $OS^kO=O$ can be derived from

$$OO = O, \qquad OS = O. \tag{3}$$

But (3) does not define the function O. Indeed, any constant function can be substituted for O in (3) and it will still be satisfied. However, the system consisting of (2) and (3) does define O and D, and from it, every equation of the forms $OS^kO=O$, $DS^kO=S^{2k}O$ can be derived.

COROLLARY 1. *A function of one variable is recursive if and only if there is a system $\Sigma(S, O, F, U_1, \cdots)$ of functional equations such that*

(i) *Σ has a unique solution in which S is the successor function and O is the zero function. In this solution, F is the given function.*

(ii) *For all natural numbers n and m, the equation $FS^nO = S^mO$ is derivable from Σ if and only if $Fn = m$.*

REMARKS. The "only if" part of (ii) follows from (i) since F satisfies Σ with S and O the successor and zero functions, it must also satisfy any equation derived from Σ. If (ii) holds, we say the *values of F can be derived from Σ*. The theorem corresponding to Corollary 1 for functions of any number of variables was proved by Kreisel and Tait [1].

PROOF. Start with the equations $OO = O$, $OS = O$.

To obtain Σ for a recursive function F, adjoin a sequence of equations corresponding to successive definitions of new functions by compositions and general recursions from O, S, and functions already defined, ending with a definition of F. This is possible by the theorem. Clearly, Σ satisfies (i) since each function introduced is uniquely determined by its defining equations from earlier functions.

We will show now by induction that the values of F can be derived from Σ. Suppose G is O or S. The values of O are derivable from the first two equations. The values of the successor function are given by identities $S(S^kO) = S^{k+1}O$.

Suppose $G = AB$ is in Σ and the values of A and B can be derived from Σ. Then we can derive the values of G since

$$GS^nO = ABS^nO = AS^{Bn}O = S^{ABn}O = S^{Gn}O.$$

Here Bn, ABn, and Gn are natural numbers indicating the number of S's in the expression.

Suppose $GA = M$ and $GB = NG$ are in Σ, every number n is B^kAt for some k and t, and the values of A, B, M, and N are derivable from Σ. Then the values of G are derivable, since

$$\begin{aligned} GS^nO &= GB^kAS^tO = N^kGAS^tO = N^kMS^tO = N^kS^{Mt}O \\ &= S^{N^kMt}O = S^{Gn}O. \end{aligned}$$

Conversely, if Σ satisfies (ii), then F is computable. We simply systematically derive all possible equations from Σ. For each n, we will eventually derive an equation of the form $FS^nO = S^mO$ and conclude that $Fn = m$. Hence by Church's Thesis, F is recursive. A mathematical proof that F is recursive can be given in the standard way.

COROLLARY 2 (POST, TURING 1948). *The word problem for semigroups (associative systems) is recursively unsolvable.*

PROOF. Let F be a recursive function whose range is not recursive and let Σ satisfy Corollary 1 for this F. Let Σ' be obtained from Σ by adjoining the single equation $GF=O$.

Then GS^nO will be derivable from Σ' if and only if n is in $\mathcal{R}(F)$. The function G_0 given by

$$\begin{aligned} G_0 n &= 0 \quad \text{if } n \in \mathcal{R}(F), \\ &= 1 \quad \text{otherwise} \end{aligned}$$

satisfies Σ' and hence $GS^nO=O$ cannot be derived from Σ' if $n \notin \mathcal{R}(F)$. Since $\mathcal{R}(F)$ is not recursive, there is no general method of telling whether or not an arbitrary equation of the form $\alpha=O$ is derivable from Σ'.

REFERENCES

1. Georg Kreisel and William W. Tait, *Finite definability of number-theoretic functions and parametric completeness of equational calculi*, Z. Math. Logik Grundlagen Math. **7** (1961), 28–38.

2. Julia Robinson, *An introduction to hyperarithmetical functions*, J. Symbolic Logic **32** (1967), 325–342.

3. Raphael M. Robinson, *Primitive recursive functions*, Bull. Amer. Math. Soc. **53** (1947), 925–942.

UNIVERSITY OF CALIFORNIA, BERKELEY

FINITE GENERATION OF RECURSIVELY ENUMERABLE SETS

JULIA ROBINSON

Suppose we wish to build up the class of recursively enumerable sets by starting with the set $\mathfrak{N}$ of natural numbers and constructing new sets from those already obtained using as little auxiliary machinery as possible. One way would be to start with a finite number of functions $F_1, \cdots, F_k$ (of one variable, from and to $\mathfrak{N}$) such that every recursively enumerable set can be obtained from $\mathfrak{N}$ by constructing new sets $F_j[\mathfrak{J}]$ where $\mathfrak{J}$ is a previously obtained set. We can think of $F_1, \cdots, F_k$ as unary operations on sets of natural numbers. Any set $\mathcal{S}$ obtained in this way is the range of a function F obtained by composition from $F_1, \cdots, F_k$. If we consider the values of $F_1, \cdots, F_k$ as given, then the number of steps needed to compute Fn does not depend on n. Hence for all $x \in \mathcal{S}$, there exists a proof that $x \in \mathcal{S}$ of bounded length in terms of $F_1, \cdots, F_k$ (just as there is a one-step proof that a composite number is composite in terms of multiplication).

We say a set of natural numbers is *generated* by $F_1, \cdots, F_k$ if it is the range of a function obtained by composition from $F_1, \cdots, F_k$. Also a class $\mathcal{C}$ of sets is *generated* by $F_1, \cdots, F_k$ if every nonempty set of $\mathcal{C}$ is generated by $F_1, \cdots, F_k$ and every set generated by $F_1, \cdots, F_k$ is in $\mathcal{C}$.

EXAMPLE. Let $G_0, G_1, \cdots$ be the primitive recursive functions listed systematically so that the function G given by

$$G(2^n(2x+1)-1) = G_n x$$

is recursive. Then

$$G_n = G(SD)^n D$$

where $Sx = x+1$ and $Dx = 2x$. Every nonempty recursively enumerable set is the range of some primitive recursive function and hence the range of $G(SD)^n D$ for some n. Since S, D, and G are recursive functions, any function obtained from them by composition will be recursive and will have a recursively enumerable range. Thus G, S, and D generate the class of recursively enumerable sets. However, they do not form an interesting set of generators since all the work of listing recursively enumerable sets is done in computing G.

Received by the editors August 22, 1967.

In this paper, we give sets of easy-to-compute generators for the classes of recursively enumerable sets and diophantine sets, i.e. sets which are existentially definable in terms of $+$ and $\cdot$. General theorems describing classes of sets which can be finitely generated are proved in [3].

Let $Ix=x$, $Ox=0$, $Tx=2^x$, $Zx=0^x$, and $E(x, y)$ be the characteristic function of equality, i.e. $E(x, y)=0^{|x-y|}$. We shall also use pairing functions. If $J(x, y)$ maps the set of ordered pairs of natural numbers onto $\mathfrak{N}$ in a one-to-one way, then J and its inverse functions K and L, given by $KJ(x, y)=x$ and $LJ(x, y)=y$, are called pairing functions. We shall always take J to be the Cantor pairing function given by

$$J(x, y) = \tfrac{1}{2}((x + y)^2 + 3x + y)$$

unless otherwise stated. Its inverse functions are probably the simplest functions which assume every natural number infinitely often. They can easily be computed recursively by the equations

$$\begin{aligned}
&K0 = L0 = 0, \\
(1)\quad &K(x + 1) = 0, \qquad L(x + 1) = Kx + 1 \quad \text{if } Lx = 0, \\
&K(x + 1) = Kx + 1, \qquad L(x + 1) = Lx - 1 \quad \text{if } Lx \neq 0.
\end{aligned}$$

EXAMPLE. Let J, K, and L be any pairing functions and let G be given by $G(J(x, n))=G_n x$. Then $G_n x=GJ(K, SL)^n J(I, O)x$. In the example above, we took $J(x, n)=2^n(2x+1)-1$, so $J(x, n+1)=2J(x, n)+1$ and $J(x, 0)=2x$.

LEMMA 1. *The functions Ox, Zx, $x+y$, $x\cdot y$, and $E(x, y)$ are all obtainable from J, K, S, and D by substitution.*

PROOF. R. M. Robinson [4, p. 665], derived the remarkable identity for all x and y with $x\geqq y$:

$$(2)\qquad x - y = KDDJ(SSSKDDJ(SSSKDDJ(y, DDx), x), Dy).$$

He then used it to define $x+y$ by substitution from J, K, S, and D. We shall let $x-y$ denote the function on the right of (2) for *all* x and y. Then $Ox=x-x$ and $x\cdot y=(J(0, x+y)-J(0, x))-J(0, y)$. To define Z, we make use of (1) to see that

$$SJ(0, 0) = J(0, 1), \qquad SJ(0, x + 1) = J(1, x).$$

Hence $KSJ(0, x)=\operatorname{sgn} x$. Since $K2=1$ and $K3=0$, we have $Zx=KSSKSJ(0, x)$. Finally, $E(x, y)=Z((x-y)+(y-x))$.

THEOREM 1. *Every nonempty recursively enumerable set is the range of a function of one variable obtained from S, D, T, K, and L by com-*

position and pairing. Conversely, the range of such a function is recursively enumerable. If T is omitted from the set of initial functions then just the nonempty diophantine sets are obtained.

PROOF. Every recursively enumerable set $\mathcal{S}$ is exponential diophantine and conversely. (See Davis, Putnam, and Robinson [1].) Hence

$$x \in \mathcal{S} \leftrightarrow \bigvee_{y_1, \cdots, y_n} F(x, y_1, \cdots, y_n) = G(x, y_1, \cdots, y_n),$$

where F and G are suitably chosen terms built up from $x, y_1, \cdots, y_n$, and particular natural numbers by means of $+$, $\cdot$, and T. (See [1, Corollary 5].) Let a be any element of $\mathcal{S}$. Then $\mathcal{S} = \mathcal{R}(H)$ where

$$H(x, y_1, \cdots, y_n) = x \cdot E(F(x, y_1, \cdots, y_n), G(x, y_1, \cdots, y_n)) \\ + a \cdot ZE(F(x, y_1, \cdots, y_n), G(x, y_1, \cdots, y_n)).$$

Let $M = H(K, KL, \cdots, KL^{n-1}, L^n)$. Then M is a function of one variable obtained from O, S, K, L, T, and Z by forming new functions F from previously obtained functions A and B by taking $F = AB$, $F = A + B$, $F = A \cdot B$, and $F = E(A, B)$. Also $\mathcal{R}(M) = \mathcal{R}(H) = \mathcal{S}$. By Lemma 1, we see that every such function M can be obtained from K, L, S, D, T, and $I = J(K, L)$ by composition and pairing. On the other hand, M is primitive recursive so $\mathcal{R}(M)$ is recursively enumerable.

The proof that every nonempty diophantine set is obtained if we omit T from the set of initial functions, is obtained from the above proof by omitting T throughout. A function F is said to be diophantine if there is a polynomial P with integer coefficients such that

$$y = Fx \leftrightarrow \bigvee_{u_1, \cdots, u_n} P(x, y, u_1, \cdots, u_n) = 0.$$

Hence if F is diophantine then $\mathcal{R}(F)$ is a diophantine set. Indeed,

$$y \in \mathcal{R}(F) \leftrightarrow \bigvee_{x, u_1, \cdots, u_n} P(x, y, u_1, \cdots, u_n) = 0.$$

Also, if F and G are diophantine then FG is diophantine. Clearly, J, K, L, S, and D are diophantine functions so the range of any function obtained from K, L, S, and D by composition and pairing is diophantine.

REMARK. It is not known whether all recursively enumerable sets are diophantine. Hence we do not know whether T is necessary to obtain all recursively enumerable sets.

Let $C = J(KL, J(LL, K))$. Then $C^3 = I$, since $CJ(x, J(y, z)) = J(y, J(z, x))$. For any function A, we define $A^* = J(K, AL)$. Then for all A and B

(3) $$(AB)^* = A^*B^*,$$

(4) $$J(A, B) = B^*J(L, K)A^*J(I, I),$$

(5) $$A^{**} = CJ(L, K)A^*J(L, K)C^2.$$

These formulas can be easily checked by carrying out the compositions indicated on the right using the fact that $J(FK, GL)J(M, N) = J(FM, GN)$ and $J(FL, GK)J(M, N) = J(FN, GM)$ etc.

LEMMA 2. *If F can be obtained from $A_1, \cdots, A_n$, K, and L by composition and pairing, then F can be obtained from $A_1^*, \cdots, A_n^*$, K, $J(L, K)$, $J(I, I)$, and $J(KL, J(LL, K))$ by composition alone. Here J, K, and L can be arbitrary pairing functions.*

PROOF. Let $\mathfrak{A}$ be the least class of functions closed under composition which contains $A_1^*, \cdots, A_n^*$, K, $J(L, K)$, $J(I, I)$, and C. We wish to show that if F is obtained from $A_1, \cdots, A_n$, K, and L by composition and pairing, then F belongs to $\mathfrak{A}$. Since $F = LF^*J(I, I)$ and $L = KJ(L, K)$, it is sufficient to show that F^* belongs to $\mathfrak{A}$, and this will be done by induction.

I. Suppose $F = A_j$ then F^* belongs to $\mathfrak{A}$ by definition. Also $K^* = LC^2$ and $L^* = J(L, K)LC$. Hence both K^* and L^* belong to $\mathfrak{A}$.

II. Suppose $F = AB$ where A^* and B^* belong to $\mathfrak{A}$. Then $F^* = A^*B^*$ so F^* belongs to $\mathfrak{A}$.

III. Suppose $F = J(A, B)$ where A^* and B^* belong to $\mathfrak{A}$. Then $F = B^*J(L, K)A^*J(I, I)$.

Hence

$$F^* = B^{**}J(L, K)^*A^{**}J(I, I)^*.$$

By (5), B^{**} and A^{**} can be obtained by composition from C, $J(L, K)$, B^*, and A^*. Hence we need only show that $J(L, K)^*$ and $J(I, I)^*$ belong to $\mathfrak{A}$. Now

$$J(L, K)^* = J(K, J(LL, KL)) = CJ(KL, J(K, LL)) = CJ(KL, L^*).$$

Hence by (4),

$$J(L, K)^* = CL^{**}J(L, K)K^*L^*J(I, I).$$

Also

$$J(I, I)^* = J(K, J(L, L)) = L^{**}CJ(I, I).$$

Finally, K^{**} and L^{**} belong to $\mathfrak{A}$ by (5) and I, hence F^* belongs to $\mathfrak{A}$.

THEOREM 2. *The class of recursively enumerable sets is generated by K, $J(K, SL)$, $J(K, DL)$, $J(K, TL)$, $J(L, K)$, $J(I, I)$, $J(KL, J(LL, K))$.*

This set of generators with $J(K, TL)$ deleted generates the class of diophantine sets.

THEOREM 2 is an immediate consequence of Lemma 2 and Theorem 1.

REMARK. By the lemma on page 714 of [2], the functions K, $J(L, K)$, and $J(I, I)$ can be replaced by the two functions $J(LK, KL)$ and $J(L, I)$. (Recall that $K=J(KK, LK)$.) Indeed, the total number of generators can be reduced to two,

$$K \quad \text{and} \quad J(L, J(F_1, J(F_2, \cdots, J(F_{n-1}, F_n))) \cdots),$$

where $F_1, \cdots, F_n$ are the generators other than K.

LEMMA 3. *If F can be obtained by composition and pairing from $A_1, \cdots, A_n$, K, and L then $F=KB$ for some function B obtained by composition from $A_1^*, \cdots, A_n^*$, $J(L, K)$, $J(I, I)$, $J(KL, J(LL, K))$, $J(L, K)^*$, $J(I, I)^*$, and $J(KL, J(LL, K))^*$. Here J, K, and L can be arbitrary pairing functions.*

PROOF. By Lemma 2,

$$F = B_0KB_1K \cdots B_{t-1}KB_t \tag{6}$$

for some $B_0, \cdots, B_t$ obtained from $A_1^*, \cdots, A_n^*$, $J(L, K)$, $J(I, I)$, and $J(KL, J(LL, K))$ by composition. We can take $t>0$ since $B_0=B_0KJ(I, I)$. (If B_i is the identity function, then $B_i=J(L, K)^2$.) Now

$$J(L, K)A^*J(L, K) = J(AK, L).$$

Hence $AK=KJ(L, K)A^*J(L, K)$. Thus each K in (6) can be brought to the front in turn. For example,

$$\begin{aligned} F &= B_0KB_1K \cdots \\ &= KJ(L, K)B_0^*J(L, K)B_1K \cdots \\ &= K^2J(L, K)J(L, K)^*B_0^{**}J(L, K)^*B_1^*J(L, K)B_2K \cdots \\ &\ \ \vdots \\ &= K^tH \end{aligned}$$

where H is obtained from the functions listed in the lemma by composition. Finally,

$$KK = KJ(KL, J(LL, K))J(L, K) = KCJ(L, K)$$

so

$$F = K^t H = K(CJ(L, K))^{t-1}H.$$

THEOREM 3. *Every nonempty recursively enumerable set is the range of a function KB where B is obtained by composition from*

(7) S^*, D^*, T^*, $J(L, K)$, $J(I, I)$, C, $J(L, K)^*$, $J(I, I)^*$, C^*.

The range of B is a primitive recursive set. If T^* is deleted from the set of functions (7) then just the nonempty diophantine sets are obtained. In this case, both the range of B and its complement are diophantine.

PROOF. In light of Lemma 3 and Theorem 1, we need only show that $\mathcal{R}(B)$ is a primitive recursive set in the first case and $\mathcal{C}\mathcal{R}(B)$ is diophantine in the second case. (Here $\mathcal{C}\mathcal{R}(B)$ is the complement of the range of B.) Notice that $Gx \geqq x$ for G equal to S^*, D^*, T^*, $J(I, I)^*$, or $J(I, I)$. This is clear for the *-functions since $J(x, y) \leqq J(x, z)$ whenever $y \leqq z$. Thus $S^*x = J(K, SL)x \geqq J(K, L)x = x$, etc. Also $J(I, I)x = J(x, x) \geqq x$. Hence if G is one of these functions, then

$$x \in \mathcal{R}(GH) \leftrightarrow \bigvee_{y \leqq x} (Gy = x \wedge y \in \mathcal{R}(H)).$$

Furthermore, if G is one of the remaining functions of (7), then G is a primitive recursive permutation such that G^{-1} is also primitive recursive. Indeed, $C^{-1} = C^2$, $(C^*)^{-1} = C^*C^*$, $J(L, K)^{-1} = J(L, K)$, and $(J(L, K)^*)^{-1} = J(L, K)^*$. Hence

$$x \in \mathcal{R}(GH) \leftrightarrow G^{-1}x \in \mathcal{R}(H).$$

Therefore by induction starting with $H = I$, we see that $\mathcal{R}(B)$ is primitive recursive.

In the diophantine case, all the functions of (7) with T^* excluded are diophantine. Hence if B is obtained from them by composition, $\mathcal{R}(B)$ is diophantine. If M is strictly monotone, then

$$x \in \mathcal{C}\mathcal{R}(M) \leftrightarrow \bigvee_{y} (My < x < M(y + 1)), \tag{8}$$

$$x \in \mathcal{R}(M^*) \leftrightarrow \bigvee_{y} (My = Lx), \tag{9}$$

$$x \in \mathcal{C}\mathcal{R}(M^*) \leftrightarrow \bigvee_{y} (My < Lx < M(y + 1)). \tag{10}$$

If G is a univalent function, then

$$\mathcal{C}\mathcal{R}(GH) = \mathcal{R}(G \restriction \mathcal{C}\mathcal{R}(H)) \cup \mathcal{C}\mathcal{R}(G). \tag{11}$$

Suppose $\mathcal{R}(H)$ and $\mathcal{C}\mathcal{R}(H)$ are diophantine, and G is one of the functions S^*, D^*, $J(I, I)$, and $J(I, I)^*$. Then $\mathcal{C}\mathcal{R}(G)$ is diophantine by (8) or (10). Hence by (11),

$$x \in \mathcal{CR}(GH) \leftrightarrow \bigvee_{y} ((y \in \mathcal{CR}(H) \wedge Gy = x) \vee x \in \mathcal{CR}(G)),$$

so $\mathcal{CR}(GH)$ is diophantine. If G is one of the permutations of (7), then

$$x \in \mathcal{CR}(GH) \leftrightarrow \bigvee_{y} (Gy = x \wedge y \in \mathcal{CR}(H)).$$

Hence by induction $\mathcal{CR}(B)$ is diophantine.

References

1. Martin Davis, Hilary Putnam and Julia Robinson, *The decision problem for exponential diophantine equations*, Ann. of Math. 74 (1961), 425–436.

2. Julia Robinson, *General recursive functions*, Proc. Amer. Math. Soc. 1 (1950), 703–718.

3. ———, *Finitely generated classes of sets*, Proc. Amer. Math. Soc. (to appear).

4. Raphael M. Robinson, *Primitive recusive functions*. II, Proc. Amer. Math. Soc. 6 (1955), 663–666.

University of California, Berkeley

DIOPHANTINE DECISION PROBLEMS

Julia Robinson

1. INTRODUCTION

In 1900, Hilbert asked for a general method to determine if an arbitrary diophantine equation has a solution. Since the work of Gödel in 1931 and the development of the theory of computability, logicians have modified Hilbert's problem. They ask instead: Is there such a method? This is the *decision problem* for diophantine equations. The answer is not known even today.

In general, the *decision problem* for a set of questions is to find out if there is a general method to answer the questions. In the case of diophantine equations, the questions are those of the form: Does the diophantine equation $P(x_1, \cdots, x_k) = 0$ have a solution? Here P is a polynomial in $x_1, \cdots, x_k$ with integer coefficients. If we consider the wider class of exponential diophantine equations, i.e., equations in which variables may occur as exponents, then the answer is negative: *There is no general method to determine if an arbitrary exponential diophantine equation has a solution* (Section 11).

These decision problems are closely related to certain diophantine problems which are turned around from the usual problems. Ordinarily, a diophantine equation with, for example, one parameter a is given and we seek a simple characterization of the values of a for which the equation has a solution. Turning the problem around, we take some interesting set $\mathcal{S}$ of natural numbers† and seek a diophantine equation with one parameter a which has a solution for exactly those values of a which belong to $\mathcal{S}$. If there is such an equation, then we say that $\mathcal{S}$ is *diophantine.*

It turns out that *if the set of primes is diophantine, then there is no general method to determine if an arbitrary diophantine equation has a solution* (Section 12). Now it may seem unlikely that there is a diophantine equation with one parameter a which is solvable if and only if a is a prime. However, *there is an exponential diophantine equation with one parameter a which has a solution if and only if a is prime* (Section 10). Actually, much more is proved in Section 11, namely, *every recursively enumerable set is exponential diophantine.*

But *if the set of primes is not diophantine, then for each diophantine equation* $P(x, y, \cdots) = 0$, *either for every n there is a solution with* $y \geqq x * n$ *or there is some m such that every solution of* $P = 0$ *with* $x > 1$ *satisfies* $y > x^m$ (Section 12). Here $x * n$ is the nth superpower of x, i.e., $*$ is the binary operation defined recursively by the equations $x * 0 = 1$ and $x * (n + 1) = x^{(x*n)}$.

Historical Remark: *Gödel's revolutionary work of* 1931 *permeates almost all of this study. The theory of computability was developed by Kleene, Turing, Post, and Church. Since this account is both an amalgamation and a specialization of their ideas, it is impossible to identify their contributions specifically and completely. The applications to diophantine equations have been made by Davis, Putnam, and myself.*

† The natural numbers are 0, 1, 2, $\cdots$.

2. PAIRS AND SEQUENCES

Cantor proved that the set of ordered pairs of natural numbers can be put in one-to-one correspondence with the set of natural numbers. This can be done in many ways, of course, but there is one particularly simple mapping function which we shall find useful. Let $J(x, y) = \frac{1}{2}((x + y)^2 + 3x + y)$. Then $J(x, y)$ is an integer-valued function which assumes each natural number as value exactly once for x and y natural numbers. In fact, $J(0, 0) = 0$ and

$$J(x, y) + 1 = \begin{cases} J(x + 1, y - 1), & \text{if } y > 0, \\ J(0, x + 1), & \text{if } y = 0. \end{cases}$$

So J maps the set of ordered pairs of natural numbers onto the natural numbers according to the following diagram.

	$J(x, y)$	y = 0	1	2	3	4
	0	0	1	3	6	10
	1	2	4	7	11	
x	2	5	8	12		
	3	9	13			
	4	14				

Two inverse functions K and L are uniquely determined by the equation $u = J(Ku, Lu)$.† The first few values of K and L which clearly indicate the general pattern of their sequence of values are given below:

u	0	1	2	3	4	5	6	7	8	9	10	11	12	13	14	15
Ku	0	0	1	0	1	2	0	1	2	3	0	1	2	3	4	0
Lu	0	1	0	2	1	0	3	2	1	0	4	3	2	1	0	5

† We usually omit parentheses in the functional notation for functions of one variable. Thus $F(x)$ is written Fx and $F(G(x))$ is written FGx.

To give an example of the use of pairing functions, suppose $\mathcal{S}$ is the range of a function $F(x, y)$ of two variables. Then $\mathcal{S}$ is also the range of $G(u) = F(Ku, Lu)$, a function of one variable. We can also recover the function F from G, since $F(x, y) = G(J(x, y))$.

We can also let a set $\mathcal{S}$ of natural numbers represent a binary relation $\mathcal{R}$ by the equivalence: $\mathcal{R}(x, y)$ if and only if $J(x, y) \in \mathcal{S}$. Thus, a sequence of sets $\mathcal{S}_0, \mathcal{S}_1, \mathcal{S}_2, \cdots$ is represented by a single set $\mathcal{S}$ through the correspondence $x \in \mathcal{S}_n$ if and only if $J(n, x) \in \mathcal{S}$.

Another example of the usefulness of pairing functions is a simple system of numbering diophantine equations. Here a diophantine equation is an equation of the form $F(x_0, x_1, \cdots) = G(x_0, x_1, \cdots)$ where F and G are terms built up from $x_0, x_1, \cdots$ and natural numbers by addition and multiplication. A diophantine equation can also be put in the form $P(x_0, x_1, \cdots) = 0$, where P is a polynomial with integer coefficients.

First we number the terms $\tau_0, \tau_1, \tau_2, \cdots$ built up from the variables and natural numbers by repeated additions and multiplications as follows:

$$\begin{aligned} \tau_{4n} &= n, \\ \tau_{4n+1} &= x_n, \\ \tau_{4n+2} &= \tau_{Kn} + \tau_{Ln}, \\ \tau_{4n+3} &= \tau_{Kn} \cdot \tau_{Ln}. \end{aligned}$$

Then we number the equations with the nth equation being $\tau_{Kn} = \tau_{Ln}$. We could also write the nth equation as $\tau_{Kn} - \tau_{Ln} = 0$. Thus, the eighth equation is $0 + 0 = x_0$ and the number of $2 \cdot x_0^2 = 1 + x_1^2$ is 7,697,614,550.

Finally, we wish to give a method of representing finite sequences of natural numbers due to Gödel. Let Rem (x, y) be the least nonnegative remainder of x divided by y.

LEMMA (GÖDEL 1931): *For every finite sequence $s_0, s_1, \cdots, s_k$ of natural numbers, there are natural numbers a and d such that*

$$s_t = \text{Rem}(a, 1 + (t + 1)d) \quad \textit{for} \quad t = 0, \cdots, k. \tag{2.1}$$

Proof: Now (2.1) is equivalent to

$$\begin{aligned} &a \equiv s_t \pmod{1 + (t + 1)d}, \\ &0 \leqq s_t < 1 + (t + 1)d \quad \text{for} \quad t = 0, \cdots, k. \end{aligned} \tag{2.2}$$

By the Chinese remainder theorem, the congruences can be satisfied for some a if the moduli are relatively prime and the inequalities will be satisfied if d is sufficiently large. Take for d a sufficiently large multiple of $k!$ so that $d > s_t$ for all $t \leqq k$. If a prime p divides both $1 + (t + 1)d$ and $1 + (t' + 1)d$, then $p \mid (t - t')d$. If $t \neq t'$, then $0 < |t - t'| \leqq k$. Hence $p|k!$ and so $p \mid d$. But $p \nmid d$ since by hypothesis $p \mid 1 + (t + 1)d$. Therefore for this choice of d, the moduli in (2.2) are relatively prime and a satisfying (2.2) exists.

This lemma is at the root of much of what follows.

3. COMPUTING AND LISTING

In Section 1 we spoke of a "general method" to tell if an arbitrary diophantine equation has a solution. What do we have in mind? We mean a set of instructions (necessarily finite) which describe in a completely deterministic way, how to start from an arbitrary diophantine equation $P(x_1, \cdots, x_k) = 0$ and to obtain after a finite number of steps the correct answer to the question: Does $P = 0$ have a solution? At no step in the process should the instructions call for either ingenuity or chance. On the other hand, we do not demand practicality of the method or place any restrictions on time or space needed to carry out the process.

Suppose we find a method to tell if an arbitrary diophantine equation has a solution. We give a proof that our method works, i.e., for every diophantine equation, it yields the correct answer to the question: Does this equation have a solution? Then if the argument is sound, everyone will presumably admit that Hilbert's problem is solved. We do not need to agree ahead of time on the exact meaning of the term "general method." But if there is no general method and we wish to prove that there is none, then we need to be precise. Hence until the notion of computability was defined, no one could ask if such a method exists.

Let us number all diophantine equations in some systematic way such as in the last section. Thus given any n, we can write down the nth equation and given an equation, we can figure out its number (or at least one of its numbers if some equations occur more than once). Let $\mathcal{S}$ be the set of numbers of equations which have solutions. Then a method to tell whether or not an arbitrary natural number n belongs to $\mathcal{S}$ would also give us a method to tell whether or not an arbitrary diophantine equation has a solution.

We call a set of numbers *computable* if there is a method of deciding whether or not an arbitrary natural number n belongs to the set. A *method* consists of a finite set of instructions which for each natural number n specifies a calculation terminating in a "yes" or "no" answer to the question: Does n belong to the set being computed?

Clearly, this is not a mathematical definition of the class of computable sets but, rather, it is a description of the intuitive concept of a computable set. In Section 5 we will give a strictly mathematical characterization of the sets which are identified with this intuitive concept of computable set.

We call a set $\mathcal{L}$ of natural numbers *listable* if there is a method of making a list of the elements of $\mathcal{L}$. A *list* is either a finite or infinite sequence, with or without repetitions. For example, the set $\mathcal{S}$ of numbers of solvable diophantine equations is listable. To make the list, we simply try the possible values of the arguments in the equations in a systematic way. Each time we find a solution of some equation, we put its number on our list.

What is a method of listing a set of numbers? It is a set of instructions which provides for a completely deterministic calculation which may or may not terminate. From time to time during the calculation, a particular number is designated as the next number on the list and we place it next on the list.

We say a function $F(x_1, \cdots, x_k)$ defined for all natural numbers and assuming natural numbers as values is *computable* if there is a method to compute the value of $F(x_1, \cdots, x_k)$ for an arbitrary k-tuple of natural numbers. Here a method is a set of instructions such that if it is applied to any k-tuple of natural numbers $x_1, \cdots, x_k$ it will yield a completely deterministic calculation of the natural number which is $F(x_1, \cdots, x_k)$.

We will take the concept of *listable set* as our fundamental intuitive notion. The following observations show that the concepts of computable set and computable function can be defined in terms of listable sets.

I. A set $\mathcal{S}$ of natural numbers is computable if and only if $\mathcal{S}$ and its complement, $\bar{\mathcal{S}}$, are both listable.

If we have instructions for computing $\mathcal{S}$, we can modify them trivially so that "yes" and "no" answers are interchanged. Thus we obtain instructions for computing $\bar{\mathcal{S}}$. Clearly we can list every computable set so both $\mathcal{S}$ and $\bar{\mathcal{S}}$ are listable. On the other hand, suppose $\mathcal{S}$ and $\bar{\mathcal{S}}$ are both infinite. If we have instructions for listing both $\mathcal{S}$ and $\bar{\mathcal{S}}$, we can make up a single list by alternately putting the next element from the list of $\mathcal{S}$ and then the next element from the list of $\bar{\mathcal{S}}$. To tell if n belongs to $\mathcal{S}$, we need only observe whether it occurs first in an even or odd position on our new list. (We know it will eventually occur and always at positions with the same parity.) Finally, if either $\mathcal{S}$ or $\bar{\mathcal{S}}$ is finite, then both are evidently computable.

In view of I and the earlier remark that the set of numbers of solvable diophantine equations is listable, the decision problem for diophantine equations is reduced to determining if the set of numbers of unsolvable diophantine equations is listable.

II. The range of a computable function is listable and conversely, every nonempty listable set $\mathcal{S}$ is the range of a computable function F.

We can list the values of a computable function Fx by computing $F0, F1, F2, \cdots$, in turn. Also, we can agree on an ordering of k-tuples in case the given function has k arguments and then list the functional values in the corresponding order. On the other hand, a nonempty listable set $\mathcal{S}$ is either finite so we can choose an ultimately constant function for F, or $\mathcal{S}$ is infinite and we take Fn equal to the nth number on a list of $\mathcal{S}$. In either case, F is computable.

III. A function Fx is computable if and only if its graph, i.e., the set $\{J(x, Fx)\}$, is listable.

If F is computable, then certainly $J(x, Fx)$ is also and by II, its range is listable. Suppose the graph of some function G is listable. To calculate Gn, we simply list the elements of the graph of G until some number u is listed with $Ku = n$. Then $Gn = Lu$.

[*Note*: Let the graph of $F(x, y)$ be $\{J(J(x, y), F(x, y))\}$. Then, in a similar way, we could prove III for functions of two variables and in fact, for any number of variables.]

We now assume that there is some mathematical language $\mathfrak{L}$ sufficiently rich that instructions can be given in $\mathfrak{L}$ for listing any listable set. Clearly we can assume that $\mathfrak{L}$ has only a finite number of symbols. Furthermore, we assume that we can recognize suitable instructions in $\mathfrak{L}$ and that the prescribed calculation consists of a sequence of steps of finite length. (This last condition means that if we decide to carry out one more step of the calculation we can eventually complete it.) In Section 5, we will justify these assumptions.

IV. There is a listable set $\mathfrak{U}$ such that $\{\mathfrak{U}_0, \mathfrak{U}_1, \cdots\}$ is the class of listable sets. [Here $k \in \mathfrak{U}_n$ if and only if $J(n, k) \in \mathfrak{U}$.]

We number the suitable sets of instructions in $\mathfrak{L}$: $\mathfrak{I}_0, \mathfrak{I}_1, \cdots$. (For example, we could first order them by the total number of symbols occurring in each set of instructions and then use a dictionary ordering among those of the same length.) Let the corresponding sets be $\mathfrak{U}_0, \mathfrak{U}_1, \cdots$. We now carry out the following sequence of calculations: We do one step of each set of instructions according to the sequence $\mathfrak{I}_0, \mathfrak{I}_0, \mathfrak{I}_1, \mathfrak{I}_0, \mathfrak{I}_1, \mathfrak{I}_2, \cdots$. When the calculation corresponding to $\mathfrak{I}_m$ yields the result that n is the next number on the list of $\mathfrak{U}_m$, we place $J(m, n)$ on our list of $\mathfrak{U}$. Hence $\mathfrak{U}$ is a listable set with the required property.

V. There is a listable set which is not computable.

Let $\mathfrak{U}$ be a set satisfying IV. Let $\mathfrak{D}$ be the diagonal set defined by $\mathfrak{D} = \{n | n \in \mathfrak{U}_n\}$. Then $\mathfrak{D}$ is certainly listable, since we can list the elements of $\mathfrak{U}$ and each time we come to a pair with equal components on the list of $\mathfrak{U}$ we can put their common value on the list of $\mathfrak{D}$. Suppose $\mathfrak{D}$ is computable; then $\overline{\mathfrak{D}}$ is listable, so $\overline{\mathfrak{D}} = \mathfrak{U}_k$ for some k. Hence for all n, $n \in \mathfrak{U}_k$ if and only if $n \notin \mathfrak{U}_n$. But this is impossible for $n = k$. Hence $\mathfrak{D}$ is not computable.

4. RECURSIVE FUNCTIONS

We now turn to the problem of defining mathematically a class of number-theoretic functions which we can identify with computable functions. We consider only functions defined for all natural numbers and whose values are natural numbers.

How do we compute familiar functions such as $[x/y]$, Rem (x, y), $\pi(x)$, $x * y$, the partition function, $\nu(n)$, etc.? We start with certain operations or initial functions, say addition, multiplication, and exponentiation, and then give formulas for computing new functions in terms of old by substitution and recursion (that is, computing the value of a function in terms of earlier values of the same function.)

A function is called *primitive recursive* if it can be obtained from certain initial functions by repeated substitution and recursion. The initial functions are: the zero function of n variables O_n with $O_n(x_1, \cdots, x_n) = 0$ for $n \geqq 0$; the successor function S with Sx equal to the next number after x; and for every n, k with $1 \leqq k \leqq n$, the identity function I_{nk} of n variables defined by

$$I_{nk}(x_1, \cdots, x_n) = x_k.$$

The function F of n variables is said to be obtained by *substitution* from a function A of m variables and m functions $B_1, \cdots, B_m$ of n variables, provided

$$F(x_1, \cdots, x_n) = A(B_1(x_1, \cdots, x_n), \cdots, B_m(x_1, \cdots, x_n)).$$

In particular, we can obtain by substitution the functions $SO_n(x_1, \cdots, x_n)$, $SSO_n(x_1, \cdots, x_n)$, $\cdots$, i.e. all constant functions. Using identity functions, we can introduce extra variables or change the order of variables so that arbitrary substitutions may be made. For example, if we want to define a function F satisfying

$$F(x, y, z) = B(z, x),$$

we can use our substitution scheme by putting

$$F(x, y, z) = B(I_{33}(x, y, z), I_{31}(x, y, z)).$$

A function F of $n + 1$ variables is said to be obtained by *recursion* from a function A of n variables and a function B of $n + 2$ variables, provided

$$F(x_1, \cdots, x_n, y) = A(x_1, \cdots, x_n),$$
$$F(x_1, \cdots, x_n, Sy) = B(x_1, \cdots, x_n, y, F(x_1, \cdots, x_n, y)).$$

For example, the functions $x + y$, $x \cdot y$, and x^y are primitive recursive as we see from the following sequence of equations:

$$x + 0 = x, \quad x + Sy = S(x + y);$$
$$x \cdot 0 = 0, \quad x \cdot Sy = x \cdot y + x;$$
$$x^0 = 1, \quad x^{Sy} = x^y \cdot x.$$

Note that sgn $x = 0^{0^x}$ can be obtained from u^v by substitution. We can also define the predecessor function P by recursion: $P0 = 0$ and $PSx = x$. Let $x \dot{-} y$ be the function given by

$$x \dot{-} y = \begin{cases} x - y, & \text{if } x - y \geqq 0, \\ 0, & \text{otherwise.} \end{cases}$$

Using the predecessor function, we can obtain $x \dot{-} y$ by recursion as follows:

$$x \dot{-} 0 = x \quad \text{and} \quad x \dot{-} Sy = P(x \dot{-} y).$$

We obtain $|x - y| = (x \dot{-} y) + (y \dot{-} x)$ by substitution. Then, $\text{Rem}(x, y)$ is defined by recursion as follows:

$$\text{Rem}\,(0, y) = 0,$$
$$\text{Rem}\,(Sx, y) = (S\,\text{Rem}\,(x, y)) \cdot \text{sgn}\,|y - S\,\text{Rem}\,(x, y)|.$$

Suppose $A(x_1, \cdots, x_k, y)$ is primitive recursive, then the functions B and C given by

$$B(x_1, \cdots, x_k, z) = \sum_{y \leqq z} A(x_1, \cdots, x_k, y),$$

and

$$C(x_1, \cdots, x_k, z) = \prod_{y \leqq z} A(x_1, \cdots, x_k, y)$$

are also primitive recursive. In fact, they are obtained by recursion as follows:

$$B(x_1, \cdots, x_k, 0) = A(x_1, \cdots, x_k, 0),$$
$$B(x_1, \cdots, x_k, Sz) = B(x_1, \cdots, x_k, z) + A(x_1, \cdots, x_k, Sz);$$

$$C(x_1, \cdots, x_k, 0) = A(x_1, \cdots, x_k, 0),$$
$$C(x_1, \cdots, x_k, Sz) = C(x_1, \cdots, x_k, z) \cdot A(x_1, \cdots, x_k, Sz).$$

We can now define the function $\nu(n)$, which is the number of divisors of n for $n > 0$ and $\nu(0) = 1$,

$$\nu(n) = \sum_{x \leqq n} 0^{\mathrm{Rem}(n,x)}.$$

Then

$$\pi(x) = \sum_{y \leqq x} 0^{|\nu(y)-2|}.$$

Actually all the functions mentioned in the second paragraph of this section can be shown to be primitive recursive. Indeed, all the functions which arise naturally in number theory turn out to be primitive recursive. It is only by deliberately making up a counter-example or defining a function metamathematically that we obtain a function which is not primitive recursive. Evidently, every primitive recursive function is computable since the definition of a function F by substitution or recursion provides a way to compute F.

Suppose we number all primitive recursive functions of one variable $F_0, F_1, \cdots$ systematically according to the way they are generated by substitution and recursion. Let G be the function defined by

$$G(t) = F_t(t) + 1.$$

Then G is not F_n for any n and hence G is not primitive recursive. But G is clearly computable.

A function F of n variables is *general recursive* if there are primitive recursive functions A of one variable and B of $n + 1$ variables such that

$$F(x_1, \cdots, x_n) = A\mu y\{B(x_1, \cdots, x_n, y) = 0\},$$

where for every n-tuple of natural numbers $x_1, \cdots, x_n$ there is some natural number y such that $B(x_1, \cdots, x_n, y) = 0$. Here $\mu y\{\cdots\}$ is the least number operator, i.e., $\mu y\{\cdots\}$ is the least number y which satisfies the condition in braces if there is such a y; otherwise $\mu y\{\cdots\}$ is undefined. This characterization of general recursive functions is due to Kleene and is usually called the Kleene normal form for general recursive functions.

Clearly, a general recursive function is computable. Namely, we compute $B(x_1, \cdots, x_n, 0), B(x_1, \cdots, x_n, 1), \cdots, B(x_1, \cdots, x_n, y)$ where $B(x_1, \cdots, x_n, y)$ is the first term equal to zero in this sequence. Since B is primitive recursive, we can compute each of these values. Then $F(x_1, \cdots, x_n) = Ay$, where A is also computable.

We cannot apply the diagonal argument to general recursive functions to show that some computable function is not general recursive because there is no method of listing all general recursive functions. More specifically, there is no method of telling which primitive recursive functions $B(x_1, \cdots, x_n, y)$ are such that for every n-tuple $x_1, \cdots, x_n$ of natural numbers there is a y with $B(x_1, \cdots, x_n, y) = 0$.

5. RECURSIVELY ENUMERABLE SETS

A nonempty set of natural numbers is *recursively enumerable* if it is the range of a primitive recursive function of one variable. The empty set is also taken to be recursive enumerable. A binary relation $\mathcal{R}$ is recursively enumerable if it is empty or if there are two primitive recursive functions F and G such that $\mathcal{R}(x, y)$ if and only if there is a natural number n such that $x = Fn$ and $y = Gn$.† Recursively enumerable relations of higher degree are defined similarly.

Every recursively enumerable set is listable by observation II of Section 3, since a primitive recursive function is computable. Conversely, we have the

CONTENTION: *Every listable set is recursively enumerable.*‡

† Equivalently, we could say $\mathcal{R}$ is recursively enumerable if the set $\{J(x, y) \mid \mathcal{R}(x, y)\}$ is recursively enumerable.

‡ This contention is equivalent to Church's thesis, i.e., every computable function is recursive.

We cannot prove this, since "listable" is an intuitive concept and not precisely defined. Why do we believe that every listable set is recursively enumerable? Not because computing successive values of a primitive recursive function is the most general method of making a list of numbers that we can think of but because of a metamathematical argument which can only be fully appreciated after some study of primitive recursive functions. Roughly, it goes like this. Any calculation can be broken down into a sequence of relatively simple basic steps. We can keep a record of the calculation so that at the end of each step, all the information needed to carry on an additional step has been recorded. Finally, if a step yields a number to be added to the list, we put the number at the end of the record of that step; if not, we put a period or other special symbol at the end. We assign numbers to the symbols being used including a number for each natural number. Next we assign numbers to all configurations of symbols in some simple systematic way according to their appearance. We can do this so that the function A given by

$$Ax = \begin{cases} n + 1, & \text{if } x \text{ is the number of a configuration} \\ & \text{which ends in the natural number } n, \\ 0, & \text{otherwise,} \end{cases}$$

is primitive recursive. Let Bx be the number of the record of the calculation after x steps. (The record at any stage is a configuration of symbols.) Now $B(x + 1)$ is completely determined by the configuration whose number is Bx. Also $B(x + 1)$ is the number of the configuration obtained by making the small changes in the record due to one additional basic step of the calculation. Thus $B(x + 1) = CBx$ for some function C. Recalling how wide the class of primitive recursive functions is, we grant that C (and hence B) is primitive recursive. Let $\mathcal{A}$ be the set being listed and a some element of $\mathcal{A}$. Then $\mathcal{A}$ is the range of the primitive recursive function F given by

$$Fu = PABu + a \cdot 0^{ABu}.$$

Hence $\mathcal{A}$ is recursively enumerable.

It is easy to see directly that the range of a general recursive function is recursively enumerable. Suppose $Fx = A\mu y\{B(x, y) = 0\}$ where A and B are primitive recursive functions. Let

$$C(x, t) = \operatorname{sgn} \prod_{z \leq t} B(x, z).$$

Then, as we have seen, C is also primitive recursive. Let $k = \mu y\{B(0, y) = 0\}$. Then the range of F is also the range of the primitive recursive function G with

$$G(x, y) = A(0^{C(x,y)} \sum_{t \leq y} C(x, t) + C(x, y) \cdot k).$$

Hence the range of F is the same as the range of $G(Ku, Lu)$, a primitive recursive function of one variable.

Actually, a much smaller class of functions than the class of primitive recursive functions is sufficient to obtain all recursively enumerable sets as ranges of functions of the class.

If the graph of a function F is recursively enumerable — say the range of A — then

$$Fx = LA\mu t\{KAt = x\}.$$

Hence F is general recursive. Conversely, the graph of a general recursive function F is the range of the general recursive function $J(x, Fx)$ and hence is recursively enumerable.

A set is *recursive* if both it and its complement are recursively enumerable.† In view of observations I–III of Section 3, recursive sets should be identified with computable sets and general recursive functions with computable functions.

There have been many different proposals for defining the class of listable sets.‡ It has turned out that the class of sets proposed has always been the same. Furthermore, the proofs that the proposed class is contained in the class of recursively enumerable sets can always be carried out according to the above sketch. Once the techniques of recursive function theory are familiar, these proofs are routine. And it is just this "routineness" which con-

† Equivalently, a set is recursive if its characteristic function is general recursive.

‡ A good account of several of these definitions is contained in Mendelson [10].

vinces us that the intuitive notion of "listable set" is correctly identified with recursively enumerable set.

6. GÖDEL'S THEOREM

We introduce logical symbols here as a necessary convenience although we do not need any knowledge of formal logic. We will use the universal quantifier $\bigwedge$ (for every), the existential quantifier $\bigvee$ (there exists), and the connectives $\wedge$ (and), $\vee$ (or), and $\neg$ (not). A formula is *arithmetical* if it is built up by means of logical symbols from equations of the form $a = b$, $a + b = c$, and $a \cdot b = c$ where a, b, and c are variables or symbols for particular numbers.

Examples The formula

$$(\bigwedge x)(\bigvee u, v, w, z)[x = u^2 + v^2 + w^2 + z^2]$$

expresses Lagrange's theorem that every natural number is the sum of four squares. The condition "p is a prime" is expressed by the formula

$$\neg\,(p^2 = p) \wedge \neg(\bigvee u, v)[p = (u + 2)(v + 2)].$$

Finally, the condition $y = [\sqrt{x}]$ is equivalent to the formula

$$(\bigvee u, v)[y^2 + u = x \wedge x + 1 + v = (y + 1)^2].$$

Relations defined by arithmetical formulas are called *arithmetical.* Gödel in 1931 proved the remarkable theorem that the relation given by an equation $F(x_1, \cdots, x_k) = y$ where F is any primitive recursive function is arithmetical. Consequently, every recursively enumerable set is arithmetical. Since as we have seen, there are some recursively enumerable sets which are not computable, there is some arithmetical formula Φ which defines a noncomputable set. Hence there is no decision method to answer questions of the sort: Does n satisfy Φ? Thus, there is no decision method to determine whether or not an arbitrary arithmetical statement is true.

In writing out arithmetical formulas, we can also use bounded quantifiers, i.e., $(\bigvee x \leqq y)[\cdots]$ (there exists an x not exceeding y such that $\cdots$) and $(\bigwedge x \leqq y)[\cdots]$ (for all x not exceeding y, $\cdots$). In fact, we could replace $(\bigvee x \leqq y)[\cdots]$ by $(\bigvee x, t)[x + t = y \wedge \cdots]$ and $(\bigwedge x \leqq y)[\cdots]$ by

$$(\bigvee x)(\bigvee t)[x = y + 1 + t \vee \cdots].$$

We shall now prove a strengthened form of Gödel's theorem† due to Martin Davis—the proof is essentially that of Gödel. In the next section, we will prove a still stronger theorem of Davis.

GÖDEL-DAVIS THEOREM: *Every relation given by* $F(x_1, \cdots, x_n) = y$, *where* F *is a primitive recursive function, can be expressed by an arithmetical formula in which the universal quantifiers are all bounded and there are no negation signs.*

For example, x is a power of 2 if and only if

$$(\bigwedge u \leqq x)(\bigwedge v \leqq x)[x \neq (2u + 3)v].$$

This formula can be put in the required form by eliminating the inequation. Namely, x is a power of 2 if and only if

$$(\bigwedge u \leqq x)(\bigwedge v \leqq x)(\bigvee w)[(x - (2u + 3)v)^2 = 1 + w].$$

Proof: We prove the theorem for functions F in the order in which they are constructed according to the definition of primitive recursive functions. That is, we check that the theorem holds if F is an initial function; then we show that the theorem holds for F if F is obtained by substitution or recursion from functions that satisfy the theorem.

I. If F is an initial function, then the theorem holds for F. In fact,

$$\begin{aligned} O_n(x_1, \cdots, x_n) = y \quad &\leftrightarrow \quad y = 0, \\ Sx = y \quad &\leftrightarrow \quad y = x + 1, \\ I_{nk}(x_1, \cdots, x_n) = y \quad &\leftrightarrow \quad y = x_k. \end{aligned}$$

Here we use $\leftrightarrow$ for "if and only if".

† This theorem is used to prove Davis' theorem (in the next section), which is needed only in Section 10. A reader unfamiliar with formal logic might do well to study the statement of Davis' theorem for the present and return to the proof later.

II. If $F(x_1, \cdots, x_n) = B(A_1(x_1, \cdots, x_n), \cdots, A_m(x_1, \cdots, x_n))$ where $A_1, \cdots, A_m$, and B satisfy the theorem, then F satisfies the theorem.

We have

$$F(x_1, \cdots, x_n) = y \leftrightarrow (\bigvee z_1, \cdots, z_m)[z_1 = A_1(x_1, \cdots, x_n) \wedge \cdots \\ \wedge z_m = A_m(x_1, \cdots, x_n) \wedge y = B(z_1, \cdots, z_m)].$$

By the inductive hypothesis, each of the equations on the right side of the equivalence can be replaced by formulas of the required sort so F satisfies the theorem.

III. If $F(x_1, \cdots, x_k, 0) = A(x_1, \cdots, x_k)$ and $F(x_1, \cdots, x_k, y+1) = B(x_1, \cdots, x_k, y, F(x_1, \cdots, x_k, y))$, where A and B satisfy the theorem, then F satisfies the theorem.

Notice that for a fixed y, the sequence $u_0, \cdots, u_{y+1}$ is completely determined by the equations:

$$(6.1)\qquad \begin{aligned} u_0 &= A(x_1, \cdots, x_k), \\ u_1 &= B(x_1, \cdots, x_k, 0, u_0), \\ u_2 &= B(x_1, \cdots, x_k, 1, u_1), \\ &\vdots \\ u_y &= B(x_1, \cdots, x_k, y-1, u_{y-1}), \\ u_{y+1} &= B(x_1, \cdots, x_k, y, u_y). \end{aligned}$$

(For convenience, we use one more equation than we need.) Hence if $u_0, \cdots, u_y, u_{y+1}$ satisfy (6.1), then $F(x_1, \cdots, x_k, t) = u_t$ for $t \leqq y+1$. Now represent this sequence according to Gödel's lemma of Section 2. Thus, for some a and d

$$u_t = \text{Rem}(a, 1 + (t+1)d) \quad \text{for} \quad 0 \leqq t \leqq y+1.$$

Finally,

$$(6.2)\qquad \begin{aligned} &F(x_1, \cdots, x_k, y) = z \leftrightarrow \\ &(\bigvee a, d)\{\text{Rem}\,(a, 1+d) = A(x_1, \cdots, x_k) \\ &\wedge z = \text{Rem}\,(a, 1 + (y+1)d) \\ &\wedge (\bigwedge t \leqq y)[\text{Rem}\,(a, 1 + (t+2)d) \\ &= B(x_1, \cdots, x_k, t), \text{Rem}\,(a, 1 + (t+1)d)]\}. \end{aligned}$$

Now

$$(6.3)\quad \mathrm{Rem}\,(u, 1 + v) = w \leftrightarrow (\bigvee r, q)[u = (1 + v)r + w \wedge w + q = v].$$

Hence each of the equations in (6.2) can be replaced by a formula of the required sort, so F satisfies the theorem.

As an example of the proof of the theorem, we can write down the following definition of $y = 2^x$:

$$\begin{aligned} y = 2^x \leftrightarrow \\ &(\bigvee a, d)(\bigwedge t \leqq x)[\mathrm{Rem}\,(a, 1 + d) = 1 \\ &\wedge \mathrm{Rem}\,(a, 1 + (t + 2)d) = 2\cdot \mathrm{Rem}\,(a, 1 + (t + 1)d) \\ &\wedge y = \mathrm{Rem}\,(a, 1 + (x + 1)d)]. \end{aligned}$$

Using (6.3), we can easily put the right side in the form required by the theorem.

7. DAVIS NORMAL FORM

In 1950, Davis considerably sharpened Gödel's result in the following theorem:

Every recursively enumerable relation can be put in the form

$$(7.1)\quad (\bigvee y)(\bigwedge u \leqq y)(\bigvee v_1 \leqq y)\cdots(\bigvee v_k \leqq y) [P(x_1, \cdots, x_n, y, u, v_1, \cdots, v_k) = 0],$$

where P is a polynomial with integer coefficients. Conversely, a relation defined by a formula of the form (7.1) *is recursively enumerable.*

We shall call this the Davis normal form of a recursively enumerable relation.

Proof: We start with the theorem of the last section. So we may assume the given recursively enumerable relation is given by a formula built up from diophantine equations by conjunctions, disjunctions, existential quantifiers, and bounded universal quantifiers. Now we show how various reductions can be made which lead to the Davis normal form.

1. If all the quantified variables are distinct, a formula of the Gödel type is equivalent to the formula obtained by putting all

the quantifiers at the beginning of the formula while maintaining their relative order and leaving the rest of the formula unchanged.

2. A formula built from equations by using conjunctions and disjunctions is equivalent to a polynomial equal to zero.

Since

$$A = 0 \wedge B = 0 \leftrightarrow A^2 + B^2 = 0$$

and

$$A = 0 \vee B = 0 \leftrightarrow A \cdot B = 0,$$

we can reduce the given formula step-wise to the form $P = 0$.

3. The relations $x > y$ and $x \geqq u$ are expressible as follows:

$$x > y \leftrightarrow (\bigvee z)[x = y + 1 + z]$$

and

$$x \geqq y \leftrightarrow (\bigvee z)[x = y + z].$$

Also we could just as well use the bounded quantifier $(\bigvee z \leqq x)$ in both formulas.

4. A formula of the form

$$(\bigwedge t \leqq x)(\bigvee u_1, \cdots, u_k)(\bigwedge z \leqq x')(\bigvee v_1, \cdots, v_l)[P = 0]$$

is equivalent to a formula in which the two universal quantifiers are adjacent:

$$(\bigvee a_1, \cdots, a_k, d_1, \cdots, d_k)(\bigwedge t \leqq x)(\bigwedge z \leqq x')(\bigvee u_1, \cdots, u_k, v_1, \cdots, v_l)$$
$$[u_1 = \mathrm{Rem}(a_1, 1 + (t+1)d_1) \wedge \cdots$$
$$\wedge\ u_k = \mathrm{Rem}\,(a_k, 1 + (t+1)d_k) \wedge P = 0].$$

Note that $a_1, \cdots, a_k, d_1, \cdots, d_k$ can be chosen so that $u_1, \cdots, u_k$ are arbitrary numbers depending on t for each t with $t \leqq x$. Actually, we could take $d_1 = d_2 = \cdots = d_k$ since the conditions on d in Gödel's lemma (Section 2) could be satisfied by a single number. Using the equivalence

$$\mathrm{Rem}\,(u, 1 + v) = w \leftrightarrow (\bigvee r, q)[u = (1 + v)r + w \wedge w + q = v],$$

and reductions 1 and 2, we can reduce the formula to one of the form

$$(\bigvee \cdots)(\bigwedge t \leqq x)(\bigwedge z \leqq x')(\bigvee \cdots)[P' = 0]$$

where P' is some polynomial with integer coefficients.

5. A formula of the form

$$(\bigwedge u \leqq x)(\bigwedge v \leqq x')(\bigvee w)[P = 0]$$

is equivalent to a formula with just one universal quantifier, namely,

$$(\bigvee z)(\bigwedge r \leqq z)(\bigvee u, v, w)$$
$$[z = J(x, x') \wedge r = J(u, v) \wedge (u > x \vee v > x' \vee P = 0)].$$

Here we use the fact that if $u \leqq x$ and $v \leqq x'$, then $J(u, v) \leqq J(x, x')$. Also $z = J(x, x')$ is equivalent to the equation $2z = 2J(x, x')$ with integral coefficients. Using reductions 2 and 3, we obtain an equivalent formula of the form

$$(\bigvee z)(\bigwedge r \leqq z)(\bigvee u, v, w, s, t)[P = 0],$$

where P is some new polynomial.

By repeated use of reductions 4 and 5, we can obtain a formula in the form

$$(\bigvee x_1, \cdots, x_k)(\bigwedge u \leqq z)(\bigvee v_1, \cdots, v_l)[P = 0].$$

6. A formula of the form

$$(\bigvee x, y)(\bigwedge u \leqq z)(\bigvee v_1, \cdots, v_k)[P = 0]$$

is equivalent to a formula with just one initial existential quantifier:

$$(\bigvee t)(\bigwedge u \leqq t)(\bigvee v_1, \cdots, v_k, x, y)$$
$$[t = J(J(x, y), z) \wedge (u > z \vee P = 0)].$$

Now by repeated applications of reduction 6, we will obtain a formula of the form

$$(\bigvee w)(\bigwedge u \leqq w)(\bigvee v_1, \cdots, v_r)[P = 0].$$

This formula is equivalent to

$$(\bigvee y)(\bigwedge u \leqq y)(\bigvee v_1 \leqq y) \cdots (\bigvee v_r \leqq y)(\bigvee w \leqq y)(\bigvee z \leqq y)(\bigvee t \leqq y)$$
$$[y = J(w, z) \wedge (u = w + 1 + t \vee P = 0)].$$

Finally, by application of reduction 2 to the last formula, we obtain the Davis normal form.

Conversely, if $\mathcal{S}$ is defined by the equivalence

$$x \in \mathcal{S} \leftrightarrow (\bigvee y)(\bigwedge u \leqq y)(\bigvee v_1 \leqq y) \cdots (\bigvee v_k \leqq y)[P = 0],$$

then $\mathcal{S}$ is recursively enumerable. Let $G(x, y)$ be defined by

$$G(x, y) = \sum_{u \leqq y} \prod_{v_1 \leqq y} \cdots \prod_{v_k \leqq y} P^2(x, y, u, v_1, \cdots, v_k);$$

then G is primitive recursive and

$$x \in \mathcal{S} \leftrightarrow (\mathbf{V}y)[G(x, y) = 0].$$

Hence if $\mathcal{S}$ is not empty, let a be some particular element of $\mathcal{S}$. Then $H(x, y)$ defined by

$$H(x, y) = 0^{G(x,y)} \cdot x + \operatorname{sgn} G(x, y) \cdot a$$

is primitive recursive. Clearly, $\mathcal{S}$ is the range of H, so $\mathcal{S}$ is recursively enumerable.

8. DIOPHANTINE RELATIONS

By a solution of a diophantine equation, we mean a solution in natural numbers. Ordinarily, integer solutions are sought. However the answer to the decision problem is the same in the two cases and it is more convenient for us to consider only natural number solutions. Indeed, $P(x_1, \cdots, x_k) = 0$ has a solution in integers if and only if one of the equations $P(\pm x_1, \cdots, \pm x_k) = 0$ given by a particular choice of signs has a solution in natural numbers. Hence if Q is the polynomial obtained by multiplying together all the 2^k polynomials $P(\pm x_1, \cdots, \pm x_k)$ obtained by different choices of signs, then $P(x_1, \cdots, x_k) = 0$ has a solution in integers if and only if $Q(x_1, \cdots, x_k) = 0$ has a solution in natural numbers. On the other hand, in view of Lagrange's theorem, $P(x_1, \cdots, x_k) = 0$ has a solution in natural numbers if and only if $P(u_1^2 + v_1^2 + w_1^2 + z_1^2, \cdots, u_k^2 + v_k^2 + w_k^2 + z_k^2) = 0$ has a solution in integers.

A set $\mathcal{S}$ of natural numbers is *diophantine* if there is a polynomial $P(n, x_1, \cdots, x_k)$ with integer coefficients such that n belongs to $\mathcal{S}$ if and only if there are natural numbers $x_1, \cdots, x_k$ with $P(n, x_1, \cdots, x_k) = 0$. Similarly, a binary relation $\mathcal{R}(m, n)$ is *diophantine* if and only if there is a polynomial $P(m, n, x_1, \cdots, x_k)$ with integer coefficients such that $\mathcal{R}(m, n)$ if and only if there are natural numbers $x_1, \cdots, x_k$ such that $P(m, n, x_1, \cdots, x_k) = 0$, and so on for relations of higher degree.

Examples. The set of perfect squares is diophantine, since n is a square if and only if $n - x^2 = 0$ has a solution in natural numbers. The set of numbers which are not squares is also diophantine. For n is not a square if and only if it lies strictly between two consecutive squares. Thus n is not a square if and only if the system of equation

$$x^2 + 1 + y = n,$$
$$n + 1 + z = (x + 1)^2,$$

has a solution for x, y, z in natural numbers. Furthermore, replacing the system of equations by a single equation, we have n is not a square if and only if the equation

$$(x^2 + 1 + y - n)^2 + (n + 1 + z - (x + 1)^2)^2 = 0$$

has a solution for x, y, z in natural numbers. A simpler equation with the same property can be obtained from Pell's equation. Namely, n is not a square if and only if the equation $(x + 2)^2 - ny^2 = 1$ has a solution for x and y in natural numbers.

The relations $m \mid n$ and $m \nmid n$ are both diophantine. Clearly, $m \mid n$ if and only if the equation $mx = n$ has a solution for x in natural numbers. Also $m \nmid n$ if and only if there are natural numbers x and y such that either $n = mx + y$ and $0 < y < m$, or $m = 0$ and $n > 0$. Hence $m \nmid n$ if and only if the equation

$$((n - mx - y)^2 + (y - 1 - z)^2 + (m - y - 1 - u)^2)(m^2 + (n - 1 - v)^2) = 0$$

has a solution for x, y, z, u, and v in natural numbers.

Finally, the relation $(m, n) = 1$ holds if and only if there are natural numbers u and v with $mu = 1 + nv$, so the relation of relative primeness is diophantine.

Another characterization of diophantine sets was found by Putnam [12].† Suppose $\mathcal{S}$ is the set of natural numbers n such that $P(n, x_1, \cdots, x_k) = 0$ has a solution for $x_1, \cdots, x_k$ in natural numbers. Let

† Putnam also shows that there is no method of telling whether or not an arbitrary diophantine equation with one parameter a is solvable for every value of a.

$$P_0(x_0, x_1, \cdots, x_k) = (x_0 + 1)(1 - P^2(x_0, x_1, \cdots, x_k)) - 1$$

Then $P_0(n, x_1, \cdots, x_k) \geqq 0$ if and only if $P(n, x_1, \cdots, x_k) = 0$. Furthermore, if $P(n, x_1, \cdots, x_k) = 0$ then $P_0(n, x_1, \cdots, x_k) = n$. Hence a natural number n belongs to $\mathcal{S}$ if and only if n belongs to the range of P_0.

9. EXPONENTIAL DIOPHANTINE RELATIONS

We have defined a diophantine equation to be an equation which can be put in the form $A = B$ where A and B are terms built from natural numbers and variables by addition and multiplication. Similarly, we define an *exponential diophantine equation* to be an equation which can be put in the form $E = F$ where E and F are terms built from natural numbers and variables by addition, multiplication, and exponentiation.

A set $\mathcal{S}$ of natural numbers is *exponential diophantine* if there is an exponential diophantine equation $E(n, x_1, \cdots, x_k) = F(n, x_1, \cdots, x_k)$ which has a solution for $x_1, \cdots, x_k$ in natural numbers if and only if n belongs to $\mathcal{S}$. Exponential diophantine relations are defined similarly.

Examples

1. The relation $m = \binom{n}{r}$ is exponential diophantine. In fact, we shall show that

$$(9.1)\quad \binom{n}{r} = [2^{nr}(1 + 2^{-n})^n] - 2^n[2^{n(r-1)}(1 + 2^{-n})^n] + 0^{n+r}.$$

Suppose $n > 0$ and $r > 0$. Then

$$2^{nr}(1 + 2^{-n})^n = 2^{nr} \sum_{t=0}^{n} \binom{n}{t} 2^{-nt} < \sum_{t=0}^{r} \binom{n}{t} 2^{n(r-t)} + \frac{(2^n - 1)}{2^n}.$$

Hence

$$[2^{nr}(1 + 2^{-n})^n] = \sum_{t=0}^{r} \binom{n}{t} 2^{n(r-t)}.$$

Similarly,

$$2^n[2^{n(r-1)}(1 + 2^{-n})^n] = \sum_{t=0}^{r-1} \binom{n}{t} 2^{n(r-t)}.$$

So (9.1) holds for $n, r > 0$. We can easily check that it also holds when either $n = 0$ or $r = 0$.

To express the relation $m = \binom{n}{r}$ by an exponential diophantine equation we use the fact that for $b > 0$, $[a/b] = c$ if and only if $bc \leqq a < b(c + 1)$. Hence $m = \binom{n}{r}$ if and only if the following conditions can be satisfied by natural numbers x and y:

$$\begin{aligned} m &= x - 2^n y + 0^{n+r}, \\ 2^{n^2} x &\leqq 2^{nr}(2^n + 1)^n < 2^{n^2}(x + 1), \\ 2^n 2^{n^2} y &\leqq 2^{nr}(2^n + 1)^n < 2^n 2^{n^2}(y + 1). \end{aligned}$$

Thus, $m = \binom{n}{r}$ if and only if the following system of equations has a solution for x, y, u, v, w, and z in natural numbers:

$$\begin{aligned} m + 2^n y &= x + 0^{n+r}, \\ 2^{n^2} x + u &= 2^{nr}(2^n + 1)^n, \\ 2^{nr}(2^n + 1)^n + 1 + v &= 2^{n^2}(x + 1), \\ 2^n 2^{n^2} y + w &= 2^{nr}(2^n + 1)^n, \\ 2^{nr}(2^n + 1)^n + 1 + z &= 2^n 2^{n^2}(y + 1). \end{aligned}$$

Now let these five equations be $E_i = F_i$ $(i = 1, \cdots, 5)$. Let

$$G = \sum_{i=1}^{5} (E_i - F_i)^2.$$

Then $G = 0$ is an exponential diophantine equation with three parameters m, n, and r which has a solution in natural numbers if and only if $m = \binom{n}{r}$.

2. The relation $m = n!$ is exponential diophantine. To see this, we use the fact that for s sufficiently large compared to n,

$$n! = \left[\frac{s^n}{\binom{s}{n}} \right]. \tag{9.2}$$

In fact, it is easy to check that (9.2) holds whenever $s > (2n)^{n+1}$. For $n > 0$,

$$n! \leqq \frac{s^n}{\binom{s}{n}} = n!\left(\frac{s}{s-1}\right)\left(\frac{s}{s-2}\right)\cdots\left(\frac{s}{s-n+1}\right),$$

$$< n!\left(\frac{1}{1-\frac{n}{s}}\right)^n.$$

We now use the elementary inequalities:

$$\frac{1}{1-\theta} < 1 + 2\theta \quad \text{for} \quad 0 < \theta < \frac{1}{2},$$

and

$$(1+\theta)^k < 1 + 2^k\theta \quad \text{for} \quad 0 < \theta < 1.$$

Thus,

$$\frac{s^n}{\binom{s}{n}} < n!\left(1 + \frac{2n}{s}\right)^n < n!\left(1 + 2^n\left(\frac{2n}{s}\right)\right)$$

$$= n! + \frac{2^n(2n)n!}{s}$$

$$< n! + 1.$$

Hence (9.2) is established for $s > (2n)^{n+1}$. Finally, $m = n!$ if and only if there are natural numbers r, s, and t such that

$$s > (2n)^{n+1},$$
$$t = \binom{s}{n},$$
$$s^n = mt + r,$$
$$r < t.$$

Since each of these conditions is exponential diophantine, the relation $m = n!$ is also exponential diophantine.

3. The set of primes is exponential diophantine.

Since p is prime if $p = x + 1$ and $p \nmid x!$ for some natural number x and both of these conditions are exponential diophantine, the set of primes is also. That is, there is an exponential diophantine equation with one parameter p which has a solution if and only if p is a prime. We could write out such an equation, as we have indicated, but it would be rather complicated. It would be interesting to find a simple equation which would define the set of primes.

4. (Davis and Putnam). The relation given by

$$m = \prod_{k=1}^{n} (a + bk)$$

is exponential diophantine.

Now for $b > 0$,

$$\prod_{k=1}^{n} (a + bk) = \binom{n + a/b}{n} b^n n!.$$

(The binomial coefficient $\binom{\alpha}{n} = \frac{\alpha(\alpha - 1)\cdots(\alpha - n + 1)}{n!}$ for rational α.) Hence it will be sufficient to show that the relation $R(s, t, p, q, r)$ which holds among the natural numbers s, t, p, q, and r if and only if

$$\text{(9.3)} \qquad \frac{s}{t} = \binom{\frac{p}{q}}{r}, \quad (s, t) = 1, \quad \text{and} \quad \frac{p}{q} > r.$$

Let α be a rational number, r a natural number less than α, and x a real number with $0 < x < 1$. Then

$$(1 + x)^\alpha = \sum_{j=0}^{r} \binom{\alpha}{j} x^j + \binom{\alpha}{r + 1} x^{r+1}(1 + \theta x)^{\alpha-r-1}$$

for some θ with $0 < \theta < 1$. Let $\alpha = p/q > r$ and $x = a^{-2}$ where a is an integer greater than 1. Then

$$\text{(9.4)} \quad a^{2r+1}(1 + a^{-2})^\alpha = \sum_{j=0}^{r} \binom{\alpha}{j} a^{2r-2j+1} + \binom{\alpha}{r + 1} a^{-1}(1 + \theta a^{-2})^{\alpha-r-1}.$$

Let

$$\text{(9.5)} \qquad S_r^\alpha(a) = \sum_{j=0}^{r} \binom{\alpha}{j} a^{2r-2j+1}.$$

Then we can define $\binom{\alpha}{r}$ explicitly in terms of these sums, since

$$\binom{\alpha}{r} = a^{-1}S_r^\alpha(a) - aS_{r-1}^\alpha(a). \tag{9.6}$$

Furthermore, since the error term in (9.4) is nonnegative for $\alpha \geqq r$, (9.4) can be rewritten as follows:

$$S_r^\alpha(a) = a^{2r+1}(1 + a^{-2})^\alpha - \theta'\alpha^{r+1}a^{-1}2^\alpha \tag{9.7}$$

for some θ' with $0 \leqq \theta' \leqq 1$. Now if $q^r r! | a$, then both $S_r^{p/q}(a)$ and $S_{r-1}^{p/q}(a)$ will be integers. Also if $a > p^{r+1}2^p$ then the error term in (9.7) is less than 1. Hence $\mathcal{R}(x, y, p, q, r)$ holds if and only if there is a natural number a so that

$$\frac{p}{q} > r,$$
$$(x, y) = 1,$$
$$a > p^{r+1}2^p,$$
$$q^r r! \mid a,$$
$$\frac{ax}{y} = [a^{2r+1}(1 + a^{-2})^{p/q}] - a^2[a^{2r-1}(1 + a^{-2})^{p/q}].$$

It is easy to see that each of these conditions is exponential diophantine,† so $\mathcal{R}$ and hence the relation $m = \prod_{k=1}^{n} (a + bk)$ are exponential diophantine. As an immediate consequence, we see that the relation $m = \prod_{k=0}^{n} (a - k)$ is exponential diophantine. Indeed, $m = \prod_{k=0}^{n} (a - k)$ if and only if there is an x such that

$$(a + x = n \wedge m = 0) \vee (a = n + x + 1 \wedge m = \prod_{k=1}^{n+1} (x + k)).$$

Since both of the alternatives are exponential diophantine, we can use the fact that $A = 0 \vee B = 0$ if and only if $AB = 0$ to see that $m = \prod_{k=0}^{n} (a - k)$ is exponential diophantine.

† Only the last condition requires simplification. We can use the fact that $u = [vw^{p/q}]$ if and only if $u^q \leqq v^q w^p$ and $v^q w^p < (u + 1)^q$.

10. EXPONENTIAL DIOPHANTINE = RECURSIVELY ENUMERABLE

It is immediately evident that every exponential diophantine relation is listable. It is enough to search systematically for solutions to the defining equation and record the successes. Hence every exponential diophantine relation is recursively enumerable. But since this is a mathematical statement, we should not rely on the contention of Section 5. Suppose n belongs to $\mathcal{S}$ if and only if $E(n, x_1, \cdots) = 0$ has a solution for $x_1, \cdots$ in natural numbers and $\mathcal{S}$ is not empty. Then $\mathcal{S}$ is the range of the primitive recursive function

$$0^{E(n,x_1,\cdots)^2} \cdot n + \operatorname{sgn}(E(n, x_1, \cdots)^2) \cdot s$$

where s is some particular element of $\mathcal{S}$.

THEOREM (DAVIS, PUTNAM, ROBINSON): *Every recursively enumerable relation is exponential diophantine.*

We shall first prove a lemma which contains the heart of the argument. Let $\mathfrak{x}$ stand for $x_1, \cdots, x_n$.

LEMMA: *Let $P(\mathfrak{x}, y, u, v_1, \cdots, v_k)$ be a polynomial with integer coefficients. Let $Q(\mathfrak{x}, y)$ be a polynomial such that*

$$Q(\mathfrak{x}, y) \geqq y \tag{10.1}$$

and

$$(\bigwedge u \leqq y)(\bigwedge v_1 \leqq y)\cdots(\bigwedge v_k \leqq y) \quad [|P(\mathfrak{x}, y, u, v_1, \cdots, v_k)| \leqq Q(\mathfrak{x}, y)]. \tag{10.2}$$

Then

$$(\bigwedge u \leqq y)(\bigvee v_1 \leqq y)\cdots(\bigvee v_k \leqq y) \quad [P(\mathfrak{x}, y, u, v_1, \cdots, v_k) = 0] \tag{10.3}$$

is equivalent to

$$\begin{aligned}(\bigvee c, t, a_1, \cdots, a_k)[t &= Q(\mathfrak{x}, y)! \\ \wedge\ 1 + (c+1)t &= \prod_{m \leqq y} (1 + (m+1)t) \\ \wedge\ 1 + (c+1)t &\mid P(\mathfrak{x}, y, c, a_1, \cdots, a_k) \\ \wedge\ 1 + (c+1)t &\mid \prod_{j \leqq y} (a_1 - j) \wedge \cdots \\ \wedge\ 1 + (c+1)t &\mid \prod_{j \leqq y} (a_k - j)].\end{aligned} \tag{10.4}$$

[Here k is a particular natural number depending only on the number of variables in P. Hence for each polynomial P, the formula (10.4) has a fixed number of terms.]

Proof of lemma: The conditions

$$(10.5) \quad t = Q(\mathfrak{x}, y)!, \quad 1 + (c+1)t = \prod_{m \leqq y} (1 + (m+1)t)$$

uniquely determine t and c. From (10.1), (10.2), and (10.5), we have for $m, m' \leqq y$ and $m \neq m'$,

$$(10.6) \qquad (1 + (m+1)t, 1 + (m'+1)t) = 1$$

and

$$(10.7) \qquad c \equiv m \pmod{1 + (m+1)t}.$$

Also if p is prime and $p \mid 1 + (m+1)t$, then $p > y$ and $p > |P(\mathfrak{x}, y, u, v_1, \cdots, v_k)|$ for all $u, v_1, \cdots, v_k$ which do not exceed y.

I. Suppose the condition in brackets in (10.4) holds. Let $u \leqq y$ and p be a prime divisor of $1 + (u+1)t$. Put $v_i = \text{Rem}\,(a_i, p)$ for $i = 1, \cdots, k$. Now since

$$1 + (c+1)t \mid \prod_{j \leqq y} (a_i - j),$$

there is some $j \leqq y$ so that $p \mid a_i - j$. Hence $v_i \leqq y$. From (10.7), $c \equiv u \pmod{p}$ and by definition, $a_i \equiv v_i \pmod{p}$. Hence

$$P(\mathfrak{x}, y, u, v_1, \cdots, v_k) \equiv P(\mathfrak{x}, y, c, a_1, \cdots, a_k) \pmod{p}.$$

But by (10.4),

$$P(\mathfrak{x}, y, c, a_1, \cdots, a_k) \equiv 0 \pmod{p}.$$

So $P(\mathfrak{x}, y, u, v_1, \cdots, v_k) \equiv 0 \pmod{p}$. We have already seen that

$$|P(\mathfrak{x}, y, u, v_1, \cdots, v_k)| < p,$$

so

$$P(\mathfrak{x}, y, u, v_1, \cdots, v_k) = 0.$$

II. Suppose (10.3) holds. Let c and t be determined by (10.5). Let $v_{ui} \leqq y$ be chosen for $u \leqq y$ and $i = 1, \cdots, k$ so that $P(\mathfrak{x}, y, u, v_{u1}, \cdots, v_{uk}) = 0$. Let $a_1, \cdots, a_k$ satisfy

$$a_i \equiv v_{ui} \pmod{1 + (u+1)t} \quad \text{for } u \leqq y.$$

There are such $a_1, \cdots, a_k$ by the Chinese remainder theorem since the moduli are relatively prime by (10.6). Hence

$$P(\mathfrak{x}, y, c, a_1, \cdots, a_k) \equiv P(\mathfrak{x}, y, u, v_{u1}, \cdots, v_{uk}) \\ \equiv 0 \pmod{1 + (u+1)t}.$$

But since the moduli are relatively prime, it follows that

$$\prod_{u \leqq y} (1 + (u+1)t) \mid P(\mathfrak{x}, y, c, a_1, \cdots, a_k),$$

i.e., $1 + (c+1)t \mid P(\mathfrak{x}, y, c, a_1, \cdots, a_k)$. Furthermore,

$$1 + (u+1)t \mid a_i - v_{ui},$$

so

$$1 + (u+1)t \mid a_i - j, \quad \text{for some } j \leqq y.$$

But since the divisors are relatively prime by (10.6), we have

$$1 + (c+1)t \mid \prod_{j \leqq y} (a_i - j), \quad \text{for } i = 1, \cdots, k.$$

Hence (10.4) is satisfied and the lemma has been established.

Proof of Theorem: Let $\mathfrak{R}(x_1, \cdots, x_n)$ be a recursively enumerable relation. Then by Davis' result, there is a polynomial $P(\mathfrak{x}, y, u, v_1, \cdots, v_k)$ so that

$$\mathfrak{R}(x_1, \cdots, x_n)$$

holds if and only if

$$(\bigvee y)(\bigwedge u \leqq y)(\bigvee v_1 \leqq y)\cdots(\bigvee v_k \leqq y)[P(\mathfrak{x}, y, u, v_1, \cdots, v_k) = 0].$$

Let $Q(\mathfrak{x}, y)$ be the polynomial obtained from P by replacing the coefficients by their absolute value, substituting y for each of the variables $u, v_1, \cdots, v_k$, and adding y to the resulting polynomial. Clearly, Q satisfies the condition of lemma. Hence $\mathfrak{R}(x_1, \cdots, x_n)$ holds if and only if there is a natural number y satisfying the relation given by (10.4). Since each of the relations occurring in (10.4) is exponential diophantine, $\mathfrak{R}$ is exponential diophantine.

We return to the intuitive theory discussed in Sections 3 and 4 for an application of the theorem.

COROLLARY: *There is a particular exponential diophantine equation $E(n, x_1, \cdots, x_k) = 0$ for which there is no general method to tell for an arbitrary natural number n whether or not $E(n, x_1, \cdots, x_k) = 0$ has a solution for $x_1, \cdots, x_k$ in natural numbers.*

Proof: We have seen that there is a listable set $\mathcal{S}$ which is not computable. By contention, every listable set is recursively enumerable and hence exponential diophantine. Thus there is an exponential diophantine equation $E(n, x_1, \cdots, x_k) = 0$ such that n belongs to $\mathcal{S}$ if and only if $E = 0$ has a solution. Hence there can be no general method of telling, for an arbitrary n, whether $E(n, x_1, \cdots, x_k) = 0$ has a solution for $x_1, \cdots, x_k$ in natural numbers.

REMARKS: *This is a stronger result than the fact that there is no general method to tell whether or not an arbitrary exponential diophantine equation has a solution. In the case of diophantine equations, it may be that every diophantine set is computable but still there is no method to tell if an arbitrary diophantine equation has a solution.*

What relation does this result have to Gödel's incompleteness theorem? Given any effective system of true axioms, Gödel showed how to construct a particular arithmetical statement which is clearly true but not provable from the given axioms. By an arithmetical statement, we mean a statement that can be expressed by an arithmetical formula (Section 6).

Suppose we number exponential diophantine equations as we did ordinary diophantine equations in Section 2, say $E_0(x_0, x_1, \cdots) = 0$, $E_1(x_0, x_1, \cdots) = 0, \cdots$. Let $\mathcal{A}$ be some effectively given axiom system. We can assume either that the axioms are true or that the system is consistent and strong enough to prove the solvability of every exponential diophantine equation which has in fact a solution.† Let $\mathcal{U}$ be the set of natural numbers n such that $E_n(n, x_1, \cdots) = 0$ can be proved to have no solution for $x_1, \cdots$ in natural numbers on the basis of the axiom system $\mathcal{A}$. The set $\mathcal{U}$ is certainly listable, since we can systematically run through the possible proofs and put n on our list each time we conclude $E_n(n, x_1, \cdots) = 0$ is unsolvable. Hence there is some natural number k such that $n \in \mathcal{U}$ if and only if $E_k(n, x_1, \cdots) = 0$

† Note that the axioms need not be limited to number theory. They can include axioms of analysis, set theory, etc.

has a solution for $x_1, \cdots$. That is, the unsolvability of $E_n(n, x_1, \cdots) = 0$ is provable from $\mathfrak{A}$ if and only if $E_k(n, x_1, \cdots) = 0$ is solvable. Putting $n = k$, we see that the unsolvability of $E_k(k, x_1, \cdots) = 0$ is provable if and only if $E_k(k, x_1, \cdots) = 0$ is solvable. If the axioms are true and $E_k(k, x_1, \cdots) = 0$ is solvable, or if the axioms are consistent and $E_k(k, x_1, \cdots) = 0$ is solvable and provably so, then $E_k(k, x_1, \cdots) = 0$ cannot be proved unsolvable. Hence $E_k(k, x_1, \cdots) = 0$ is unsolvable but cannot be proved to be unsolvable.

Now given $\mathfrak{A}$ effectively, we can actually calculate k. Briefly, Gödel showed how to obtain a primitive recursive function whose range is $\mathfrak{U}$.† We can then use the reductions of Sections 6 and 7 to obtain $\mathfrak{U}$ in the Davis normal form. Finally, the theorem of this section yields the required exponential diophantine equation and we can calculate its number k. Actually to carry out this program of obtaining the required exponential diophantine equation for a nontrivial axiom system would hardly be worthwhile since the complexity of the resulting equation would outweigh its interest.

Thus we have sketched the proof that *to any effectively given system of true axioms, there is a method for finding an exponential diophantine equation which has no solution but for which there is no proof from the given axioms that the equation is unsolvable.*

11. REDUCTION OF EXPONENTIAL DIOPHANTINE EQUATIONS

Is the relation $r = s^t$ diophantine? If so, then every exponential diophantine equation could be replaced by an equivalent diophantine equation in more variables. Also every recursively enumerable relation would be diophantine and hence Hilbert's problem would be unsolvable. At present, we don't know the answer to this question. We shall show now that if any infinite set of primes is diophantine or if any relation of roughly expo-

† Essentially, his argument is the one sketched in Section 5 showing that any listable set is recursively enumerable.

nential order of growth is diophantine, then $r = s^t$ is indeed diophantine.

We need a few simple lemmas. In the following, let $a > 1$ and a_n, a'_n be defined by the equation

$$a_n + a'_n\sqrt{a^2 - 1} = (a + \sqrt{a^2 - 1})^n. \tag{11.1}$$

Thus,

$$\begin{aligned} a_0 &= 1 & a'_0 &= 0, \\ a_1 &= a, & a'_1 &= 1, \\ a_{n+2} &= 2a \cdot a_{n+1} - a_n, & a'_{n+2} &= 2a \cdot a'_{n+1} - a'_n. \end{aligned} \tag{11.2}$$

LEMMA 1: *Natural numbers x and y satisfy the equation $x^2 - (a^2 - 1)y^2 = 1$ if and only if there is a natural number n such that $x = a_n$ and $y = a'_n$.*

This is a special case of the familiar result concerning Pell's equation $x^2 - dy^2 = 1$, where d is not a perfect square. Namely, x and y satisfy $x^2 - dy^2 = 1$ if and only if for some n, $x + \sqrt{d}y = (x_1 + \sqrt{d}y_1)^n$, where x_1, y_1 is the least positive solution of the equation which always has a nontrivial solution if d is not a square (cf. LeVeque [9], p. 116.) In our case, $a > 1$ so that $a^2 - 1$ is not a square and $x = a$, $y = 1$ is clearly the least positive solution.

LEMMA 2: *For $a > 1$,*

$$a'_n \equiv n \pmod{a - 1} \tag{11.3}$$

and

$$a_n - a'_n(a - s) \equiv s^n \pmod{2as - s^2 - 1}. \tag{11.4}$$

Proof: The proof is by induction on n. From (11.2), we see that both (11.3) and (11.4) hold for $n = 0$ and $n = 1$. Assume (11.3) and (11.4) have been proved for n and $n + 1$, then

$$a'_{n+2} \equiv 2a \cdot (n + 1) - n \equiv n + 2 \pmod{a - 1}$$

and

$$\begin{aligned} a_{n+2} - a'_{n+2}(a - s) &\equiv 2a(a_{n+1} - a'_{n+1}(a - s)) - (a_n - a'_n(a - s)), \\ &\equiv 2as^{n+1} - s^n \equiv s^{n+2} \pmod{2as - s^2 - 1}. \end{aligned}$$

LEMMA 3. *There is a diophantine relation $Q(a, u)$ such that*

(i) *if* $\mathrm{Q}(a, u)$ *then* $u \geqq a^a$,
(ii) *if* $a > 1$ *and* $u \geqq a^{2a}$ *then* $\mathrm{Q}(a, u)$.

In fact, we may take for $\mathrm{Q}(a, u)$ the relation between a and u which holds if and only if there are x and y such that

$$(11.5) \qquad \begin{aligned} &x^2 - (a^2 - 1)(a - 1)^2y^2 = 1, \\ &x > 1, \\ &u \geqq ax. \end{aligned}$$

Proof: Suppose $\mathrm{Q}(a, u)$ holds. Since $x > 1$, $a > 1$. Hence there is an $n > 0$ such that $x = a_n$ and $(a - 1)y = a_n'$. By Lemma 2, $n \equiv 0 \pmod{a - 1}$. Therefore $n \geqq a - 1$. Hence $u \geqq a \cdot a_{a-1}$. But $a_n \geqq a^n$ by (11.1), so $u \geqq a^a$.

Now for $a > 1$, the conditions (11.5) will be satisfied by $x = a_{a-1}$, $y = a_{a-1}'/(a - 1)$, and $u \geqq a \cdot a_{a-1}$. Since a_n is the rational part of $(a + \sqrt{a^2 - 1})^n$, we see that $a_n \leqq (2a)^n \leqq a^{2n}$. Hence $a \cdot a_{a-1} \leqq a \cdot a^{2a-2} < a^{2a}$. Therefore, if $u \geqq a^{2a}$ and $a > 1$ then $\mathrm{Q}(a, u)$.

THEOREM: *The relation* $r = s^t$ *is diophantine if any infinite set of primes is diophantine.*

Proof: We shall show that $r = s^t$ with $s > 0$ if and only if there are natural numbers a, x, y, and p such that

$$(11.6) \qquad \begin{aligned} &s > 0,\ a > 1, \\ &\mathrm{Q}(s + t + 1, p), \\ &p \text{ is a prime}, \\ &p \mid 2as - s^2 - 1,\ p - 1 \mid a - 1. \\ &x^2 - (a^2 - 1)y^2 = 1, \\ &y \equiv t \pmod{a - 1}, \\ &\mathrm{Rem}\,(x - (a - s)y, p) = r, \end{aligned}$$

Here Q is the relation of Lemma 3 and all the conditions are clearly diophantine except for "p is a prime." In order to obtain the theorem, we need to prove slightly more. It is sufficient to show that if (11.6) is satisfied, then $r = s^t$, and if $r = s^t$ and $s > 0$, then (11.6) can be satisfied for any sufficiently large prime p. Hence we could replace the condition "p is a prime" by the condition $p \in \mathcal{P}$ where $\mathcal{P}$ is any infinite diophantine set of primes.

Of course, if $r = s^t \wedge s > 0$ is diophantine, then $r = s^t$ is diophantine.

Suppose (11.6) holds. Then $s^t < p$, $0 < s < p$, and for some n, $x = a_n$ and $y = a'_n$. So by Lemma 2,

$$x - (a - s)y \equiv s^n \pmod{2as - s^2 - 1}$$

and

$$y \equiv n \pmod{a - 1}.$$

Hence by (11.6),

$$r \equiv s^n \pmod{p}$$

and

$$t \equiv n \pmod{p - 1}.$$

Since $s^{p-1} \equiv 1 \pmod{p}$ by Fermat's theorem,

$$r \equiv s^n \equiv s^t \pmod{p}.$$

But $s^t < p$ and $r < p$, so $r = s^t$.

Now suppose $r = s^t$ and $s > 0$. Let p be a prime so large that $\mathcal{Q}(s + t + 1, p)$ holds. We need to choose $a > 1$ so that

$$p - 1 \mid a - 1,$$
$$p \mid 2as - s^2 - 1. \tag{11.7}$$

If a satisfies (11.7), then $x = a_t$ and $y = a'_t$ will satisfy the other conditions of (11.6) and we are through. But (11.7) amounts to

$$a \equiv 1 \pmod{p - 1},$$
$$2sa \equiv s^2 + 1 \pmod{p},$$

and since $(2s, p) = 1$, we can solve these congruences simultaneously for a. Hence all the conditions of (11.6) can be satisfied.

As a consequence of this theorem, we obtain the result that *if an infinite set of primes is diophantine, then every recursively enumerable set is diophantine and there is no general method to tell if an arbitrary diophantine equation has a solution.*

REMARK: In order to satisfy (11.6) for a given r, s, t with $r = s^t$, it is sufficient to take for p any sufficiently large prime. Hence we can simultaneously define a finite set of relations

$$r_1 = s_1^{t_1}, \cdots, r_k = s_k^{t_k}$$

in terms of a single prime. We need only repeat the conditions of (11.6) with $r_i, s_i, t_i, a_i, x_i, y_i$ replacing r, s, t, a, x, y for $i = 1, \cdots, k$, leaving p fixed throughout. Hence *to every exponential diophantine set* $\mathcal{S}$, *i.e. every recursively enumerable set* $\mathcal{S}$, *there is a diophantine equation* $P(x, p, y_1, \cdots) = 0$ *which has a solution with* p *prime for a given* x *if and only if* $x \in \mathcal{S}$. This argument is similar to one used by Davis.

COROLLARY: *There is no general method of deciding whether or not*

$$\text{Range } A(x_1, \cdots) \subseteqq \text{Range } ((u + (v + 2)(w + 2))^2 + u)$$

where A is an arbitrary polynomial with positive integer coefficients.†

Proof: From the above remark, there is no general method to tell whether or not

$$F(p, y_1, \cdots) = G(p, y_1, \cdots)$$

has a solution with p prime where F and G are arbitrary polynomials with positive integer coefficients. Hence there is no method to tell whether the polynomial

$$(z + 2)(1 + (F(z + 2, y_1, \cdots) - G(z + 2, y_1, \cdots))^2)$$

represents a prime, i.e., assumes a prime number as value.

Let $T(x, y) = (x + y)^2 + x$, so that $T(x, y) = T(u, v)$ implies that $x = u$ and $y = v$. Then the polynomial $A(z, y_1, \cdots)$ given by

$$T((z + 2) + 2FG, (z + 2)(1 + (F - G)^2))$$

has positive coefficients. Furthermore,

$$\text{Range } A(z, y_1, \cdots) \subseteqq \text{Range } T(u, (v + 2)(w + 2))$$

if and only if $(z + 2)(1 + (F - G)^2)$ does not represent a prime. Hence there is no general method of telling whether or not

$$\text{Range } A(z, y_1, \cdots) \subseteqq \text{Range } T(u, (v + 2)(w + 2)),$$

where A is an arbitrary polynomial with positive coefficients.

Next we use the same lemmas and a similar argument to prove another theorem.

---

† **Open problem:** Is there a decision method to tell whether or not an arbitrary polynomial with positive integer coefficients represents all natural numbers?

THEOREM: *If there is a diophantine relation* $\mathcal{S}(p, q)$ *such that*

(11.8) *if* $\mathcal{S}(p, q)$, *then* $p > 1$ *and* $q \leqq p^p$,

and

(11.9) *for every* k, *there are* p *and* q *with* $\mathcal{S}(p, q)$ *and* $q > p^k$,

then the relation $r = s^t$ *is diophantine.*

Proof: We shall show that $r = s^t$ with $s, t > 0$ if and only if there are natural numbers a, x, y, and z such that

$$
\begin{aligned}
& s > 0, \quad t > 0, \\
& \mathcal{Q}(s + t + 1, 2as - s^2 - 1), \\
& \mathcal{S}(a, z), \\
(11.10) \qquad & x < z, \quad y > 0, \quad x^2 - (a^2 - 1)y^2 = 1, \\
& \text{Rem } (y, a - 1) = t, \\
& \text{Rem } (x - (a - s)y, 2as - s^2 - 1) = r.
\end{aligned}
$$

Suppose (12.10) is satisfied. Then $s^t < 2as - s^2 - 1$, $x < a^a$ and $y > 0$. Hence $x = a_n$ and $y = a'_n$ for some positive $n < a$. Also Rem $(y, a - 1) = t \equiv n \pmod{a - 1}$, so $n = t$. Finally, Rem $(x - y(a - s), 2as - s^2 - 1) = r \equiv s^t (\text{mod } 2as - s^2 - 1)$ so $r = s^t$.

Suppose $r = s^t$ with $s, t > 0$. The conditions (11.8) and (11.9) on $\mathcal{S}$ insure that for every k there are infinitely many p and q with $\mathcal{S}(p, q)$ and $q > p^k$. Hence we can choose a so large that

$$
\begin{aligned}
& \mathcal{Q}(s + t + 1, 2as - s^2 - 1), \\
& t < a - 1,
\end{aligned}
$$

and such that for some z, $\mathcal{S}(a, z)$ and $z > a^{2t}$. Then for this choice of a, $z > a_t$. Hence if we take $x = a_t$ and $y = a'_t$, all conditions of (11.10) will be satisfied.

THEOREM: *If there is a diophantine relation* $\mathcal{R}(p, q)$ *such that*

(11.11) *there is some* n *such that if* $\mathcal{R}(p, q)$ *then* $q \leqq p * n$

and

(11.12) *for every* k, *there are* p *and* q *with* $\mathcal{R}(p, q)$, $p > 1$, *and* $q > p^k$, *then there is a diophantine relation* $\mathcal{S}(p, q)$ *which satisfies* (11.8) *and* (11.9).

Proof: Case A. Suppose $\mathcal{R}(u, v)$ implies $v \leqq u^{mu}$ for some m. Let $\mathcal{S}(p, q)$ be the diophantine relation which holds if and only if there is a v such that

$$(11.13) \qquad p > 1, \quad \mathcal{R}(p, v), \quad \text{and} \quad q^m \leqq v.$$

If $\mathcal{S}(p, q)$, then for some v, $q^m \leqq v \leqq p^{mp}$ so $q \leqq p^p$. Hence $\mathcal{S}$ satisfies (11.8). On the other hand, for every k there is a $p > 1$ and a v so that $\mathcal{R}(p, v)$, $v > p^{2mk}$, and hence $\mathcal{S}(p, p^{2k})$. Since $p > 1$, $p^{2k} > p^k > p^0$ for $k > 0$. Therefore $\mathcal{S}$ satisfies (11.9).

Case B. Suppose $\mathcal{R}(u, v)$ implies $v \leqq u * n$ for some $n > 2$ and there is no m such that $\mathcal{R}(u, v)$ implies $v \leqq u^{mu}$. Let $\mathcal{R}_1(u, v)$ if and only if there is a natural number a such that $\mathcal{Q}(a, u)$ and $\mathcal{R}(a, v)$. Hence if $\mathcal{R}_1(u, v)$ holds, there is an a with

$$a^a \leqq u,$$
$$v \leqq a * n \leqq a^a * (n - 1) \leqq u * (n - 1).$$

Furthermore, for every k, there is an $a > 1$ and v with $\mathcal{R}(a, v)$ and $v > a^{2ka}$. By Lemma 3, $\mathcal{Q}(a, a^{2a})$. Hence for $u = a^{2a}$, we have $\mathcal{R}_1(u, v)$ and $v > u^k$. Thus we obtain the theorem, since we can continue to apply Case B until either n is reduced to 2 or we obtain a relation which falls under Case A.

Putting the last two theorems together with the fact that every recursively enumerable relation is exponential diophantine, we obtain the following.

THEOREM: *If there is a diophantine relation* $\mathcal{R}(p, q)$ *which satisfies* (11.11) *and* (11.12), *then every recursively enumerable relation is diophantine.*

12. CONCLUSIONS

Exactly one of the following two statements is true:

I. Every recursively enumerable relation is diophantine.

II. For every diophantine equation $P(p, q, x_1, \cdots, x_k) = 0$ either (1) for every n there is a solution with $q \geqq p * n$ or (2) there is an m such that every solution satisfies $q \leqq p^m$.

If I is true, then all the familiar number-theoretic sets such as the set of primes would be diophantine and as in the case of exponential diophantine equations, there could be no method of telling whether an arbitrary diophantine equation is solvable.

On the other hand, if II is true, there would be rather surprising bounds on solutions of diophantine equations. We see that (1) holds whenever there is a p such that there are infinitely many q for which $P(p, q, x_1, \cdots) = 0$ has a solution or if there is a solution with $p = 0$ or 1 and $q > 1$. Does (1) hold in any other case? There is no example known. Hence it may be that if for all $p > 1$ there are at most a finite number of q's for which $P = 0$ has a solution, then there is an m such that all solutions with $p > 1$ satisfy $q < p^m$. If this is the case, then a positive solution to the decision problem for diophantine equations would appear more likely although by no means assured. We would still need a test to tell if for some p there are infinitely many q's for which the given equation has a solution and some bound on m in terms of the coefficients of the equation.

Is it I or II that is true? We shall probably find out by disproving one of them. For example, if we could show that some recursively enumerable set is not diophantine, then I is false. Or it would be sufficient to show that the class of recursively enumerable sets differs in some regard from the class of diophantine sets. On the other hand, to disprove II, all we need is a single diophantive equation which is a counter-example. Davis and Putnam have recently made progress in this direction. (See [2], [3].)

We now put these results together.

THEOREM: *Consider diophantine equations $P(p, q, x_1, \cdots) = 0$ such that to every p there are at most a finite number of values of q for which $P = 0$ is solvable. If to every such equation $P = 0$, there is an n such that every solution of $P = 0$ with $p > 1$ satisfies $q \leqq p * n$, then to every such equation there is an n' such that each solution with $p > 1$ satisfies $q \leqq p^{n'}$.*

Proof: Suppose some diophantine relation $\mathcal{R}(p, q)$ satisfies (11.11) and (11.12). Then every recursively enumerable relation is diophantine. In particular, the relation given by $q = p * p$ is diophantine. Hence there is an equation $P_1(p, q, x_1, \cdots) = 0$ which is solvable for exactly one q for each p and for which there is no n such that all solutions with $p > 1$ of $P_1 = 0$ satisfy $q \leqq p * n$. So the theorem is true in this case. On the other hand, if all the equations considered define relations which satisfy (11.11) then the relations do not satisfy (11.12), i.e., to each of these equations, there is an n' such that every solution with $p > 1$ satisfies $q \leqq p^{n'}$.

BIBLIOGRAPHY

1. Davis, Martin, *Computability and Unsolvability*. McGraw-Hill, New York, 1958.
2. ———, "Applications of recursive function theory to number theory," *Recursive Function Theory, Proceedings of Symposia in Pure Mathematics, Volume V. Amer. Math. Soc.* (1962), 135–138.
3. ———, "Unsolvable problems: a review," *Proceedings of the Symposium on Mathematical Theory of Automata, Microwave Res. Inst. Symposia Series*, Vol. 12, Polytechnic Press, Brooklyn, 1963, 15–22.
4. ———; Putnam, Hilary; and Robinson, Julia, "The decision problem for exponential diophantine equations," *Ann. of Math.*, 74 (1961), 425–436.
5. Gödel, Kurt, "Über formal unentscheidbare Sätze der Principia Mathematica und verwandter Systeme I," *Monatsh. Math. und Physik*, 38 (1931), 173–198. For an English translation of this article, see Kurt Gödel, *On Formally Undecidable Propositions*, Basic Books, 1962.
6. Grzegorczyk, Andrzej, *Fonctions Récursives*. Gauthier-Villars, Paris, 1961.
7. Hermes, Hans, *Aufzählbarkeit, Entscheidbarkeit, Berechenbarkeit*. Springer-Verlag, Berlin, 1961.
8. Kleene, Stephen Cole, *Introduction to Metamathematics*. Van Nostrand, Princeton, N.J., 1952. Chapters 11–13.

9. LeVeque, William J., *Elementary Theory of Numbers*. Addison-Wesley, Reading, Mass., 1962.

10. Mendelson, Elliott, *Introduction to Mathematical Logic*. Van Nostrand, Princeton, N.J., 1964. Chapter 5.

11. Post, Emil, "Recursively enumerable sets of positive integers and their decision problems," *Bull. Amer. Math. Soc.* 50 (1944), 284–316.

12. Putnam, Hilary, "An unsolvable problem in number theory," *J. Symb. Logic*, 25 (1960), 220–232.

13. Robinson, Julia, "Existential definability in arithmetic," *Trans. Amer. Math. Soc.*, 72 (1952), 437–449.

FINITELY GENERATED CLASSES OF SETS OF NATURAL NUMBERS

JULIA ROBINSON

We say that a set S of natural numbers is generated by a class $\mathfrak{F}$ of functions if S is the range of a function F obtained by composition from the functions of $\mathfrak{F}$. Here we consider the identity function I to be obtained by the empty composition. We consider only the case in which $\mathfrak{F}$ consists of functions of one variable on and to the set of natural numbers N. All sets considered are subsets of N.

A class $\mathfrak{C}$ of sets is generated by $\mathfrak{F}$ if every nonempty set of $\mathfrak{C}$ is generated by $\mathfrak{F}$ and every set generated by $\mathfrak{F}$ is in $\mathfrak{C}$. $\mathfrak{C}$ is finitely generated if there is a finite class of functions $\mathfrak{F}$ which generates $\mathfrak{C}$.

A function F is compatible with $\mathfrak{C}$ if $F(S)$ belongs to $\mathfrak{C}$ for every S in $\mathfrak{C}$.

Clearly, if $\mathfrak{F}$ generates $\mathfrak{C}$ then every F in $\mathfrak{F}$ is compatible with $\mathfrak{C}$. Also N belongs to $\mathfrak{C}$. Furthermore to show that $\mathfrak{F}$ generates $\mathfrak{C}$ it is sufficient to prove that every $F \in \mathfrak{F}$ is compatible with $\mathfrak{C}$, $N \in \mathfrak{C}$, and every nonempty set in $\mathfrak{C}$ is generated by $\mathfrak{F}$.

In [1] finite sets of relatively simple functions are given which generate the classes of recursively enumerable sets and diophantine sets. Here we ask: What classes of sets can be finitely generated? In particular, we will show that if $\mathfrak{C}$ is a denumerable field of sets such that all finite sets belong to $\mathfrak{C}$ and at least one infinite set having an infinite complement belongs to $\mathfrak{C}$, then $\mathfrak{C}$ is finitely generated.

LEMMA 1. *If $\mathfrak{I}$ is an infinite class of infinite sets then there is a set M such that $M \cap T$ and $\overline{M} \cap T$ are infinite for every T belonging to $\mathfrak{I}$.*

PROOF. Let each T in $\mathfrak{I}$ be listed infinitely often in the sequence $T_0, T_1, \cdots$. We will construct M and $\overline{M}$ simultaneously. At the nth step put the least unassigned number of T_n in M and then the least remaining unassigned number of T_n in $\overline{M}$.

LEMMA 2. *Given infinite sets $T_0, T_1, \cdots$ and nonempty sets $S_0, S_1, \cdots$ such that $S_k = S_l$ whenever $T_k \cap T_l$ is not empty. Then there is a function F such that $F(T_n) = S_n$.*

PROOF. Make a list of natural numbers such that each number occurs infinitely often. We will use the list to define F at one argument for each step of the construction. Namely, if at a given step n is the

Received by the editors April 29, 1968.

next number in the list and if x is the least unassigned number in T_n and y is the least number in S_n not yet in $F(T_n)$, put $Fx=y$; while if there is no such y, put $Fx=$ the least number of S_n. Finally, in case $\bigcup T_k \neq N$, put $Fx=0$ for every x not in any T_k.

THEOREM 1. HYPOTHESES. (1) *$\mathcal{C}$ is a denumerable class of sets.* (2) *If A and B belong to $\mathcal{C}$ then $A\cup B$ belongs to $\mathcal{C}$.* (3) *N belongs to $\mathcal{C}$.* (4) *Either* (i) *$\mathcal{C}$ contains all finite sets or* (ii) *$\mathcal{C}$ contains only infinite sets and $\mathcal{C}$ contains the set obtained from a member set by adjoining a single element.* (5) *There is an infinite partition $\mathcal{P}\subset\mathcal{C}$ of N into infinite sets all of which are generated by a finite class $\mathfrak{F}$ of functions compatible with $\mathcal{C}$.*

CONCLUSION. *$\mathcal{C}$ is finitely generated.*

PROOF. Let $\mathcal{P}=\{P_0, P_1, \cdots\}$. We say that I is the index set of a set S if $I=\{n: S\cap P_n\neq\varnothing\}$. Let M be the set given by Lemma 1 such that every infinite index set of a set belonging to $\mathcal{C}$ contains infinitely many elements of both M and $\overline{M}$. (Concerning $\overline{M}$, we shall use only that $\overline{M}$ is infinite.) List the nonempty sets of $\mathcal{C}$ in a sequence T_0, $T_1, \cdots$ so that $T_m=N$ for every m in M. We will define F in such a way that (a) $F(T_n\cap P_k)=T_k$ whenever $T_n\cap P_k$ is infinite, and (b) $F(T_n)=N$ whenever T_n has an infinite index set. Let us check that this is sufficient. In the first place, every nonempty set of $\mathcal{C}$ will be generated by $\{F\}\cup\mathfrak{F}$ since $F(P_k)=T_k$. Also F is compatible with $\mathcal{C}$, i.e. for every n, $F(T_n)$ belongs to $\mathcal{C}$. To see this, let I be the index set of T_n. If I is infinite then $F(T_n)=N$ and N belongs to $\mathcal{C}$ by (3). If I is finite, then

$$F(T_n) = \bigcup_{k\in I} F(T_n\cap P_k),$$

where $F(T_n\cap P_k)$ is either finite or T_k according as $T_n\cap P_k$ is finite or infinite. Hence if $\mathcal{C}$ contains all finite sets then $F(T_n)$ is a nonempty finite union of sets of $\mathcal{C}$ and so $F(T_n)$ belongs to $\mathcal{C}$. If $\mathcal{C}$ consists only of infinite sets, then $T_n\cap P_k$ is infinite for some k. Thus $F(T_n)$ is the union of a nonempty finite union of sets of $\mathcal{C}$ and a finite set. Hence $F(T_n)$ belongs to $\mathcal{C}$.

Therefore to complete the proof, we need only define F satisfying (a) and (b). Now make a list of pairs (n, k) so that each pair occurs infinitely often. We will use this list to index the steps in the definition of F. At a step corresponding to (n, k), if there is a number in $T_n\cap P_k$ at which F has not yet been assigned, let the least such number be x. Put Fx equal to the least number in T_k which is not already in $F(T_n\cap P_k)$ if there is such a number; otherwise put Fx equal to the least number in T_k. Also if the index set of T_n is infinite and y is the least number unassigned in $T_n \cap P_m$ for any m in M, define $Fy=k$. Then F will satisfy both (a) and (b).

EXAMPLES. The classes of recursive sets, recursively enumerable sets, arithmetical sets, hyperarithmetical sets, are all finitely generated. We can let $P_k=(SD)^kD(N)$ since S, D are compatible with each of these classes. Here S is the successor function and D the double function.

By the usual diagonal argument it is easy to see that the class of arithmetical sets cannot be generated by a finite class of arithmetical functions. (Similarly hyperarithmetical sets cannot be generated by a finite class of hyperarithmetical functions.) I do not know if the class of recursive sets can be generated by a finite number of recursive functions.

An interesting class of sets which is not covered by Theorem 1 is the class of polynomial sets and I do not know whether it is finitely generated. By a polynomial set, I mean the range of a polynomial with positive integer coefficients and any number of variables.

If we assume that $\mathcal{C}$ is a field of sets which contains all finite sets, then condition (5) in Theorem 1 can be weakened to the mere existence of $\mathcal{P}$. In order to prove this, we shall need the following combinatorial result.

LEMMA 3. *Let $\mathcal{C}$ be a denumerable field of sets which contains all finite sets. Suppose further that there is an infinite partition $\mathcal{P}_0$ of N into infinite sets of $\mathcal{C}$. Then there is an infinite partition $\mathcal{P}$ of N into infinite sets $P_0, P_1, \cdots$ of $\mathcal{C}$ such that for every S in $\mathcal{C}$ either S or $\bar{S}$ is included in a finite union of the P's.*

REMARK. This lemma is stronger than the one I originally proved and was suggested to me by Dana Scott. His proof was obtained by considering the Boolean algebra of sets of $\mathcal{C}$ mod finite sets.

PROOF. Let $S_0, S_1, \cdots$ be the sets of $\mathcal{C}$. We will define $P_0, P_1, \cdots$ by means of a sequence of infinite partitions $\mathcal{P}_0, \mathcal{P}_1, \cdots$ of N into infinite sets of $\mathcal{C}$ such that for every k and n with $k<n$, P_k belongs to $\mathcal{P}_n$ and either S_k or $\bar{S}_k$ is included in $\bigcup_{j<n} P_j$. Assume that these conditions hold for all n up through t. We shall define P_t and $\mathcal{P}_{t+1}$ so that they also hold for $n=t+1$. Let S be S_t if $\bar{S}_t$ has infinitely many infinite parts under $\mathcal{P}_t$; otherwise let S be $\bar{S}_t$. In either case, $\bar{S}$ has infinitely many infinite parts under $\mathcal{P}_t$. Let those infinite parts of $\bar{S}$ which are not contained in one of $P_0, \cdots, P_{t-1}$ be $I_0, I_1, \cdots$. Then put

$$P_t = I_0 \cup \left(S - \bigcup_{k<t} P_k\right).$$

Let $a_1, a_2, \cdots$ be the numbers which do not belong to

$$\bigcup_{k\leq t} P_k \cup \bigcup_{j\geq 1} I_j,$$

if there are any. Put $T_k = I_k \cup \{a_k\}$ if there is an a_k; otherwise let $T_k = I_k$. Finally, put

$$\mathcal{P}_{t+1} = \{P_0, P_1, \cdots, P_t, T_1, \cdots\}.$$

Clearly $P_t \in \mathcal{P}_{t+1}$, $S \subseteq \bigcup_{k\leq t} P_k$, and $\mathcal{P}_{t+1}$ is an infinite partition of N into infinite sets of $\mathcal{C}$.

Now let $\mathcal{P} = \{P_0, P_1, \cdots\}$. By construction each P_k is an infinite set of $\mathcal{C}$, $\mathcal{P}$ is disjoint, and for every k, either S_k or $\bar{S}_k$ is contained in $\bigcup_{j\leq k} P_j$. Finally, every n belongs to some P_j. In fact, $\{n\} = S_l$ for some l, so n belongs to some P_j with j not exceeding l. Hence $\mathcal{P}$ satisfies the lemma.

THEOREM 2. *If a denumerable field of sets $\mathcal{C}$ contains all finite sets and includes an infinite disjoint class $\mathcal{P}$ of infinite sets $P_0, P_1, \cdots$, then $\mathcal{C}$ is finitely generated.*

PROOF. Without loss of generality, we may suppose that $\mathcal{P}$ is a partition of N; otherwise adjoin the missing numbers to the parts P_k with at most one to each P_k. By Lemma 3, we may also assume that $\mathcal{P}$ has the further property that for every S in $\mathcal{C}$ either S or $\bar{S}$ is included in a finite union of P's.

By Lemma 2, we can define a function H so that $H(S) = P_0$ for every infinite set S in $\mathcal{C}$. Hence H is compatible with $\mathcal{C}$.

Let G be defined so that $G(S) = P_{n+1}$ for every infinite set S of $\mathcal{C}$ which is included in P_n. The existence of G also follows from Lemma 2. Then $G^k H(N) = P_k$ for all k. Furthermore, if S is included in a finite union of P's, then $G(S)$ is the union of a finite set and a finite union of P's; if $\bar{S}$ is included in a finite union of P's, then $G(S)$ is the union of a finite set and the complement of a finite union of P's. In both cases $G(S)$ belongs to $\mathcal{C}$. Hence G is compatible with $\mathcal{C}$.

We can now finish the proof by noticing the hypotheses of Theorem 1 hold. We can also construct F more directly as follows: Let $S_0, S_1, \cdots$ be the nonempty sets of $\mathcal{C}$. By Lemma 2, there is a function F such that $F(T) = S_k$ if T is an infinite subset of P_k which belongs to $\mathcal{C}$. Then F is compatible with $\mathcal{C}$. Indeed, if S is contained in a finite union of P's then $F(S)$ is the union of a finite number of sets of $\mathcal{C}$ and if $\bar{S}$ is contained in a finite union of P's then $F(S)=N$. Every nonempty set in $\mathcal{C}$ is generated by F, G, and H since $S_k=FG^kH(N)$.

THEOREM 3. *Let $\mathcal{C}$ be a denumerable field of sets which contains all finite sets. Suppose $A_0, \cdots, A_n$ is a partition of N into infinite sets of $\mathcal{C}$ none of which is the union of two disjoint infinite sets of $\mathcal{C}$. Then $\mathcal{C}$ is finitely generated if and only if $n>0$.*

PROOF. If $n=0$ then $\mathcal{C}$ is the field of all finite and cofinite sets. Suppose $\mathcal{C}$ were generated by $F_1, \cdots, F_k, G_1, \cdots, G_l$ where the range of each F is finite and the range of each G is cofinite. If G is obtained by composition from the G's then the range of G is cofinite. Hence if H is obtained by composition from the F's and G's then H has a finite range only if some F, say F_j, is used to obtain H. Then the number of elements in $H(N)$ is not more than the number of elements in $F_j(N)$. Hence the number of elements in the finite sets generated by $F_1, \cdots, F_k, G_1, \cdots, G_l$ is bounded which gives a contradiction. Thus for $n=0$, $\mathcal{C}$ is not finitely generated.

Suppose $n>0$. For $K\subseteq\{0, \cdots, n\}$, let $S_K=\bigcup_{k\in K} A_k$. Every set of $\mathcal{C}$ differs from some S_K by a finite number of elements. We will first show that every S_K can be generated by three functions A, B, and C which are compatible with $\mathcal{C}$. By Lemma 2, there is a function A such that $A(S)=A_0$ for every infinite set S in $\mathcal{C}$. A is compatible with $\mathcal{C}$ since $A(S)$ is either finite or A_0 for every S in $\mathcal{C}$.

Let B be a permutation which maps A_k onto A_{k+1} for $k=0, \cdots, n-1$ and maps A_n onto A_0. B is compatible with $\mathcal{C}$ since B maps any set differing from S_K by a finite set onto a set differing from S_M by a finite set where M is obtained from K by adding 1 to each element of K other than n and replacing n by 0. Finally, let C be a function which is one-to-one on each A_i and maps A_i onto $A_0\cup A_i$. Again C is compatible with $\mathcal{C}$ since C maps any set differing from S_K by a finite set onto a set differing from S_M by a finite set where $M=\emptyset$ if $K=\emptyset$ and $M=K\cup\{0\}$ if $K\neq\emptyset$.

Suppose t is the least element of a nonempty set $K\subseteq\{0, \cdots, n\}$. If K has just one element then $S_K=B^tA(N)$. If K has more than one element, let

$$K' = \{k: k + t \in K \wedge k > 0\}.$$

Then $S_K=B^tC(S_{K'})$. Hence by induction on the number of elements of K, for every nonempty set K, there is a function H obtained from B and C by composition such that $S_K=HA(N)$.

Now if S differs from S_K (K nonempty) by a finite set, then there is a set E which differs from A_0 by a finite set such that for some H

obtained from B and C by composition $H(\boldsymbol{E})=\boldsymbol{S}$. Indeed, we can choose H as above so that $H(\boldsymbol{A}_0)=\boldsymbol{S}_K$; since $H(\boldsymbol{N})=\boldsymbol{N}$ and $H^{-1}(j)$ is finite for all j, we can obtain $\boldsymbol{E}$ by modifying $\boldsymbol{A}_0$ by a finite set.

Thus, it remains to show that there are a finite number of functions compatible with $\mathcal{C}$ which generate every finite set and every set differing from $\boldsymbol{A}_0$ by a finite set.

Order the natural numbers in a sequence of the order type of the integers, say $\cdots, a_{-1}, a_0, a_1, \cdots$, so that $\boldsymbol{A}_0=\{a_0, a_{-1}, a_{-2}, \cdots\}$ and for $j>0$, $a_j\in A_k$ if $j\equiv k \pmod n$ for $k=1, \cdots, n$. Now define three functions F, G, and H as follows:

$$
\begin{aligned}
F(a_k) &= a_{k+1}, \\
G(a_k) &= a_k \quad \text{if } k \leqq 0, \\
&= a_{k+1} \quad \text{if } k > 0, \\
H(a_k) &= a_0 \quad \text{if } k \leqq 0, \\
&= a_k \quad \text{if } k > 0.
\end{aligned}
$$

We will show that $\mathcal{C}$ is generated by A, B, C, F, F^{-1}, G and H. It is easy to check that F, F^{-1} G and H are all compatible with $\mathcal{C}$.

Now for $j, k, l, \cdots$ greater than 0,

$$F(\boldsymbol{A}_0 \cup \{a_j, a_k, a_l, \cdots\}) = \boldsymbol{A}_0 \cup \{a_1, a_{j+1}, a_{k+1}, a_{l+1}, \cdots\}$$

and

$$G(\boldsymbol{A}_0 \cup \{a_j, a_k, a_l, \cdots\}) = \boldsymbol{A}_0 \cup \{a_{j+1}, a_{k+1}, a_{l+1}, \cdots\}.$$

Hence by repeatedly applying F and G, we can obtain a function which maps $\boldsymbol{A}_0$ onto the union of an arbitrary finite set and $\boldsymbol{A}_0$. Next to obtain a set which differs from $\boldsymbol{A}_0$ by an arbitrary finite set, we apply a suitable power of F^{-1} to a set which is the union of a finite set and $\boldsymbol{A}_0$. On the other hand, if we first apply H to the union of $\boldsymbol{A}_0$ and a finite set and then apply a suitable power of F or F^{-1}, we can obtain an arbitrary nonempty finite set.

REMARK. This proof is due to R. M. Robinson. In the original proof, the number of generators depended on n.

LEMMA 4. *If $\boldsymbol{A}$ can be partitioned into arbitrarily many infinite sets of $\mathcal{C}$ and $\boldsymbol{A}=\boldsymbol{B}\cup\boldsymbol{C}$ where $\boldsymbol{B}$ and $\boldsymbol{C}$ are infinite sets of $\mathcal{C}$, then at least one of $\boldsymbol{B}$ and $\boldsymbol{C}$ can be partitioned into arbitrarily many infinite sets of $\mathcal{C}$.*

PROOF. Let $\boldsymbol{A}$ be partitioned into $2n$ infinite sets $\boldsymbol{S}_1, \cdots, \boldsymbol{S}_{2n}$ of $\mathcal{C}$. Then $\boldsymbol{B}=\bigcup(\boldsymbol{B}\cap\boldsymbol{S}_k)$ and $\boldsymbol{C}=\bigcup(\boldsymbol{C}\cap\boldsymbol{S}_k)$. Either $\boldsymbol{B}$ or $\boldsymbol{C}$ has at least n infinite parts under this partition of $\boldsymbol{A}$ and hence can be partitioned into n infinite sets of $\mathcal{C}$. Hence B or C can be partitioned into arbitrarily many infinite sets of $\mathcal{C}$.

THEOREM 4. *If $\mathcal{C}$ is a denumerable field of sets which contains all finite sets then either $\mathcal{C}$ is the field of all finite and cofinite sets or $\mathcal{C}$ is finitely generated.*

PROOF. Suppose there is a set A in $\mathcal{C}$ such that both A and $\bar{A}$ are infinite. If N can be partitioned into at most n infinite sets of $\mathcal{C}$ for some n, then Theorem 3 applies. Otherwise N can be partitioned into arbitrarily many infinite sets of $\mathcal{C}$ and by Lemma 4 so can A or $\bar{A}$. Now construct $A_0, A_1, \cdots$ in turn so that each A_j is an infinite set of $\mathcal{C}$, and $N - \bigcup_{j \leq k} A_j$ can always be partitioned into arbitrarily many infinite sets of $\mathcal{C}$. Of course, we must choose A_j to be disjoint from the earlier A's. Hence Theorem 2 applies and $\mathcal{C}$ is finitely generated.

REFERENCE

1. Julia Robinson, *Finite generation of recursively enumerable sets*, Proc. Amer. Math. Soc. **19** (1968), 1480–1486.

UNIVERSITY OF CALIFORNIA, BERKELEY

UNSOLVABLE DIOPHANTINE PROBLEMS

JULIA ROBINSON

We shall show that there is no general method of telling whether an arbitrary polynomial $P(x_1, \cdots, x_k)$ with integer coefficients is ever a power of 2 for $x_1, \cdots, x_k$ natural numbers. At present there is no general method known even in the special case with $k=1$.

Actually we shall show directly that the relation given by $r=2^t$ is diophantine in the set $\mathfrak{J}$ of powers of 2. Hence every recursively enumerable set is diophantine in $\mathfrak{J}$ by Corollary 5 of Davis, Putnam, and Robinson [4].

We call a relation $\rho(x_1, \cdots, x_n)$ among natural numbers *diophantine* if there is a polynomial P with integer coefficients such that $\rho(x_1, \cdots, x_n)$ if and only if there are natural numbers $y_1, \cdots, y_k$ with $P(x_1, \cdots, x_n, y_1, \cdots, y_k)=0$. A set $\mathcal{S}$ (function F) is *diophantine* if $x \in \mathcal{S}$ (the graph of F) is diophantine. Also, ρ *is diophantine in a set* $\mathcal{S}$ if there is a polynomial P with integer coefficients such that $\rho(x_1, \cdots, x_n)$ if and only if there are natural numbers $y_1, \cdots, y_k$, $z_1, \cdots, z_l$ with both $P(x_1, \cdots, x_n, y_1, \cdots, y_k, z_1, \cdots, z_l)=0$ and $z_1, \cdots, z_l \in \mathcal{S}$. In [6], the term "existentially definable" was used instead of "diophantine". The definitions given there are easily seen to be equivalent to these. In both [1] and [4], relations over the positive integers were considered but the definitions and theorems hold for natural numbers with only the obvious modifications.

The logical symbols $\bigvee$ (there exists), $\bigwedge$ (and), $\leftrightarrow$ (if and only if) occur in this paper. The variables always range over the natural numbers.

Let $a>1$ and a_n, a_n' be defined by

$$a_n + a_n'(a^2-1)^{1/2} = (a+(a^2-1)^{1/2})^n.$$

LEMMA 1. *Natural numbers x and y satisfy the Pell equation $x^2-(a^2-1)y^2=1$ if and only if there is a natural number n such that $x=a_n$ and $y=a_n'$.*

This is a standard result of elementary number theory.

LEMMA 2. *For $a>1$,*

$$a_n' \equiv n \pmod{a-1}, \qquad a_n - a_n'(a-2) \equiv 2^n \pmod{4a-5}.$$

For proof, see Lemmas 5 and 7 of [6].

Presented to the Society, April 26, 1969; received by the editors September 18, 1968.

LEMMA 3. *There is a binary diophantine relation* $\mathfrak{A}$ *such that* $\mathfrak{A}(t, m)$ *implies* $m > 2^t$ *and if m is sufficiently large with respect to t then* $\mathfrak{A}(t, m)$ *holds.*

PROOF. Let ψ be the relation of Lemma 8 of [6]. Then the binary relation $\mathfrak{A}$ defined by

$$\mathfrak{A}(t, m) \leftrightarrow (\vee u)(\psi(t+2, u) \wedge m > u)$$

clearly satisfies the requirements.

LEMMA 4. *A natural number n is the sum of two squares if and only if no number of the form* $4t+3$ *divides n to an odd power. Hence if x and y are relatively prime then xy is the sum of two squares if and only if x and y are each the sum of two squares.*

This is again a standard result of elementary number theory.

THEOREM 1. *Let* $\mathfrak{M}$ *be an infinite set of natural numbers and H be a diophantine function such that for all* $m \in \mathfrak{M}$,

$$(m, H(m)) = 1, \qquad 2^{H(m)} \equiv 1 \pmod{m}.$$

Then $r = 2^t$ *is diophantine in* $\mathfrak{M}$.

Before proving the theorem, we give two corollaries.

COROLLARY 1. *Every recursively enumerable set is diophantine in any infinite set of numbers m such that* $2^{m-1} \equiv 1 \pmod{m}$. *In particular,* $\mathfrak{M}$ *can be any infinite set of primes.*

PROOF. From Theorem 1 with $H(m) = m-1$ and Corollary 5 of [4]. This result is also proved in [7].

COROLLARY 2. *Every recursively enumerable set is diophantine in the set of numbers of the form* 2^{2^n}.

PROOF. Let $\mathfrak{M} = \{2^{2^n} - 1\}$ and $H(m) = m+1$. Now

$$2^{2^{2n}} \equiv 1 \pmod{2^{2^n} - 1}$$

since 2^n divides 2^{2^n}. Hence for $m \in \mathfrak{M}$,

$$2^{m+1} \equiv 1 \pmod{m}.$$

Hence the corollary follows from Theorem 1 as before.

REMARK. R. M. Robinson pointed out this application of the theorem to me.

PROOF OF THEOREM 1. We shall show that $r = 2^t$ if and only if there are natural numbers a, u, v, and m such that

(1) $a>1$, $m\in\mathfrak{M}$, $m\mid(4a-5)$, $H(m)\mid(a-1)$, $m>1$,
(2) $\mathfrak{A}(t, m)$,
(3) $u^2-(a^2-1)v^2=1$,
(4) $\operatorname{Rem}(u-v(a-2), m)=r$,
(5) $\operatorname{Rem}(v, H(m))=t$.

Suppose that a, u, v, and m satisfy (1)–(5). By Lemma 1 and (3), there is a natural number n such that $u=a_n$ and $v=a_n'$. Then by Lemma 2,

$$
\begin{aligned}
u-v(a-2) &\equiv 2^n \pmod{4a-5}\\
&\equiv 2^n \pmod{m}
\end{aligned}
$$

since $m\mid(4a-5)$ by (1). Hence by (4),

(6) $r\equiv 2^n \pmod{m}$, $r<m$.

Also by Lemma 2,

$$
\begin{aligned}
v &\equiv n \pmod{a-1}\\
&\equiv n \pmod{H(m)}
\end{aligned}
$$

since $H(m)\mid(a-1)$ by (1). Hence by (5),

$$t\equiv n \pmod{H(m)},\qquad t<H(m).$$

Thus, $n=t+q\cdot H(m)$ for some $q\geqq 0$. But by the hypothesis of the theorem $2^{H(m)}\equiv 1 \pmod{m}$ and by (2), $A(t, m)$ holds so that

$$2^n\equiv 2^t \pmod{m},\qquad 2^t<m.$$

Hence by (6), $r=2^t$.

On the other hand, suppose $r=2^t$. We need to find a, u, v, and m which satisfy (1)–(5). Choose m in $\mathfrak{M}$ sufficiently large so that $\mathfrak{A}(t, m)$ holds. This is possible by Lemma 3 since $\mathfrak{M}$ is infinite by hypothesis. Now m is odd since $2^{H(m)}\equiv 1 \pmod{m}$ and $H(m)>0$. Hence by the Chinese Remainder Theorem, we can choose $a>1$ so that

$$4a\equiv 5 \pmod{m},\qquad a\equiv 1 \pmod{H(m)}$$

since m and $H(m)$ are relatively prime. Hence (1) and (2) are satisfied. Now let $u=a_t$ and $v=a_t'$ so that (3) holds. By Lemma 2, we have

$$
\begin{aligned}
u-v(a-2) &\equiv 2^t \pmod{4a-5}\\
&\equiv 2^t \pmod{m}.
\end{aligned}
$$

Also since $A(t, m)$ holds, $2^t<m$. Hence $\operatorname{Rem}(u-v(a-2), m)=2^t=r$ so that (4) holds. By Lemma 2,

$$\begin{aligned} v &\equiv t \pmod{a-1} \\ &\equiv t \pmod{H(m)} \end{aligned}$$

since $H(m) \mid (a-1)$. Now $t < H(m)$ since $2^t < m$ and $2^{H(m)} > m$. Hence $\operatorname{Rem}(v, H(m)) = t$ so (5) holds. Since all of the relations occurring in (1)–(5) except $m \in \mathfrak{M}$ are diophantine, the theorem follows.

The following existential definition of the set of numbers of the form 2^{2^n} in terms of $\mathfrak{I}$, the set of powers of 2, was suggested by the work of Davis [3].

THEOREM 2. *$x = 2^{2^n}$ for some $n > 0$ if and only if x is a power of 2 and there are natural numbers u and v such that $x = 1 + 3(u^2 + v^2)$.*

PROOF. Let m be positive and k be odd. Then

$$2^{2^m \cdot k} - 1 = (2^{2^{m-1} \cdot k} + 1)(2^{2^{m-2} \cdot k} + 1) \cdots (2^{2k} + 1)(2^{2k} - 1).$$

Hence $(2^{2^m \cdot k} - 1)/3$ is the sum of two squares if and only if $(2^k - 1)(2^k + 1)/3$ is the sum of two squares by Lemma 4. (Note that $3 \mid (2^k + 1)$ for k odd.) Now for $k \geq 3$, $2^k - 1$ is of the form $4t + 3$ and is prime to $(2^k + 1)/3$ so the product is not the sum of two squares; while for $k = 1$, it is. In the case $m = 0$ and k odd, $2^k - 1$ is not divisible by 3. Hence for odd k, $2^{2^m \cdot k} = 1 + 3(u^2 + v^2)$ has a solution for u and v if and only if $m > 0$ and $k = 1$.

THEOREM 3. *To every recursively enumerable set $\mathcal{S}$, there is a polynomial P with integer coefficients such that*

$$\mathcal{S} = \{x \colon P(x, y_1, \cdots, y_k) = 2^t \textit{ for some } y_1, \cdots, y_k, \text{ and } t\}.$$

PROOF. Since $\mathcal{S}$ is diophantine in $\mathfrak{I}$ by Corollary 2 and Theorem 2, there is a polynomial F such that

$$x \in \mathcal{S} \leftrightarrow$$
$$(\exists u_1, \cdots, u_m, v_1, \cdots, v_n)\,(F(x, u_1, \cdots, u_m, 2^{v_1}, \cdots, 2^{v_n}) = 0).$$

Hence

$$\begin{aligned} x \in \mathcal{S} \leftrightarrow {} & (\exists u_1, \cdots, u_m, w_1, \cdots, w_n, v) \\ & (F(x, u_1, \cdots, u_m, w_1, \cdots, w_n) = 0 \wedge w_1 \mid 2^v \wedge \cdots \wedge w_n \mid 2^v). \end{aligned}$$

We can now combine the conditions into a single polynomial in more variables in the usual way (see Davis [1, p. 104]). Thus,

$$x \in \mathcal{S} \leftrightarrow (\exists z_1, \cdots, z_r, v)(G(x, z_1, \cdots, z_r, 2^v) = 0)$$

for a suitable choice of G. Finally, we see that

$$x \in \mathcal{S} \leftrightarrow (\exists z_1, \cdots, z_r, w, v)(w(1 + 2G(x, z_1, \cdots, z_r, w)^2) = 2^v).$$

Hence we can take $w(1+2G^2)$ for P.

REMARK. The arguments here are derived from those of Putnam [5] and Davis [2].

Since not every recursively enumerable set is recursive, Theorem 3 shows that there can be no general method of telling whether an arbitrary polynomial assumes a power of 2 as value.

OPEN QUESTIONS. Is every recursively enumerable set diophantine in the set of powers of 3? of n? Is there a general method of telling whether an arbitrary polynomial P with positive integer coefficients assumes a power of 2 as value? (In this case, there is an obvious method of telling whether P assumes a particular power of 2.)

BIBLIOGRAPHY

1. Martin Davis, *Computability and unsolvability*, McGraw-Hill, New York, 1958.

2. ———, *Extensions and corollaries of recent work on Hilbert's tenth problem*, Illinois J. Math. **7** (1963), 246–250.

3. ———, *One equation to rule them all*, Memorandum RM-5494-PR, The RAND Corporation, Santa Monica, Calif., 1968.

4. Martin Davis, Hilary Putnam and Julia Robinson, *The decision problem for exponential diophantine equations*, Ann. of Math. **74** (1961), 425–436.

5. Hilary Putnam, *An unsolvable problem in number theory*, J. Symbolic Logic **25** (1960), 220–232.

6. Julia Robinson, *Existential definability in arithmetic*, Trans. Amer. Math. Soc. **72** (1952), 437–449.

7. ———, "Diophantine decision problems" in *Studies in number theory*, Studies in Math., vol. 6, Math. Assoc. of America; Prentice-Hall, Englewood Cliffs, N. J., 1969, pp. 76–116.

UNIVERSITY OF CALIFORNIA, BERKELEY

HILBERT'S TENTH PROBLEM

JULIA ROBINSON[1]

Ju. V. Matijasevič [4] has recently shown that Hilbert's tenth problem is unsolvable—there is no general method of telling whether a (polynomial) diophantine equation has a solution. Hilbert had asked for such a method. However in his address at the 1900 International Congress when he proposed his list of problems for the twentieth century, Hilbert spoke of "the conviction (which every mathematician shares, but which no one has yet supported by a proof) that every definite mathematical problem must necessarily be susceptible of an exact settlement, either in the form of an actual answer to the question asked, or by the proof of the impossibility of its solution and therewith the necessary failure of all attempts".[2] So Hilbert would be satisfied with Matijasevič's solution.

Matijasevič actually proved the stronger theorem: *Every recursively enumerable (listable) relation is diophantine.* A relation $R(m_1, \ldots, m_j)$ is *diophantine* if there is a polynomial $P(x_1, \ldots, x_j, y_1, \ldots, y_k)$ with integer coefficients such that $R(m_1, \ldots, m_j)$ holds if and only if $P(m_1, \ldots, m_j, y_1, \ldots, y_k) = 0$ has a solution for $y_1, \ldots, y_k$ in natural numbers.

Since there are recursively enumerable sets which are not recursive (computable), Hilbert's problem is unsolvable. For suppose there is a general method for telling whether a diophantine equation has a solution. Let S be a recursively enumerable set which is not recursive. Choose P so that m belongs to S if and only if $P(m, y_1, \ldots, y_k) = 0$ has a solution for $y_1, \ldots, y_k$. For each m, we could then tell whether m belongs to S by using the assumed method to tell whether the

[1]My lectures at Stony Brook were an expository account of earlier work looking towards a negative solution of Hilbert's tenth problem. They are covered by the article *Diophantine decision problems* in the MAA volume, Studies in number theory (referred to here as Studies). The original paper submitted to these Proceedings was a short review written with the hope of stimulating number theorists to work on the question of the existence of a "roughly exponential" diophantine relation R (see I below). I learned with great delight of Matijasevič's ingenious answer in February, 1970. These notes are minor comments on the consequences of Matijasevič's theorem. There is also an appendix containing a simpler proof of one of the results in Studies.

[2]For an account of the circumstances of Hilbert's address, see Reid [5].

corresponding equation has a solution. But this is impossible, since S is not computable.

BACKGROUND. Matijasevič's proof is based on two earlier results:

I. (J. Robinson, 1952, [7]). The relation given by $r = s^t$ is diophantine if there is a diophantine relation $R(u, v)$ and an exponential function $E(u)$ such that $R(u, v)$ implies $v < E(u)$ but there is no polynomial for which $R(u, v)$ implies $v < P(u)$.

II. (M. Davis, H. Putnam, and J. Robinson, 1961, [2]). Every recursively enumerable relation is exponential diophantine. (A relation is *exponential diophantine* if it can be defined by a diophantine equation in which the variables may occur as exponents in the same way as a diophantine relation is defined by a polynomial diophantine equation.)

Matijasevič has found a diophantine relation $R(u, v)$ satisfying I. Namely,

III. (Matijasevič, 1970). The relation given by $v = \phi_{2u}$ is diophantine, where ϕ_k is the kth Fibonacci number.

Putting I and III together, we obtain that exponentiation is diophantine. Thus every exponential diophantine equation can be reduced to a polynomial diophantine equation in more variables. Hence Matijasevič's theorem follows by II. Proofs of I and II are in Studies. The entire proof uses only very elementary number theory—the binomial theorem, Chinese Remainder Theorem, solutions of Pell's equations, and Lucas' sequences.[3]

N. K. Kosovskiĭ, G. V. Čudnovskiĭ, Martin Davis, K. Schütte and Matijasevič have since given direct diophantine definitions of exponentiation based on Matijasevič's work.

The starting point for the proof of II was Gödel's theorem that every primitive recursive function is arithmetical. His proof showed more, namely, every recursively enumerable set S can be put in the form

$$m \in S \leftrightarrow \mathbf{Q}_1 y_1 \ldots \mathbf{Q}_k y_k [P(m, y_1, \ldots, y_k) = 0] \tag{1}$$

where P is a polynomial with integer coefficients and each $\mathbf{Q}_i$ is either an existential quantifier or a bounded universal quantifier. Martin Davis [1] showed in 1950 that it was sufficient to have just one bounded universal quantifier and any number of existential quantifiers. This was essential for the proof of II. Indeed, the key lemma is that a bounded universal quantifier applied to a diophantine relation can be transformed into an exponential diophantine relation. Gödel's form is not enough, since we could not handle a bounded universal quantifier applied to an exponential diophantine relation. S. Kochen pointed out that now we can remove the bounded universal quantifiers one at a time, obtaining first an exponential diophantine relation and then converting it to a diophantine relation by Matijasevič's theorem. Hence §7 of Studies is no longer needed.

CONSEQUENCES. What are the practical consequences of Matijasevič's theorem to the business of solving diophantine equations? None at all. Because for any particular diophantine equation of reasonable size (the only ones we are interested in), we may still with ingenuity and perhaps new insight be able to tell if it has a solution. At least, Matijasevič has not destroyed our conviction that this is so. What we cannot do, is write out instructions once and for all and then turn over the testing of equations for solvability to a machine. We will have to add constantly to our tools as we take on new equations.

---

[3]Both the numbers a'_n in Studies and Matijasevič's $\psi_{m,n}$ are Lucas' sequences [3].

Still in a sense Hilbert's problem lives on. Hilbert asked for a method for telling whether an arbitrary diophantine equation has a solution and there is none. So now we ask for what classes of diophantine equations, is there a method, and seek to find it. For example, Alan Baker, these Proceedings, has made great advances on equations in two variables.

One of the corollaries to Matijasevič's theorem, which he mentions [4, p. 279] is the existence of a universal diophantine equation. This follows immediately by well-known properties of recursively enumerable sets. But since it is the most interesting purely number theoretical consequence, it seems worthwhile to give a direct proof. The standard argument is especially simple in this case. (See IV, p. 83, Studies for an intuitive argument.)

Number diophantine equations as on p. 79, Studies. Then, there is a polynomial $U(v, y_0, y_1, \dots, y_k)$ with integer coefficients such that m belongs to the nth diophantine set D_n if and only if $U(n, m, y_1, \dots, y_k) = 0$ has a solution for $y_1, \dots, y_k$. Here $D_n = \{m$: the nth diophantine equation has a solution with $x_0 = m\}$. We call $U = 0$ a *universal diophantine equation*, since every diophantine set is obtained for some n. We will use the same notation as in §2 of Studies. Let $s_k = \operatorname{Rem}(K(s), 1 + (1 + k)L(s))$ for all s and k. Notice that the relation given by $t = s_k$ is diophantine, since K, L, and $\operatorname{Rem}(x, 1 + y)$ are. By Gödel's lemma, to every finite sequence of natural numbers $a, a', a'', \dots, a^{(n)}$, there is an s such that $a^{(i)} = s_i$ for all $i \leqq n$. Number the terms built up from natural numbers and the variables $x_0, x_1, \dots$ as follows:

$$\tau_{4n} = n, \qquad \tau_{4n+1} = x_n, \qquad \tau_{4n+2} = \tau_{Kn} + \tau_{Ln}, \qquad \tau_{4n+3} = \tau_{Kn} \cdot \tau_{Ln}.$$

Let the nth diophantine equation be $\tau_{Kn} = \tau_{Ln}$. Also let $\tau_n(s)$ be the value of τ_n when $s_0, s_1, \dots$ are substituted for $x_0, x_1, \dots$ respectively. We can then write

$$\begin{aligned} m \in D_n \leftrightarrow \bigvee_{s,t} \Big[\bigwedge_{k \leqq n} (t_{4k} = k \wedge t_{4k+1} = s_k \wedge t_{4k+2} \\ = t_{Kk} + t_{Lk} \wedge t_{4k+3} = t_{Kk} \cdot t_{Lk}) \wedge t_{Kn} = t_{Ln} \wedge m = s_0 \Big]. \end{aligned} \tag{2}$$

Notice that if t satisfies the expression in parentheses for all $k \leqq n$, then $t_j = \tau_j(s)$ for all $j \leqq 4n + 3$. Hence $t_{Kn} = t_{Ln}$ insures that $x_0 = s_0$, $x_1 = s_1, \dots$ is a solution of the nth diophantine equation. On the other hand, if m belongs to D_n, then such t and s must exist by Gödel's lemma. Then the right side of (2) can be put in the Davis normal form, reduced to an exponential diophantine relation by the lemma on p. 103 of Studies, and then to a diophantine relation by Matijasevic's theorem. Clearly, each of these reductions can be carried out explicitly but the universal equation obtained would be very complicated.

If m is composite, there is a proof that m is composite consisting of one multiplication. Until now it was not known if there was a similar proof consisting of a bounded number of additions and multiplications to show that a prime p is prime. Let c be the number of additions and multiplications necessary to check that $U(n, m, y_1, \dots, y_k) = 0$. Then if p is prime there is a proof that p is prime using just c additions and multiplications. The work of finding the right values of $y_1, y_2, \dots$ to use in the computation is of course scratch work. The same c can be used for any recursively enumerable set.

OPEN PROBLEMS. I think the most exciting problem is to find some interesting bound on the number of variables needed in a universal diophantine equation.

A bound can be obtained by starting with the result of R. M. Robinson [8] that every recursively enumerable set can be put in the Davis form with just four inner existential quantifiers, reducing it to an exponential diophantine equation in more variables, and then to a diophantine equation in still more variables. A bound obtained this way would be too large to be interesting. We might define an interesting equation to be one written in the latin alphabet without subscripts.

Another natural problem is the decision problem for nontrivial solutions of homogeneous diophantine equations, or equivalently, for rational solutions of arbitrary diophantine equations. The only result in the negative direction is that the set of natural numbers is arithmetically definable in the field of rational numbers [6]. This implies that there is no general method of telling whether an arithmetical statement is true in the rational field.

APPENDIX. The result of Davis and Putnam that $m = \prod_{k=1}^{n} (a + bk)$ is exponential diophantine can be proved more simply than on pp. 98–101 of Studies. We may assume $b \neq 0$. Then $m = \prod_{k=1}^{n} (a + bk)$ if and only if there are t and x such that

$$(3) \quad t > (a + bn)^n, \qquad bx \equiv a \pmod{t}, \qquad \operatorname{Rem}\left(\binom{x+n}{n} n!b^n, t\right) = m.$$

If (3) holds, then

$$\begin{aligned}\binom{x+n}{n} n!b^n &= (bx + b)(bx + 2b) \cdots (bx + nb) \\ &\equiv (a + b)(a + 2b) \cdots (a + nb) \pmod{t}.\end{aligned}$$

Since the product on the right is less than t, it must be the remainder m. On the other hand, suppose m, n, a, b satisfy the relation. Choose $t > (a + bn)^n$ and prime to b. Then there is an x such that $bx \equiv a \pmod{t}$. For this x and t, the third condition of (3) holds. Since all the relations of (3) are exponential diophantine (see pp. 98–100 of Studies), the result is obtained.

REFERENCES

1. Martin Davis, a) *On the theory of recursive unsolvability*, Doctoral Dissertation, Princeton University, Princeton, N.J., 1950; b) *Arithmetical problems and recursively enumerable predicates*, J. Symbolic Logic **18** (1953), 33–41. MR **14**, 1052.

2. Martin Davis, Hilary Putnam and Julia Robinson, *The decision problem for exponential diophantine equations*, Ann. of Math. (2) **74** (1961), 425–436. MR **24** #A3061.

3. E. A. Lucas, *Théorie des fonctions numérique simplement périodique*, Amer. J. Math. **1** (1878), 184–239, 289–321.

4. Ju. V. Matijasevič, *Enumerable sets are diophantine*, Dokl. Akad. Nauk SSSR **191** (1970), 279–282. = Soviet Math. Dokl. **11** (1970).

5. Constance Reid, *Hilbert. With an appreciation of Hilbert's mathematical work by H. Weyl*, Springer-Verlag, New York, 1970.

6. Julia Robinson, *Definability and decision problems in arithmetic*, J. Symbolic Logic **14** (1949), 98–114. MR **11**, 151.

7. ———, *Existential definability in arithmetic*, Trans. Amer. Math. Soc. **72** (1952), 437–449. MR **14**, 4.

8. R. M. Robinson, *Arithmetical representation of recursively enumerable sets*, J. Symbolic Logic **21** (1956), 162–186. MR **18**, 272.

UNIVERSITY OF CALIFORNIA
BERKELEY, CALIFORNIA

SOLVING DIOPHANTINE EQUATIONS[1]

JULIA ROBINSON

University of California, Berkeley, California, USA

MATIJASEVIČ (1970) showed that every recursively enumerable set is diophantine and hence that there is no algorithm for telling whether an arbitrary diophantine equation has a solution. Thus the problem of solving diophantine equations is not trivial. Now it seems to me we should turn the problem around. Instead of asking whether a given diophantine equation has a solution, ask for what equations do known methods yield the answer? A successful investigation of this problem will need the tools of both logic and number theory.

Since the set of solvable (in N) diophantine equations is recursively enumerable but not recursive, the problem is to find ways of proving equations are unsolvable and of characterizing classes of equations for which we have a decision method. Matijasevič's theorem shows that we will ultimately be thwarted in our attempts to find a method of solving diophantine equations but it in no way decrees that we cannot make progress. We should keep in mind that by the usual fixed-point argument, we can construct other unsolvable equations from any given recursively enumerable set of unsolvable diophantine equations. This method, at least, has no parallel in classical number theory.

Let S be the set of solvable diophantine equations. Suppose a given method M generates a set U' of unsolvable equations, that is, U' is recursively enumerable. Let U'' be the complementary set of unsolvable equations. Then $S \cup U'$ is not recursive. Otherwise we would have a method of testing diophantine equations for solvability. Namely, if E is in $S \cup U'$, test by simultaneously looking for a solution of E and using M to generate U' watching out for E (one of these searches will end); if E is in U'', then

[1] This paper is a modified version of the talk given at the Congress with the title *Existential decision problems.*

E is unsolvable. Hence the best we can do is to find useful recursive subsets of $S \cup U'$ and for equations in these sets, we will have a decision method.

In this note, we will look at a generalization of the ancient method of showing a diophantine equation is unsolvable by checking that for some m, it has no solution mod m. The major advance in this method was made by Ax (1967) where he showed that there is a decision method for telling whether a diophantine equation has a solution mod m for every m. (The result was not stated but is an immediate consequence of the decision method for solvability of an equation in p-adic integers for all p.)

A diophantine equation is an equation in $L(+, \cdot, 0, 1)$. Let D be the set of variable-free equations true in N. Then a diophantine equation E has a solution in N iff $D \vdash (\bigvee x) E$ where x stands for all the variables in E. If M is a model of D and E has no solution in M, then E has no solution in N. Thus, if $\mathrm{Th}(M)$ is decidable, we have a method of showing certain equations are unsolvable in N. It is enough if the *diophantine theory* of M, i.e., the set of diophantine equations solvable in M, is recursive. In this case, we say that M has a *decidable diophantine theory*. We will write $\mathrm{DTh}(M)$ for the diophantine theory of a model, a class of models, or a theory M. Notice that the set of equations solvable in an extension of D is the diophantine theory of some model.

One beautiful classical example of this type of argument is the Hasse–Minkowski theorem: A quadratic form F represents 0 nontrivially in the rational field Q iff F represents 0 nontrivially in every p-adic field and in the real field. An easy algorithm is known for checking the truth of the right side for a given quadratic form and hence of the left side. The 'only if' part of the theorem is apparent as above while the 'if' part is a success story.

It is a common plaint that the 'trouble' with number theory is that it has only one simple model. But D has a host of models and besides, we need only work with their diophantine theories. We might hope for a nonconstructive Hasse principle: A diophantine equation has a solution in N iff it has a solution in every model of D with a decidable diophantine theory. However this is not true as we see from Matijasevič's theorem and the following familiar argument. Let A and B be disjoint recursively enumerable sets which cannot be separated by a recursive set. By Matijasevič's theorem, there are diophantine equations $A_1 = A_2$ and $B_1 = B_2$ such that

$$n \in A \leftrightarrow (\bigvee x)[A_1(\boldsymbol{n}, x) = A_2(\boldsymbol{n}, x)]$$
$$n \in B \leftrightarrow (\bigvee y)[B_1(\boldsymbol{n}, y) = B_2(\boldsymbol{n}, y)],$$

where N is the range of the variables and $\boldsymbol{n}$ is the sum of n 1's. (As before, x and y stand for the variables in the equations.) Let $J(u_1, u_2) = \{(u_1+u_2) \cdot (u_1+u_2)\}+u_1$. Then $J(u_1, u_2) = J(v_1, v_2)$ implies $u_1 = v_1$ and $u_2 = v_2$ in N. Hence the equation

$$J(A_1(z, x), B_1(z, y)) = J(A_2(z, x), B_2(z, y)) \tag{1}$$

has no solution in N since A and B are disjoint. (We say that (1) is the equation corresponding to the given diophantine definitions of A and B.) On the other hand, (1) is solvable in every decidable diophantine theory. For suppose (1) is unsolvable in T (an extension of D). Then

$$A_1(\boldsymbol{n}, x) = A_2(\boldsymbol{n}, x) \tag{2}$$

has a solution in T for all $n \in A$ since it does in N but has no solution for $n \in B$ since then (1) would have a solution in T contrary to assumption. Hence the set

$$\{n: A_1(\boldsymbol{n}, x) = A_2(\boldsymbol{n}, x) \text{ has a solution in } T\}$$

separates A and B so DTh(T) is undecidable.

Thus, the set of equations unsolvable in N splits into two sets, U_1 the set of equations unsolvable in some decidable diophantine theory and U_2 the set of equations solvable in every decidable diophantine theory. We do not know whether U_1 is recursively enumerable so we cannot use our earlier argument to show that U_2 is not recursively enumerable. However this is the case. Suppose we have a list of equations in U_2. We will sketch a proof that this would lead to the recursive separation of A and B. We build up two sequences of sets starting with $A_{-1} = A$ and $B_{-1} = B$ as follows: $A_n = A_{n-1} \cup \{n\}$, $B_n = B_{n-1}$ if the equation corresponding to $A_{n-1} \cup \{n\}$ and B_{n-1} occurs on the list of equations in U_2 before the equation corresponding to A_{n-1} and $B_{n-1} \cup \{n\}$ does; otherwise $A_n = A_{n-1}$ and $B_n = B_{n-1} \cup \{n\}$. Here we use the equation $(M_1(x, y) \cdot x)+(M_2(x, y) \cdot \boldsymbol{n}) = (M_2(x, y) \cdot x)+(M_1(x, y) \cdot \boldsymbol{n})$ to define $M \cup \{n\}$ where $M_1 = M_2$ defines M. Notice that at least one of the equations is in U_2 and that A_n, B_n are disjoint recursively inseparable recursively enumerable sets. Hence we can continue. Also $\bigcup A_n$ and $\bigcup B_n$ are complementary sets and hence recursive. Thus $\bigcup A_n$ would give a recursive separation of A and B which is contrary to assumption. Hence U_2 is not recursively enumerable.

Problem. Is U_1 recursively enumerable? recursive?

Finally, we turn to the decidability results which lead to the smallest decidable diophantine theory which is known. Let $R_t = \{x : x \in R \wedge (x = 0 \vee x = 1 \vee \ldots \vee x \geqq t)\}$ for $t = 0, 1, \ldots$. Clearly, R_t is decidable by Tarski's theorem on the decidability of real closed fields since each variable v occurring in a sentence S can be restricted by the formula

$$(\bigvee z)\,[v = 0 \vee \ldots \vee v = t + z^2].$$

Then $S \in \mathrm{Th}(R_t)$ iff the corresponding relativized sentence belongs to $\mathrm{Th}(R)$.

Furthermore, $R' = \bigcap \mathrm{DTh}(R_t)$ is also decidable. To see this, let $E(x_1, \ldots, x_n)$ be a given diophantine equation. If

$$R \vdash (\bigvee t)(\bigwedge x_1 \ldots x_n)[E(x_1, \ldots, x_n) \wedge x_1 \geqq 0 \wedge \ldots \wedge x_n \geqq 0 \rightarrow x_1 \leqq t \vee \ldots \vee x_n \leqq t] \tag{3}$$

is false, then E has a solution in R'. If (3) is true, let t_0 be the least integer for which it holds. Consider the equations in one less variable obtained by assigning in turn $x_i = j$ for $j = 0, 1, \ldots, t_0$ and $i = 1, \ldots, n$. Then E has a solution in R' iff one of these $n(t_0+1)$ equations in $n-1$ variables has a solution in R'. By repeating the argument, we eventually determine whether E has a solution in R'.

Ax (1967) proved that there is a decision method for the solvability of a diophantine equation mod m for all m. From this it follows easily that $\mathrm{DTh}(F)$ is decidable where F is the class of all finite models of D. We need only look at the standard models, i.e., those in which every element is a sum of 1's. Every standard model is determined by the values of k and l where k is the least integer such that $k = k+t$ for some $t \neq 0$ and l is the least such t. For any model with $l > 1$, an equation E has a solution iff E has a solution mod l. For $l = 1$, E has a solution if E has a variable on each side or in case one side is variable-free, E has a solution in N. Hence Ax's theorem can be extended to give a decision method for $\mathrm{DTh}(F)$.

Remark. The elementary theory of finite models of D is undecidable. Indeed, if S is a sentence of L not containing $+$, then

$$F \models (\bigvee x)[x \neq x+0] \rightarrow S'$$

where S' is obtained from S by relativizing all quantifiers in S to $v \neq v+0$ if and only if S is valid in all finite structures with one binary operation. Hence $\mathrm{Th}(F)$ is undecidable by the theorem of Trakhtenbrot (1950) that the elementary theory of finite models of a binary operation is undecidable.

Thus, the least decidable diophantine theory that we have obtained is the intersection of R' and DTh(F). I doubt that this is *in fact* the least decidable diophantine theory but, if so, then U_1 is recursive.

References

AX, J., 1967, *Solving diophantine problems modulo every prime*, Annals of Mathematics, vol. 85, pp. 161–183

MATIJASEVIČ, Y. V., 1970, *Enumerable sets are diophantine* (Russian), Doklady Akademii Nauk SSSR, vol. 191, pp. 279–282, Engl. transl., Soviet Mathematics Doklady, vol. 11, pp. 354–358

TRAKHTENBROT, B. A., 1950, *The impossibility of an algorithm for the decision problem for finite domains* (Russian), Doklady Akademii Nauk SSSR, vol. 70, pp. 569–572.

USA
University of California, Berkeley
Julia ROBINSON

AXIOMS FOR NUMBER THEORETIC FUNCTIONS

To the memory
of a friend and a leader in mathematics

We shall give a finite set of axioms in terms of one binary operation (composition) and three designated elements O, S and I for number theoretic functions of one variable. The theory $\mathcal{T}$ based on these axioms is sufficiently strong so that the first-order Peano axioms of arithmetic can be derived for the set of „constant" functions.

Ryll-Nardzewski [1] showed that Peano arithmetic is not finitely axiomatizable. Rabin [2] showed that it is not a subtheory of any finitely axiomatizable theory of arithmetic. On the other hand, it is well-known that the Peano axioms can be derived from a finite number of axioms of set theory.

We shall use the following logical symbols: $=$ (equals), $\neg$(not), $\wedge$ (and), $\vee$ (or), $\rightarrow$ (if, then), $\longleftrightarrow$ (if and only if), $\bigvee$(there exists), $\overset{1}{\bigvee}$ (there exists exactly one), and $\bigwedge$ (for every). We will usually drop the initial universal quantifiers.

The axioms

1. $F(GH) = (FG)H$ (Associative axiom)
2. $IF = FI = F$ (Identity axiom)
3. $OF = O$ (Axiom of constants)
4. $(\bigwedge X)[F(XO) = G(XO)] \rightarrow F = G$ (Axiom of extension)
5. $SF \neq O$ (Zero axiom)
6. $FO = O \vee (\bigvee X[SX = FO]$ (Predecessor axiom)
7. $FO = O \wedge FS = SF \rightarrow F = I$ (Induction axiom)
8. $(\bigvee K, L)(\bigwedge F, G)(\overset{1}{\bigvee} X)[KX = F \wedge LX = G]$ (Axiom of pairing)

9. $(\bigvee X, Y)[UX = VY] \vee (\bigvee F)[FU = AU \wedge FV = BV]$ (Axiom of piecing)

10. $(\bigvee F)[FO = AO \wedge FS = BF]$ (Axiom of iteration)

11. $(\bigwedge X)(\bigvee Y)[FY = XO] \rightarrow (\bigvee G)[FG = I]$ (Axiom of choice)

12. $(\bigvee G)(\bigwedge X)[G(XO) = O \longleftrightarrow (\bigvee Y)[FY = XO]]$ (Axiom of the characteristic)

The first four axioms hold for any class of functions from and to an arbitrary domain provided it is closed under composition and contains all constant functions as well as the identity function.

A *standard model* is a class of number theoretic functions of one variable closed under composition (indicated by juxtaposition) and containing O, S, I representing the zero, successor, and identity functions. Notice that the first seven axioms are true in all standard models.

Every axiom is true in the model of all number theoretic functions. The last five axioms insure that the class of functions satisfies certain closure properties. We shall show that *the least standard model which satisfies all the axioms is the class of arithmetical functions.*

We have not investigated the independence of the axioms. However the pairing axiom cannot be dropped since the class of number theoretic functions which assume at most a finite number of values infinitely often satisfies all the other axioms. Another version of the axiom of piecing is

$$9'. \quad (\bigwedge X)[(\bigvee T)[UT = XO] \longleftrightarrow \neg(\bigvee T)[VT = XO]] \rightarrow (\bigvee F)[FU = AU \vee FV = BV].$$

This can be substituted for 9 without changing $\mathcal{T}$ (see the corrollary to Theorem 2). The class of recursive functions satisfies 8, 9′, 10 and 11 but not 12.

Informally, we shall identify the ,,constant" functions with numbers, i. e. X is a number if and only if $X = XO$. We shall use small letters only for numbers. Thus, $(\bigwedge x)[\varphi(x)]$ stands for $(\bigwedge X)[X = XO \rightarrow \varphi(X)]$, etc. Also x is in the range of F if $(\bigvee t)[Ft = x]$.

Pairing functions

We shall select one pair of functions K and L which satisfy the axiom of pairing and keep them fixed throughout the paper.

D e f i n i t i o n. *$J(A, B)$ denotes the unique function X such that $KX = A$ and $LX = B$.*

254

Of course J is not a function of the theory $\mathcal{T}$ but just a suggestive abbreviation.

L e m m a 1. $J(A, B)C = J(AC, BC)$.

P r o o f. Let $X = J(A, B)$. Then $K(XC) = (KX)C = AC$ and $L(XC) = (LX)C = BC$.

L e m m a 2. $J(K, L) = I$.

P r o o f. Since $KI = K$ and $LI = L$, it follows from the uniqueness of $J(K, L)$ that $J(K, L) = I$.

D e f i n i t i o n. $J_1(X_1) = X_1$ *and* $J_{n+1}(X_1, \ldots, X_{n+1}) = J(X_1, J_n(X_2, \ldots, X_{n+1}))$ for $n > 0$. (*Here* n *is an index and not an individual of* $\mathcal{T}$).

L e m m a 3. *For every* $X_1, \ldots, X_n$, *there is exactly one function* F *such that* $F = J_n(X_1, \ldots, X_n)$.

Proof by induction on n.

L e m m a 4. *For every* F *and index* $n > 0$, *there are* $X_1, \ldots, X_n$ *such that* $F = J_n(X_1, \ldots, X_n)$. *Indeed,* $X_1, \ldots, X_n$ *are determined by*

(a) $X_n = L^{n-1} F$, (b) *if* $1 \leqq i < n$, *then* $X_i = KL^{i-1} F$.

P r o o f b y i n d u c t i o n. It is clear for $n = 1, 2$. Suppose $n > 1$. Then $F = J(X_1, G)$ where $X_1 = KF$ and $G = LF$. By the inductive hypothesis, $G = J_n(X_2, \ldots, X_{n+1})$ where

$$X_{i+1} = KL^{i-1} G = KL^i F \text{ and } X_{n+1} = L^{n-1} G = L^n F.$$

L e m m a 5. $(\bigwedge x)[(\bigvee t)[J(K, O)t = x] \longleftrightarrow \neg (\bigvee t)[J(K, SL)t = x]]$.

P r o o f. $LJ(K, O)t = Ot \neq SLt' = LJ(K, SL)t'$. Hence no x is in the range of both $J(K, O)$ and $J(K, SL)$.

On the other hand, every x is either in the range of $J(K, O)$ or in the range of $J(K, SL)$. Indeed,

(a). $Lx = O$. Then $J(K, O)x = J(Kx, O) = x$,

(b). $Lx \neq O$, say $Lx = Sy$. Then $J(K, SL)J(Kx, y) = J(Kx, Sy) = x$.

R e m a r k. These lemmas depend only on the axiom of pairing and the basic theory of axioms 1—7.

Development of the theory

T h e o r e m 1. $(\bigvee x)[Gx = O] \rightarrow (\bigvee F)(\bigwedge x)[Gx = O \longleftrightarrow (\bigvee t)[Ft = x]]$.

P r o o f. Suppose $Gx = O$. Let M satisfy the piecing $MJ(K, O) = KJ(K, O) = K$, $MJ(K, SL) = xJ(K, SL) = x$. M exists by Lemma 5 and the weak piecing axiom 9′. Then

255

$F = MJ(I, G)$ will satisfy the theorem. Indeed, if $Gy = O$, then $Fy = MJ(y, O) = MJ(K, O)J(y, O) = y$; if $Gy \neq O$, say $Gy = Sz$, then $Fy = MJ(K, SL)J(y, z) = x$.

T h e o r e m 2. $(\bigvee x)(\bigwedge y)[Gy \neq x] \rightarrow$

$$\rightarrow (\bigvee F)(\bigwedge t)[(\bigvee u)[Fu = t] \longleftrightarrow \neg (\bigvee u)[Gu = t]].$$

P r o o f. Let H be a characteristic function of F and Z be defined by the weak piecing axiom, $ZO = SO$, $ZS = O$. Then ZH is a characteristic function of the complement of the range of G. If x satisfies the hypothesis of the theorem, then $ZHx = O$. Hence by Theorem 1, there is a function F for which ZH is a characteristic function.

C o r o l l a r y . *Axioms* 1—8, 9′, 10 *and* 12 *imply Axiom* 9.

P r o o f . Suppose $UX \neq VY$. Let U' be a function whose range is the complement of the range of U by Theorem 2. Then by 9′ there is a function F such that $FU = AU$ and $FU' = BU'$. Since Vx is not in the range of U by hypothesis, there is a z such that $U'z = Vx$. Hence $FVx = FU'z = BU'z = BVx$ and by extension $FV = BV$. Thus Axiom 9 is satisfied by F.

T h e o r e m 3. $(\bigvee X)[GX = O] \rightarrow$

$$\rightarrow (\bigvee F)(\bigwedge X)[GX = O \longleftrightarrow (\bigvee T)[FT = X]].$$

P r o o f. Suppose O is in the range of G. Then by Theorem 1, there is an H such that $Gx = O \longleftrightarrow (\bigvee t)[Ht = x]$. If H assumes all values, the theorem is satisfied. Indeed, by the axiom of choice, there is a function M such that $HM = I$ so $HMX = X$. Thus F can be put equal to HM. Otherwise, suppose there is a number not in the range of H. By Theorem 2, there is a function H' whose range is the complement of the range of H. Define M by the piecing

$$MJ(K, O) = HKJ(K, O),$$
$$MJ(K, SL) = H'KJ(K, SL).$$

M assumes all values, so by the axiom of choice, let T satisfy $MT = I$. Suppose $GX = O$. For all z, $GXz = O$. Hence Xz is in the range of H, i. e. $MTXz$ is in the range of H. Hence TXz is in the range of $J(K, O)$. So $MTXz = HKTXz$ and by extension $(MT)X = HKTX = X$. Hence F equal to HKT satisfies the theorem.

T h e o r e m 4. $(\bigvee X)[FX = GX] \rightarrow$

$$\rightarrow (\bigvee H)(\bigwedge X)[FX = GX \longleftrightarrow (\bigvee T)[HT = X]].$$

P r o o f. Since $J(I, I)X \neq J(O, SO)$, there is a function M whose range is the complement of the range of $J(I, I)$. Let C

be defined by the piecing

$$CJ(I, I) = OJ(I, I) = O,$$
$$CM = SOM = SO.$$

If $FX = GX$ then $CJ(F, G)X = CJ(FX, FX) = CJ(I, I)FX = O$. If $FX \neq GX$ then for some t, $FXt \neq GXt$ by the axiom of extension. Then $CJ(F, G)Xt = CJ(FXt, GXt) = CMy = SO \neq O$ for some y. Hence $FX = GX$ if and only if $CJ(F, G)X = O$. Now by Theorem 3, there is a function H such that

$$CJ(F, G)X = O \longleftrightarrow (\bigvee T)[HT = X]$$

so H will satisfy Theorem 4.

Theorem 5. $(\bigvee F)(\bigwedge x)[(\bigvee t)[Gt = x \bigvee Ht = x] \longleftrightarrow$

$$\longleftrightarrow (\bigvee t)[Ft = x]].$$

Proof. F can be defined by the piecing

$$FJ(K, O) = GKJ(K, O),$$
$$FJ(K, SL) = HKJ(K, SL).$$

Since $FJ(t, O) = Gt$ and $FJ(t, SO) = Ht$, the range of F includes the ranges of G and H. On the other hand, if $Lt = O$, then $Ft = FJ(Kt, O) = GKt$; if $Lt = Sy$, then $Ft = HKJ(K, SL) \cdot J(Kt, y) = HKt$. Hence the range of F is the union of the range of G and the range of H.

Remark. Notice that up to now, we have not used iteration nor will we use it in proving the following metatheorem.

Metatheorem. *Let* $\varphi(A_1, \ldots, A_m, x_1, \ldots x_n)$ *be a formula with at most the indicated free varlables and any number of bound numerical vaviables but no other bound variables. Then*

$$(\bigwedge A_1, \ldots, A_m)[(\bigvee x_1, \ldots, x_n)[\varphi] \rightarrow$$
$$\rightarrow (\bigvee F)(\bigwedge x_1, \ldots, x_n)[\varphi \longleftrightarrow (\bigvee t)[Ft = J_n(x_1, \ldots, x_n)]]]$$

is a theorem of $\mathcal{T}$.

Proof. We can take formulas of the sort $Ax = y$ as the atomic formulas of $\mathcal{T}$. Of course x and y may be the same variable and A may be O, S, or I. Indeed,

$$AB = C \longleftrightarrow (\bigwedge x)(\bigvee y, z)[Ay = z \bigwedge Bx = y \bigwedge Cx = z]$$

by the associative and extension axioms.

It is sufficient to prove the theorem for atomic formulas and for $\theta \bigvee \psi$, $\neg \theta$ and $(\bigvee x)[\theta]$ assuming the theorem holds for θ and ψ.

I. Suppose φ is $Av_r = v_s$ with $r \neq s$. Then take F equal to $J_n(K_1, \ldots, K_{s-1}, AK_r, K_{s+1}, \ldots, K_n)$ where $I = J_n(K_1, \ldots, K_n)$. If $x = J_n(x_1, \ldots, x_{s-1}, Ax_r, x_{s+1}, \ldots, x_n)$ then $Fx = x$. Also if $Fy = J_n(y_1, \ldots, y_n)$ then $y_s = K_s Fy = AK_r y = Ay_r$.

II. Suppose φ is $Av_r = v_r$. If there is no u with $Au = u$ then we have nothing to prove. Otherwise, there is a function B such that

$$(\bigwedge u)[Au = u \longleftrightarrow (\bigvee t)[Bt = u]]$$

by Theorem 4. Take F equal to $J_n(K_1, \ldots, K_{r-1}, BK_r, K_{r+1}, \ldots, K_n)$. Then $K_r Fx = BK_r x = ABK_r x = AK_r Fx$. Also if $x = J_n(x_1, \ldots, x_n)$ with $x_r = Ax_r$, then $(\bigvee t)[Bt = x_r]$. Hence $FJ_n(x_1, \ldots, x_{r-1}, t, x_{r+1}, \ldots, x_n) = x$.

In the following two cases, there is no difficulty if either θ or ψ cannot be satisfied. Hence we can assume that there are functions G and H such that

$$\theta(x_1, \ldots, x_n) \longleftrightarrow (\bigvee t)[Gt = J_n(x_1, \ldots, x_n)],$$

$$\psi(x_1, \ldots, x_n) \longleftrightarrow (\bigvee t)[Ht = J_n(x_1, \ldots, x_n)].$$

III. Suppose φ is $\theta \bigvee \psi$. Then F satisfying the theorem exists by Theorem 5.

IV. Suppose φ is $\rceil\, \theta$. If every number is in the range of G then φ cannot be satisfied; if θ cannot be satisfied then take F equal to I. Otherwise by Theorem 2, there is a function F whose range is the complement of the range of G and this F satisfies the theorem for φ.

V. Suppose φ is $(\bigvee x)[\theta]$. If θ can be satisfied so can φ. By hypothesis there is a function G such that $\theta(x, x_1, \ldots, x_n) \longleftrightarrow (\bigvee t)[Gt = J_{n+1}(x, x_1, \ldots, x_n)]$. Take F equal to LG since

$$J_{n+1}(x, x_1, \ldots, x_n) = J(x, J_n(x_1, \ldots, x_n)).$$

Theorem 6. $FX = O \bigwedge FY = SF \rightarrow (\bigvee G)[FG = I]$.

Proof. By iteration, there is a function G such that $GO = XO$ and $GS = YG$. Then $(FG)O = FXO = O$ and $(FG)S = FYG = S(FG)$. Hence by the axiom of induction, $FG = I$.

Corollary. $(\bigvee x)[Fx = O] \bigwedge (\bigwedge t)(\bigvee y)[Fy = SFt] \rightarrow (\bigwedge x)(\bigvee t)[Ft = x]$.

Proof. By the metatheorem, there is a function H whose range is $\{J(y, t) : Fy = SFt\}$. By the axiom of choice, choose T so that $LHT = I$. Then Y equal to KHT satisfies the hypothesis of Theorem 6.

Principle of Induction. *Let* φ *be a formula with at least one free numerical variable* x *and no bound function variables. Then*

$$\varphi(O) \wedge (\bigwedge x)[\varphi(x) \to \varphi(Sx)] \to (\bigwedge x)[\varphi(x)].$$

Proof. Any free numerical variables can be considered as free function variables in the metatheorem putting in the usual condition that $V = VO$. Hence there is a function F such that $\varphi(x) \longleftrightarrow (\bigvee t)[Ft = x]$. Hence by the corollary to Theorem 6, the Principle of Induction follows.

Remark. While the corollary and the Principle of Induction are true in all standard models, Theorem 6 is not. In the least standard model containing H the function given by $Hx = [x/2]$, we have $HO = O$ and $HSS = SH$ but there is no G with $HG = I$.

Theorem 7. $(\bigvee P)[PS = I]$.

Proof. Define H by the piecing

$$HO = J(O, O)O,$$
$$HS = J(SK, K)S.$$

Then KH assumes all values. Namely, $KHO = O$ and if $J(y, z) = St$ then $KHSt = KJ(SK, K)St = Sy$. Hence there is a function T such that $KHT = I$. Then by extension, we can take P equal to LHT since $LHTSx = x$.

Theorem 8. $SA = SB \to A = B$.

Proof. Let P satisfy Theorem 7. Then $A = PSA = PSB = B$

Theorem 9. $(\bigwedge x)(\bigvee y)[Fx = Gy] \to (\bigvee H)[F = GH]$.

Proof. By Theorem 4, there is a function M such that $FKX = GLX \longleftrightarrow (\bigvee T)\ [MT = X]$. If $Fx = Gy$ then there is a z such that $Mz = J(x, y)$. Hence KM assumes all values and by the axiom of choice, there is a function N such that $KMN = I$. Then $FKMN = F = GLMN$ and H equal to LMN satisfies the theorem.

Theorem 10. $FO = O \wedge FS = F \to F = O$.

Proof. The set $\{x : Fx = O\}$ contains O and $Fx = O \to FSx = O$. Hence by the Principle of Induction $Fx = O$ for all x and by extension $F = O$.

Recursion

Next we will show that we can also define functions by simultaneously carrying out infinitely many iterations.

Recursion theorem. *There is a unique function* F *such that* $FU = AU$ *and* $FV = BF$ *where* $U \quad J(K, O)$ *and* $V = J(\quad, SL)$.

Note that $J(r, s) = J(K, SL)^s J(r, O)$. Hence $FJ(r, s) = B^s AJ(r, O)$. Thus the values of F are given by A along the x-axis and are obtained elsewhere by iteration along vertical lines.

Proof. The idea is to define a function H such that $Ht = J_3(x_1, x_2, x_3)$ implies that $x_2 = Fx_1$ and x_3 is an index which insures that KH assumes all values. As t runs through consecutive values, x_1 will traverse the xy-plane by repeatedly starting up vertical lines and then returning to the x-axis. Let $K_1 = K$, $K_2 = KL$ and $K_3 = LL$.

By the metatheorem there are functions W and W' such that

$$(\bigvee t)[Wt = J_3(x_1, x_2, x_3)] \longleftrightarrow Lx_1 = Lx_3,$$
$$(\bigvee t)[W't = J_3(x_1, x_2, x_3)] \longleftrightarrow Lx_1 \neq Lx_3.$$

Clearly, the range of W and the range of W' are complementary sets. Hence M can be defined by the piecing

$$MW = J_3(USK_3, AUSK_3, SK_3)W,$$
$$MW' = J_3(VK_1, BK_2, K_3)W'.$$

Let H be defined by the iteration

$$HO = J_3(UO, AUO, O),$$
$$HS = MH.$$

Now we shall establish some properties of H.

$$\text{(A)}\ (\bigvee y)[K_3Hy = x \wedge K_1Hy = Ux] \rightarrow (\bigvee z)[K_3Hz = x \wedge LK_1Hz = Lx].$$

Proof. Given x and y such that $K_3Hy = x$ and $K_1Hy = Ux$. Suppose there is no z satisfying the conclusion of (A). If $Lx = O$, then z equal to y satisfies the conclusion of (A). Hence under the assumption, $Lx \neq O$. Suppose $K_3Hw = x$ and $LK_1Hw = u$. Since w is in the range of W', $HSw = J_3(VK_1, BK_2, K_3)Hw$. So $K_3HSw = x$ and $LK_1HSw = Su$. Hence by the induction principle, LK_1Hw assumes all values with $K_3Hw = x$. This contradicts the assumption that the conclusion of (A) is false, proving (A).

$$\text{(B)}\ (\bigvee y)[K_3Hy = x \wedge K_1Hy = Ux].$$

Proof. For $x = O$, we can take $y = O$. Suppose now that (B) holds for x. Thus for some t, $K_3Ht = x$ and $K_1Ht = Ux$. Then by (A), there is a z with $K_3Hz = x$ and $LK_1Hz = Lx$. Hence z is in the range of W so $K_3HSz = Sx$ and $K_1HSz = USx$. Hence by the induction principle, (B) holds for all x.

$$\text{(C)}\ KK_1H = KK_3H.$$

260

P r o o f b y i n d u c t i o n.

(a) $KK_1HO = KJ(KO, O) = KO = KK_3HO$.

(b) $y = Wt \wedge KK_1Hy = KK_3Hy \rightarrow$
$$\rightarrow KK_1HSy = KUSK_3Hy = KK_3HSy.$$

(c) $y = W't \wedge KK_1Hy = KK_3Hy \rightarrow KK_1HSy =$
$$= KVK_1Hy = KK_1Hy = KK_3Hy = KK_3HSy.$$

(D) $(\bigvee t)[K_1Ht = x]$.

P r o o f. Choose z by (A) and (B) so that $K_3Hz = x$ and $LK_1Hz = Lx$. By (C), $KK_1Hz = KK_3Hz = Kx$. Hence $K_1Hz = x$

(E) $K_1Ht = Uz \rightarrow K_2Ht = AUz$.

P r o o f. For $t = O$, $K_1HO = UO$ and $K_2HO = AUO$ by definition of H. Suppose $t \neq O$, then $HPt = Wz$ for some z where P satisfies Theorem 7. Hence $K_1Ht = USK_3HPt$ and $K_2Ht = AUSK_3HPt = AUK_1Ht$ since $U^2 = U$.

(F) $K_1Ht = K_1Ht' \rightarrow K_2Ht = K_2Ht'$.

P r o o f. The proof is by induction on LK_1Ht. If $LK_1Ht = O$, then (E) implies (F). Suppose (F) holds whenever $LK_1Ht = y$. Now $K_1Hu = K_1Hu'$ and $LK_1Hu = Sy$ implies that u and u' are not O. Hence
$$K_1Hu = VK_1HPu \wedge LK_1HPu = y,$$
$$K_1Hu' = VK_1HPu' \wedge LK_1HPu' = y.$$
So $K_1HPu = K_1HPu'$ and $K_2HPu = K_2HPu'$. But $K_2Hu = BK_2HPu = BK_2HPu' = K_2Hu'$. Hence (F) follows by induction.

(G) $K_1Ht = VK_1Ht' \rightarrow K_2Ht = BK_2Ht'$.

P r o o f. For the hypothesis to hold $t \neq O$. Hence $K_1Ht = VK_1HPt = VK_1Ht'$. Also $K_2Ht = BK_2HPt = BK_1Ht'$ by (F) since $K_1HPt = K_1Ht'$.

We now return to the proof of the theorem. Since K_1H assumes all values, there is a function T with $K_1HT = I$. Put F equal to K_2HT. Then $FU = K_2HTU = AU$ by (E). Also $K_1HTV = VK_1HT$ so $K_2HTV = BK_2HT$ by (G). Hence $FV = FVK_1HT = FK_1HTV = BK_2HT = BF$.

Suppose F' also satisfies the recursion equations. Then $FJ(x, O) = AJ(x, O) = F'J(x, O)$. Suppose $FJ(x, y) = F'J(x, y)$. Then $FJ(x, Sy) = BFJ(x, y) = F'J(x, Sy)$. Hence by induction on y, $FJ(x, y) = F'J(x, y)$ for all x and y. But $u = J(Ku, Lu)$ so by extension $F = F'$.

The Peano Axioms

In this section, we shall show that the first order Peano axioms hold for the numbers of $\mathcal{T}$. Let $\mathcal{P} = \langle N, S, +, \cdot, 0\rangle$ be Peano arithmetic, i. e. the theory of the axioms

P1. $Sx = Sy \to x = y$.

P2. $0 \neq Sx$.

P3. $\varphi(0) \wedge (\bigwedge x)[\varphi(x) \to \varphi(Sx)] \to (\bigwedge x)[\varphi(x)]$.

P4. $x + 0 = x$.

P5. $x + Sy = S(x + y)$.

P6. $x \cdot 0 = 0$.

P7. $x \cdot Sy = (x \cdot y) + x$.

In the axiom scheme P3, φ is any formula of $\mathscr{P}$ with at least one free variable.

We shall now define a subtheory of $\mathscr{P}$ which satisfies P1—P7. Let $\mathscr{N} = \{X : X = XO\}$, $0 = O$, and S be the unary operation defined by $Sx = y \longleftrightarrow SXO = YO$. By the Recursion Theorem, there is a function A such that

$$AJ(K, O) = KJ(K, O);$$
$$AJ(K, SL) = SA.$$

D e f i n i t i o n. $x + y = z \longleftrightarrow (\bigvee t)[Kt = x \wedge Lt = y \wedge At = z]$, i. e. $x + y = AJ(x, y)$.

For a particular choice of K and L satisfying the pairing axiom, there is a unique t such that $Kt = x$ and $Lt = y$. Hence $x + y$ is uniquely determined for each x and y. We will show that in fact $+$ is uniquely determined independent of the choice of K and L. We prove by induction on y that $+$ satisfies P4 and P5

$$x + 0 = AJ(x, 0) = AJ(K, 0)\, J(x, 0) = KJ(x, 0) = x,$$
$$x + Sy = AJ(K, SL)J(x, y) = SAJ(x, y) = S(x + y).$$

Now by induction, we see that

$$FO = x \wedge (\bigwedge z)[FSz = SFz] \wedge GO = x \wedge (\bigwedge z)[GSz = SGz] \to F = G.$$

Hence $x + y$ is uniquely determined for all x and y without regard to the choice of pairing functions.

Let W be defined by the recursion

$$WJ(K, O) = J(O, K)J(K, O)$$
$$WJ(K, SL) = J(A, L)W$$

where A is the addition function defined above. Notice that $(LW)J(K, O) = KJ(K, O)$ and $(LW)J(K, SL) = LW$. Hence LW satisfies the same recursion equations as K so $LW = K$.

D e f i n i t i o n. $x \cdot y = KWJ(x, y)$.

We now check that $x \cdot y$ satisfies P6 and P7.

$$x \cdot 0 = KWJ(x, 0) = KJ(O, K)J(x, 0) = 0.$$

$$x \cdot Sy = KWJ(K, SL)J(x, y) = KJ(A, L)WJ(x, y) = $$
$$= KJ(A, L)J(KWJ(x, y), LWJ(x, y)) = KJ(A. L)J(x \cdot y, x) =$$
$$= (x \cdot y) + x.$$

As before W is uniquely determined in terms of K and L. Furthermore, we can show by induction that

$$FO = O \wedge (\wedge z)[FSz = Fz + x] \wedge GO = O \wedge (\wedge z)[GSz = Gz + x] \rightarrow$$
$$\rightarrow F = G.$$

Hence $x \cdot y$ does not depend on the choice of K and L and does satisfy P6 and P7.

Finally, P1 and P2 are special cases of T8 and the zero axiom. P3 is simply the Principle of induction applied to arithmetical formulas, i. e. formulas which can be expressed in the language of $\mathcal{P}$.

T h e o r e m. *The class of arithmetical functions is a model of $\mathcal{T}$ and is contained in every standard model of $\mathcal{T}$.*

P r o o f. Any arithmetical function F can be defined in the form $Fx = y \longleftrightarrow \varphi(x, y)$ where φ is a formula with two free variables in the language of $\mathcal{P}$. By the metatheorem, there is a function G whose range is $\{J(x, y) : \varphi(x, y)\}$. Since KG assumes all values, there is a function T with $KGT = I$. Then $F = LGT$. Hence every arithmetical function is contained in any standard model of $\mathcal{T}$.

The class $\mathcal{A}$ of arithmetical functions does indeed satisfy the axioms of $\mathcal{T}$. We can take K and L corresponding to $J(x, y) = ((x + y)^2 + 3x + y)/2$. There is no difficulty in showing that $\mathcal{A}$ is closed under piecing. Gödel's classical argument shows that it is also closed under iteration. It is easy to give an arithmetical definition of the function G satisfying the axiom of choice such that $Gx = \mu z\{Fz = x\}$. Finally, the characteristic function which is 0 on the range of F and 1 elsewhere is easily defined arithmetically.

References

1. R y l l - N a r d z e w s k i C. The role of the axiom of induction in elementary arithmetic. Fund. Math., 1952, 39, 239—263.
2. R a b i n, M i c h a e l O. Non-standard models and the independence of the induction axiom. Essays on the foundations of mathematics. Jerusalem, 1961; Amsterdam, 1962, 287—299.

Поступило 11 июля 1969 г.

УДК 51.01

Юрий Матиясевич, Джулия Робинсон
(Ленинград) (Беркли, США)

ДВА УНИВЕРСАЛЬНЫХ ТРЕХКВАНТОРНЫХ ПРЕДСТАВЛЕНИЯ ПЕРЕЧИСЛИМЫХ МНОЖЕСТВ

1. Условимся о следующих обозначениях. Строчные латинские буквы от a до n включительно с индексами и без них будут использоваться в качестве переменных для натуральных, т.е. целых неотрицательных чисел, остальные строчные латинские буквы – в качестве переменных для целых чисел. Аналогично строчные греческие буквы от α до ν будут использоваться в качестве метапеременных для натуральных чисел, а остальные – в качестве метапеременных для целых чисел.

Заглавные латинские буквы будут использоваться для обозначения полиномов. При этом под полиномами понимаются только полиномы, имеющие лишь целочисленные коэффициенты; это свойство полиномов не будет далее оговариваться специально.

2. Мы говорим, что множество $\mathfrak{M}$ натуральных чисел представимо арифметической формулой $\mathfrak{F}$, имеющей лишь одну свободную переменную a, если справедлива эквивалентность $a \in \mathfrak{M} \Leftrightarrow \mathfrak{F}$.

Как показал Курт Гедель [1], любое перечислимое множество представимо некоторой арифметической формулой. Этот результат можно усилить, наложив те или иные ограничения на вид формул. Исследования такого рода проводились в работах [2–13]. Цель настоящей статьи – показать, что любое перечислимое множество представимо формулами каждого из следующих двух видов:

$$\exists b \exists c \mathop{\&}_{i=1}^{\varepsilon} \exists d [P_i(a, b, c) < D_i(a, b, c) d < Q_i(a, b, c)], \qquad (1)$$

$$\exists b \exists c \forall f [f \leq F(a, b, c) \Rightarrow W(a, b, c, f) > 0]. \qquad (2)$$

3. Пусть $\mathfrak{R}$ – произвольное перечислимое множество натуральных чисел. Найдем формулу вида

$$\exists h_1 \ldots \exists h_\delta [R(a, h_1, \ldots, h_\delta) = 0], \qquad (3)$$

представляющую множество $\mathfrak{R}$ (существование такой формулы показано, например, в [6-9, 14]).

Обозначим степень полинома R посредством λ. Не ограничивая общности, мы будем считать, что $\lambda \geq 1$.

Если формула (2) эквивалентна формуле (3), то пара чисел $\langle b, c \rangle$, существование которой утверждается в (2), должна нести полную информацию о наборе из δ чисел $\langle h_1, \ldots, h_\delta \rangle$, существование которого утверждается в (3).

Известно много способов кодирования наборов натуральных чисел посредством одного или пары натуральных чисел. Мы выберем один нестандартный способ, который позволит нам проверять справедливость равенства

$$R(a, h_1, \ldots, h_\delta) = 0 \qquad (4)$$

непосредственно по коду без нахождения самих чисел $h_1, \ldots, h_\delta$.

Обозначим через $B(h_1, \ldots, h_\delta, k)$ полином

$$\sum_{i=1}^{\delta} h_i k^{(\lambda+1)^i}.$$

"Геометрический смысл" этого полинома таков: если k превосходит каждое из чисел $1, h_1, \ldots, h_\delta$, то $h_1, \ldots, h_\delta$ – это цифры в записи числа $B(h_1, \ldots, h_\delta, k)$ в k-ичной системе счисления, стоящие на $(\lambda+1)$-м, ..., $(\lambda+1)^\delta$-м местах; при этом на остальных местах стоят нули.

Не трудно проверить, что имеет место тождество вида

$$(1 + ak + B(h_1, \ldots, h_\delta, k))^\lambda =$$

$$= \sum_{\alpha_0 + \ldots + \alpha_\delta \leq \lambda} \kappa_{\alpha_0, \ldots, \alpha_\delta} a^{\alpha_0} h_1^{\alpha_1} \ldots h_\delta^{\alpha_\delta} k^{N(\alpha_0, \ldots, \alpha_i)}, \qquad (5)$$

где

$$N(l_0, \ldots, l_\delta) = \sum_{i=0}^{\delta} l_i (\lambda + 1)^i,$$

а $\kappa_{\alpha_0, \ldots, \alpha_\delta}$ – какие-то целые положительные числа. Легко видеть, что полином N обладает следующим свойством:

$$\left\{\mathop{\&}_{i=0}^{\delta}((l_i' \le \lambda)\&(l_i'' \le \lambda))\&N(l_0', \dots, l_\delta') = N(l_0'', \dots, l_\delta'')\right\} \Rightarrow \mathop{\&}_{i=0}^{\delta}(l_i' = l_i''). \quad (6)$$

Это свойство становится очевидным из "геометрического смысла" полинома N: если $l_i \le \lambda$, то $l_\delta, \dots, l_0$ – это в точности все цифры в записи числа $N(l_0, \dots, l_\delta)$ в системе счисления с основанием $\lambda + 1$. Именно для получения свойства (6) мы выбрали в качестве B такой полином, что в k-ичной записи числа $B(h_1, \dots, h_\delta, k)$ отличные от нуля цифры расставлены редко. Свойство (6) позволяет дать "геометрическое истолкование" тождеству (5): в k-ичной записи числа $(1 + ak + B(h_1, \dots, h_\delta, k))^\lambda$ цифрами являются всевозможные числа вида

$$\kappa_{\alpha_0, \dots, \alpha_\delta} a^{\alpha_0} h_1^{\alpha_1} \dots h_\delta^{\alpha_\delta} \quad (7)$$

при условии, что k превосходит каждое из них.

Не ограничивая общности, мы будем считать, что полином R является линейной комбинацией одночленов (7):

$$R(a, h_1, \dots, h_\delta) = \sum_{\alpha_0 + \dots + \alpha_\delta \le \lambda} \rho_{\alpha_0, \dots, \alpha_\delta} \kappa_{\alpha_0, \dots, \alpha_\delta} a^{\alpha_0} h_1^{\alpha_1} \dots h_\delta^{\alpha_\delta}. \quad (8)$$

Очевидно, что если $l_0 + \dots + l_\delta \le \lambda$, то

$$N(l_0, \dots, l_\delta) \le \lambda(\lambda + 1)^\delta.$$

Обозначим число $\lambda(\lambda + 1)^\delta$ посредством ν, посредством $V(k)$ обозначим полином

$$\sum_{\alpha_0 + \dots + \alpha_\delta \le \lambda} \rho_{\alpha_0, \dots, \alpha_\delta} k^{\nu - N(\alpha_0, \dots, \alpha_\delta)}.$$

Легко видеть, что имеет место тождество вида

$$V(k)(1 + ak + B(h_1, \dots, h_\delta, k))^\lambda = \sum_{i=0}^{2\nu} T_i(a, h_1, \dots, h_\delta) k^i, \quad (9)$$

где $T_0, \dots, T_{2\nu}$ – какие-то полиномы, степени которых не превосходят λ.

Тождеству (9) также можно дать естественную интерпретацию, если рассмотреть k-ичную систему счисления, в которой имеются и отрицательные цифры; например, можно потребовать, чтобы

– 114 –

$$k > |2T_\iota(a, h_1, \dots, h_\delta)| \quad (\iota = 0, \dots, 2\nu) \tag{10}$$

и рассмотреть систему с цифрами от $[-(k-1)/2]$ до $[(k-1)/2]$.

Нетрудно проверить, что из (5), (6), (8) и (9) следует, что имеет место тождество

$$T_\nu(a, h_1, \dots, h_\delta) = R(a, h_1, \dots, h_\delta). \tag{11}$$

Таким образом, если

$$b = B(h_1, \dots, h_\delta, k) \tag{12}$$

и k столь велико, что выполнены неравенства (10), то равенство (4) имеет место тогда и только тогда, когда стоящая на ν-ом месте цифра в k-ичной записи числа $V(k)(1+ak+b)^\lambda$ есть 0; как будет сейчас показано, последнее условие легко может быть записано с помощью одного квантора существования.

Лемма 1. *Каковы бы ни были числа $a, b, h_1, \dots, h_\delta$, k, удовлетворяющие условиям (10) и (12), для того чтобы было выполнено равенство (4), необходимо и достаточно, чтобы существовало целое число z такое, что*

$$-k^\nu < 2(V(k)(1+ak+b)^\lambda - zk^{\lambda+1}) < k^\nu. \tag{13}$$

Необходимость. Положим

$$z = \sum_{\iota=\nu+1}^{2\nu} T_\iota(a, h_1, \dots, h_\delta)k^{\iota-\nu-1}.$$

Согласно (9), (12) и (4),

$$V(k)(1+ak+b)^\lambda - zk^{\nu+1} = \sum_{\iota=0}^{\nu-1} T_\iota(a, h_1, \dots, h_\delta)k^\iota.$$

Из (10) получаем, что

$$\begin{aligned} \left|2\sum_{\iota=0}^{\nu-1} T_\iota(a, h_1, \dots, h_\delta)k^\iota\right| &\le \\ \le \sum_{\iota=0}^{\nu-1} |2T_\iota(a, h_1, \dots, h_\delta)k^\iota| &\le \\ \le \sum_{\iota=0}^{\nu-1}(k-1)k^\iota = k^\nu - 1 < k^\nu&, \end{aligned} \tag{14}$$

следовательно, неравенства (13) выполнены.

Достаточность. Нетрудно видеть, что существует не более одного целого числа y такого, что

$$-k^\nu < 2(V(k)(1+ak+b)^\lambda - yk^\nu) < k^\nu.$$

С одной стороны, таким числом y является, согласно (13), число zk, с другой стороны, – число

$$\sum_{i=\nu}^{2\nu} T_i(a, h_1, \ldots, h_\delta) k^{i-\nu},$$

поскольку, согласно (9) и (12),

$$V(k)(1 + ak + b)^\lambda - \left(\sum_{i=\nu}^{2\nu} T_i(a, h_1, \ldots, h_\delta) k^{i-\nu}\right) k^\nu =$$

$$= \sum_{i=0}^{\nu-1} T_i(a, h_1, \ldots, h_\delta) k^i$$

и имеют место неравенства (14). Таким образом,

$$\sum_{i=\nu}^{2\nu} T_i(a, h_1, \ldots, h_\delta) k^{i-\nu} = zk.$$

Переходя от равенства к сравнению, мы получаем, что

$$T_\nu(a, h_1, \ldots, h_\delta) \equiv 0 \pmod{k}.$$

Это вместе с (10) дает равенство

$$T_\nu(a, h_1, \ldots, h_\delta) = 0. \qquad (15)$$

Из (11) и (15) следует (4).

Лемма доказана.

4. Займемся преобразованием неравенств (10). Обозначим через γ какое-либо положительное число, превосходящее удвоенную сумму абсолютных величин всех коэффициентов любого из полиномов $T_0, \ldots, T_{2\nu}$. Очевидно, что справедливы неравенства

$$|2\, T_i(a, h_1, \ldots, h_\delta)| < \gamma(\max\{1, a, h_1, \ldots, h_\delta\})^\lambda.$$

Таким образом, если

$$c \geq \max\{h_1, \ldots, h_\delta\}$$

и

$$k = K(a, c) = \gamma(2 + a + c)^\lambda,$$

то неравенства (10) выполнены.

С помощью леммы 1 можно легко показать, что формула (3) эквивалентна формуле

$$\exists b \exists c [\mathfrak{I}_1 \,\&\, \exists z \mathfrak{I}_2],$$

$\mathfrak{I}_1$ здесь и ниже обозначает формулу

$$\exists h_1 \ldots \exists h_\delta [c \geq \max\{h_1, \ldots, h_\delta\} \;\&$$
$$b = B(h_1, \ldots, h_\delta, K(a, c))],$$

а $\mathfrak{I}_2$ – формулу

$$-(K(a,c))^\nu < 2\,(V(K(a,c))(1 + aK(a,c) + b)^\lambda -$$
$$- z(K(a,c))^{\nu+1}) < (K(a,c))^\nu.$$

Лемма 2. Формула $\mathfrak{J}_1$ эквивалентна формуле

$$\mathfrak{J}_3 \& \mathfrak{J}_4 \& \mathfrak{J}_5, \tag{16}$$

$\mathfrak{J}_3$ здесь и ниже обозначает формулу

$$\exists d [b = d(K(a, c))^{\lambda+1}],$$

$\mathfrak{J}_4$ – формулу

$$\mathop{\&}_{i=1}^{\delta-1} \exists d \exists e [b = d(K(a,c))^{(\lambda+1)^{i+1}} + e \,\& \, e < (c+1)(K(a,c))^{(\lambda+1)^i}],$$

$\mathfrak{J}_5$ – формулу

$$b < (c+1)(K(a,c))^{(\lambda+1)^\delta}.$$

Доказательство леммы становится особенно наглядным, если заметить, что обе формулы $\mathfrak{J}_1$ и (16) означают, что в записи числа b в системе счисления с основанием $K(a,c)$ отличные от нуля цифры могут стоять лишь на $(\lambda+1)$-м, ..., $(\lambda+1)^\delta$-м местах, причем эти цифры не могут превосходить c.

5. Как показывают леммы 1 и 2, формула (3) эквивалентна формуле

$$\exists b \exists c [\exists z \mathfrak{J}_2 \& \mathfrak{J}_3 \& \mathfrak{J}_4 \& \mathfrak{J}_5]. \tag{17}$$

Теорема 1. Каждое перечислимое множество натуральных чисел представимо формулой вида (1).

Доказательство. Искомая формула получается из формулы (17) посредством несложных алгебраических преобразований.

В формулу $\mathfrak{J}_2$ входит переменная z, допустимыми значениями которой являются все целые числа. Однако из $\mathfrak{J}_2$ следует, что

$$2z(K(a,c))^{\nu+1} > -(K(a,c))^\nu + 2V(K(a,c))(1 + aK(a,c) + b)^\lambda,$$

и потому

$$z \geq V(K(a,c))(1 + aK(a,c) + b)^\lambda.$$

Обозначим через $\mathfrak{J}_6$ формулу, которая получается из формулы $\mathfrak{J}_2$ подстановкой вместо z полинома

$$d + V(K(a,c))(1 + aK(a,c) + b)^\lambda$$

и разнесением членов, содержащих и не содержащих d, по разные стороны от знаков неравенств. Очевидно, что формула $\exists z \mathfrak{J}_2$ эквивалентна формуле $\exists d \mathfrak{J}_6$.

– 117 –

Формулу $\mathfrak{I}_3$ мы преобразуем в эквивалентную ей формулу

$$\exists d \, [\, b - 1 < (K(a,c))^{\lambda+1} d < b + 1 \,]. \qquad (18)$$

Равенства, входящие в каждый конъюнктивный член формулы $\mathfrak{I}_4$, позволяют явно выразить e через a, b, c, d и, следовательно, дают возможность исключить эту переменную; при этом необходимо добавить неравенство, соответствующее неотрицательности e. Окончательно получаем формулу

$$\mathop{\&}_{i=1}^{\delta-1} \exists d \, [\, b - (c+1)(K(a,c))^{(\lambda+1)^i} < (K(a,c))^{(\lambda+1)^{i-1}} d < b + 1 \,]. \qquad (19)$$

Нам осталось заменить формулу $\mathfrak{I}_5$ эквивалентной ей формулой

$$\exists d \, [\, b - 1 < bd < (c+1)(K(a,c))^{(\lambda+1)^\delta} \,].$$

Теорема доказана.

6. Перейдем к построению формулы вида (2), представляющей множество $\mathfrak{N}$. Для этого мы предварительно покажем, что для любого ε существуют полиномы F_ε и W_ε от 2ε и $3\varepsilon + 1$ неизвестной соответственно такие, что если числа $g_1, \ldots, g_\varepsilon$, $s_1, \ldots, s_\varepsilon$, $t_1, \ldots, t_\varepsilon$ удовлетворяют неравенствам

$$0 < g_i, \quad t_i - s_i \le g_i \qquad (i = 1, \ldots, \varepsilon), \qquad (20)$$

то формула

$$\mathop{\&}_{i=1}^{\varepsilon} \exists z \, [\, s_i < z g_i < t_i \,] \qquad (21)$$

эквивалентна формуле

$$\forall f \, [\, f \le F_\varepsilon(s_1, \ldots, s_\varepsilon, t_1, \ldots, t_\varepsilon) \Rightarrow W_\varepsilon(g_1, \ldots, g_\varepsilon, s_1, \ldots, s_\varepsilon, t_1, \ldots, t_\varepsilon, f) > 0 \,].$$

Мы начнем с рассмотрения случая $\varepsilon = 1$ и найдем сперва полиномы X и Y такие, что если $g > 0$, то формула

$$\exists z \, [\, s < zg < t \,] \qquad (22)$$

эквивалентна формуле

$$\forall y \, [\, -s^2 - t^2 - 2 < y \le s^2 + t^2 + 2 \Rightarrow X(g, s, t, y) > 0 \vee Y(g, s, t, y) > 0 \,]. \qquad (23)$$

Формула (23), не содержащая кванторов существования, должна тем не менее содержать полную информацию о некотором числе z, удовлетворяющем неравенствам

$$s < zg < t. \tag{24}$$

Обоснование эквивалентности формул вида (23) и (22) мы будем проводить с помощью следующей очевидной леммы, которую можно рассматривать как дискретный аналог теоремы Коши об обращении в нуль непрерывной функции, принимающей на концах интервала значения разных знаков.

Пусть p и q – целые числа, такие что $p<q$, Φ и Ψ – одноместные предикаты, определенные на всех целых числах от p до q включительно. Если имеет место $\Phi(p)\,\&\,\Psi(q)$ и для каждого w, такого что $p<w<q$, имеет место $\Phi(w)\vee\Psi(w)$, то существует число r, такое что $p\le r<q$ и $\Phi(r)\,\&\,\Psi(r+1)$.

Лемма 3. Если

$$g>0, \tag{25}$$

то формула (22) эквивалентна формуле

$$\forall y[-s^2-t^2-2<y\le s^2+t^2+2\Rightarrow$$
$$(y-1)g-s>0\vee t-yg>0]. \tag{26}$$

Доказательство. Пусть g, s, t – числа, удовлетворяющие условиям (25) и (26). Покажем, что они удовлетворяют и условию (22).

Согласно (25),

$$t-(-s^2-t^2-1)g\ge t+s^2+t^2+1>0,$$
$$(s^2+t^2+1)g-s\ge s^2+t^2+1-s>0.$$

По сформулированному выше дискретному аналогу теоремы Коши получаем, что существует число z такое, что

$$t-zg\quad 0\,\&\,zg-s>0.$$

Таким образом, условие (22) выполнено.

Пусть теперь g, s, t – числа, удовлетворяющие условиям (25) и (22). Найдем число z, удовлетворяющее неравенствам (24). Допустим, что условие (26) не выполнено, и пусть y – такое число, что

$$(y-1)g-s\le 0\,\&\,t-yg\le 0. \tag{27}$$

Из (24) и (27) получаем, что

$$(y-1)g\;\_\;s\quad zg,\quad zg\quad t\;\_\;yg,$$

следовательно,

$$y-1<z<y.$$

Полученное противоречие завершает доказательство эквивалентности формул (22) и (26).

Заметим, что если

$$t-s\le g, \tag{28}$$

то два неравенства, входящие в формулу (26), несовместимы. Более того, если $(y-1)g-s>0$, то $t-yg<0$, и, наоборот, если $t-yg>0$, то $(y-1)g-s<0$.

Это позволяет преобразовать дизъюнкцию двух неравенств в одно:

$$(y-1)g-s>0 \vee t-yg>0 \Leftrightarrow ((y-1)g-s>0 \,\&\, t-yg<0) \vee$$
$$\vee (t-yg>0 \,\&\, (y-1)g<0)) \Leftrightarrow ((y-1)g-s)(yg-t)>0.$$

Таким образом, если выполнены неравенства (25) и (28), то формула (22) эквивалентна формуле

$$\forall y\,[-s^2-t^2-2<y\le s^2+t^2+2 \Rightarrow Z(g,\ s,\ t,\ y)>0],$$

здесь и ниже $Z(g,s,t,y)$ обозначает полином

$$((y-1)g-s)(yg-t).$$

Отметим, что если $g>0$, то

$$\forall y\,[\,y\le -s^2-t^2-2 \vee y>s^2+t^2+2 \Rightarrow Z(g,\ s,\ t,\ y)>0]. \tag{29}$$

7. Рассмотрим теперь произвольную формулу вида (21). Если числа $g_1,\dots,g_\varepsilon,\ s_1,\dots,\ s_\varepsilon,\ t_1,\dots,t_\varepsilon$ удовлетворяет неравенствам (20), то, как показано выше, формула (21) эквивалентна формуле

$$\mathop{\&}_{\iota=1}^{\varepsilon} \forall y\,[-s_\iota^2-t_\iota^2-2<y\le s_\iota^2+t_\iota^2+2 \Rightarrow Z(g_\iota,\ s_\iota,\ t_\iota,\ y)>0]. \tag{30}$$

Введем следующие обозначения:

$$F_\iota(s_1,\dots,\ s_\iota,\ t_1,\dots,\ t_\iota)=\sum_{\mu=1}^{\iota}(2s_\mu^2+2t_\mu^2+4)\quad (\iota=0,\dots,\varepsilon),$$

$$Z_\iota(g_\iota,\ s_1,\dots,\ s_\iota,\ t_1,\dots,\ t_\iota,\ y)=Z(g_\iota,\ s_\iota,\ t_\iota,$$
$$y-F_{\iota-1}(s_1,\dots,\ s_{\iota-1},\ t_1,\dots,\ t_{\iota-1})-s_\iota^2-t_\iota^2-2)\quad (\iota=1,\dots,\varepsilon).$$

Очевидно, что формула (30) эквивалентна формуле

$$\mathop{\&}_{\iota=1}^{\varepsilon} \forall y\,[F_{\iota-1}(s_1,\dots,\ s_{\iota-1},\ t_1,\dots,\ t_{\iota-1})<y\le$$
$$\le F_\iota(s_1,\dots,\ s_\iota,\ t_1,\dots,\ t_\iota)\Rightarrow$$
$$\Rightarrow Z_\iota(g_\iota,\ s_1,\dots,\ s_\iota,\ t_1,\dots,\ t_\iota,\ y)>0]. \tag{31}$$

- 120 -

Обозначим через $W_\varepsilon(g_1, \ldots, g_\varepsilon, s_1, \ldots, s_\varepsilon, t_1, \ldots, t_\varepsilon, y)$ полином

$$\prod_{\iota=1}^{\varepsilon} Z_\iota(g_\iota, s_1, \ldots, s_\iota, t_1, \ldots, t_\iota, y).$$

Лемма 4. Если числа $g_1, \ldots, g_\varepsilon, s_1, \ldots, s_\varepsilon, t_1, \ldots$ $\ldots, t_\varepsilon$ удовлетворяют неравенствам (20), то формула (21) эквивалентна формуле

$$\forall f[f \leq F_\varepsilon(s_1, \ldots, s_\varepsilon, t_1, \ldots, t_\varepsilon) \Rightarrow$$
$$\Rightarrow W_\varepsilon(g_1, \ldots, g_\varepsilon, s_1, \ldots, s_\varepsilon, t_1, \ldots, t_\varepsilon, f) > 0].$$

Доказательство леммы легко проводится на основе свойства (29).

Теорема 2. Каждое перечислимое множество натуральных чисел представимо формулой вида (2).

Доказательство. Преобразуем формулу $\exists z \mathfrak{J}_2 \& \mathfrak{J}_3 \& \mathfrak{J}_4 \& \mathfrak{J}_5$ к виду, аналогичному (21).

В формуле $\mathfrak{J}_2$ нам достаточно разнести члены, содержащие и не содержащие z, по разные стороны от знаков неравенств. Получающуюся в результате формулу мы обозначим через $\mathfrak{J}_7$.

В формуле $\mathfrak{J}_3$ заменим переменную d, допустимыми значениями которой являются целые неотрицательные числа, на переменную z, допустимыми значениями которой являются целые числа. Поскольку всегда

$$b \geq 0, \quad (K(a,c))^{\lambda+1} > 0,$$

то получившаяся формула эквивалентна формуле $\mathfrak{J}_3$. Перепишем получившуюся формулу в виде, аналогичном (18), и обозначим новую формулу посредством $\mathfrak{J}_8$.

Аналогично в каждом конъюнктивном члене формулы $\mathfrak{J}_4$ мы заменим переменную d на z. Поскольку всегда

$$b \geq 0, \quad (K(a,c))^{(\lambda+1)^{\iota+1}} > (c+1)(K(a,c))^{(\lambda+1)^{\iota}} > 0,$$

то получившаяся формула эквивалентна формуле $\mathfrak{J}_4$. Проделаем с получившейся формулой такие же преобразования, какие были проделаны с формулой $\mathfrak{J}_4$ в доказательстве теоремы 1. В результате мы получим формулу $\mathfrak{J}_9$, аналогичную формуле (19).

Формулу $\mathfrak{J}_5$ мы заменим эквивалентной ей формулой

$$\exists z[b-(c+1)(K(a,c))^{(\lambda+1)^{\delta}} < 2(c+1)(K(a,c))^{(\lambda+1)^{\delta}} z <$$
$$< (c+1)(K(a,c))^{(\lambda+1)^{\delta}} - b],$$

которую обозначим через $\mathfrak{J}_{10}$.

Формула

$$\exists z\, \mathfrak{I}_7 \,\&\, \mathfrak{I}_8 \,\&\, \mathfrak{I}_9 \,\&\, \mathfrak{I}_{10} \qquad (32)$$

имеет вид, аналогичный (21). Отличие состоит лишь в том, что вместо переменных g_1, s_1, t_1 в (32) стоят некоторые полиномы от параметров a, b, c. Легко проверить, что при всех значениях параметров имеют место неравенства, аналогичные (20). Это позволяет найти с помощью леммы 4 искомые полиномы F и W.

Теорема доказана.

Литература

1. Gödel K. Über formal unentscheidbare Sätze der Principia Mathematica und verwandter Systeme, I. Monatsh. Math. und Phys., 1931, 38, 173–198.

2. Davis M. Arithmetical problems and recursively enumerable predicates. J. Symbol. Log., 1953, 18, № 1, 33–41 (в сб. пер. иностр. период. лит. Математика, 1964, 8, № 5, 15–22).

3. Robinson R.M. Arithmetical representation of recursively enumerable sets. J. Symbol. Log., 1956, 21, № 2, 162–186 (в сб. перев. ин. период. лит. Математика, 1964, 8, № 5, 23–47).

4. Putnam H. An unsolvable problem in number theory. J. Symbol. Log., 1960, 25, № 3, 220–232 (в сб. пер. иностр. период. лит. Математика, 1964, 8, № 5, 55–67).

5. Adler A. Existential formulas in arithmetic. Doct. diss. Univ. Toronto. 1967. Dissert. Abstrs, 1969, 29, № 8, 2962–2963.

6. Матиясевич Ю.В. Диофантовость перечислимых множеств. Докл. АН СССР, 1970, 191, № 2, 279-282.

7. Чудновский Г.В. Диофантовы предикаты. Успехи матем. наук, 1970, 25, № 4, 185-186.

8. Косовский Н.К. О диофантовых представлениях последовательности решении уравнения Пелля. Зап. науч. семинаров ЛОМИ АН СССР, 1971, 20, 49-59.

9. Davis M. An explicit Diophantine definition of the exponential function. Communic. Pure and Appl. Math., 1971, 24, № 2, 137–145.

10. Robinson R.M. An undecidable diophantine problem. A talk given to the IV Int. Congr. for Logic Meth. and Phil. of Science, Bucharest, 1971.

11. Matijasevič Yu. On recursive unsolvability of Hilbert's 10–th problem. Proc. IV Int. Congr. for Log., Meth. and Phil. of Science (в печати).

12. Матиясевич Ю.В. Арифметические представления перечислимых множеств с небольшим числом кванторов. Зап. науч. семинаров ЛОМИ АН СССР, 1972, 32, 77-84.

13. Robinson R.M. Some representations of diophantine sets. J. Symbol. Log., 1972, 37, № 3, 572–578.

14. Матиясевич Ю.В. Диофантовы множества. Успехи матем. наук, 1972, 27, № 5, 185-222.

ACTA ARITHMETICA
XXVII (1975)

Reduction of an arbitrary diophantine equation to one in 13 unknowns

by

YURI MATIJASEVIČ (Leningrad) and JULIA ROBINSON (Berkeley, Calif.)

A diophantine equation is a polynomial equation in some *parameters* $a_1, \ldots, a_\mu$ and some *unknowns* $z_0, \ldots, z_\nu$. That is, an equation of the form

$$P(a_1, \ldots, a_\mu, z_0, \ldots, z_\nu) = 0 \tag{1}$$

where P is a polynomial with integer coefficients (μ and ν are particular natural numbers). Both parameters and unknowns will be restricted to the natural numbers $\boldsymbol{N} = \{0, 1, \ldots\}$. The distinction between parameters and unknowns will be clear if we consider Mordell's equation

$$y^2 = x^3 + a. \tag{2}$$

Here x and y are unknowns and a is a parameter. The problem is to find the values of the parameter a for which (2) has a solution for the unknowns x and y. A trivial problem would result if x and a were unknowns and y the parameter. (For every y, we can find x and a satisfying (2).) Thus, it is convenient to think of P as a polynomial in $z_0, \ldots, z_\nu$ with coefficients which are polynomials in $a_1, \ldots, a_\mu$ with integer coefficients. We then write $P(z_0, \ldots, z_\nu)$ and speak of $a_1, \ldots, a_\mu$ as the *parameters of* P and $z_0, \ldots, z_\nu$ as the *variables of* P.

Given a polynomial P, we shall construct another polynomial $\bar{P}$ such that

$$P(a_1, \ldots, a_\mu, z_0, \ldots, z_\nu) = 0 \tag{3}$$

for some $z_0, \ldots, z_\nu$ if and only if

$$\bar{P}(a_1, \ldots, a_\mu, b, c, d, e, f, g, h, i, j, k, l, m, n) = 0 \tag{4}$$

for some $b, c, d, e, f, g, h, i, j, k, l, m, n$.

We say that (3) is a *diophantine definition* with $\nu+1$ unknowns of the relation among $a_1, \ldots, a_\mu$ which holds if and only if (3) has a solution, while (4) is a diophantine definition of the same relation with 13 unknowns.

We shall also speak of the equation $P = 0$ being *reduced* to the equation $\bar{P} = 0$. Two equations with the same parameters are *equivalent* if they are solvable for exactly the same values of the parameters.

In 1970, it was shown that every recursively enumerable relation has a diophantine definition [7]. From this and the existence of a universal recursively enumerable relation, it follows that for some δ and ν, there is a polynomial U of degree δ in $z_0, \ldots, z_\nu$ such that

$$U(a, n, z_0, \ldots, z_\nu) = 0$$

for some $z_0, \ldots, z_\nu$ if and only if $a \in \mathscr{D}_n$, where $\mathscr{D}_0, \mathscr{D}_1, \ldots$ is a list of all diophantine sets, i.e. sets which have a diophantine definition. By the use of pairing functions, this result can be extended to the case of any number of parameters leaving δ and ν unchanged. Hence every diophantine equation is equivalent to an equation with the same parameters of degree δ in $\nu+1$ unknowns. Before 1970, the existence of such a ν was widely disbelieved.

Our proof that the number of unknowns needed for a diophantine definition is $\leqslant 13$ is virtually self-contained, using only familiar or easily proved facts from elementary number theory.

Skolem [11] showed that every diophantine equation is equivalent to an equation of total degree 4. To see this we can replace (1) by an equation of the form $A = B$ where A and B are polynomials with positive integer coefficients. We then make up a system of equations $A_i = B_i$ of the forms $\alpha+\beta = \gamma$ and $\alpha\cdot\beta = \gamma$ where each α, β, γ is either a natural number, a parameter or variable of P, or a new variable introduced to represent an expression in the stepwise calculation of A and B, making sure to represent A and B by the same variable. Then $P = 0$ for some $z_0, \ldots, z_\nu$ if and only if $\sum(A_i-B_i)^2 = 0$ has a solution. Since A_i-B_i is of at most the second degree, the resulting equation will be of at most the fourth degree. Skolem's reduction increases the number of unknowns according to the number of steps needed to calculate P. Neither his result nor ours shows that both the degree (with respect to the unknowns) and the number of unknowns can be bounded simultaneously, although this is clear from the existence of U.

Now a diophantine equation $F(a, u_1, \ldots, u_\nu) = 0$ has a solution for $u_1, \ldots, u_\nu$ if and only if

$$(u_0+1)\big(1-F^2(u_0, \ldots, u_\nu)\big)-1 = a$$

has a solution for $u_0, \ldots, u_\nu$ (see Putnam [9]). Thus, our theorem shows that *every diophantine set is the non-negative part of the range on* $\boldsymbol{N}$ *of a polynomial with* 14 *variables.*

On the other hand, if we consider the range on $\boldsymbol{N}$ of a polynomial with positive coefficients, we cannot put a bound on the number of variables needed as the following example of R. M. Robinson shows. Let

$$F = x_1^n + 2x_2^n + 4x_3^n + \ldots + 2^{k-1}x_k^n.$$

Then F assumes the values $0, 1, \ldots, 2^k-1$ and no other values less than 2^n. We shall show that for n sufficiently large, the range (on $\boldsymbol{N}$) of F is not the same as the range of any polynomial with positive coefficients and fewer variables.

Suppose G is a polynomial with positive integer coefficients in $k-1$ variables with the same range as F. The G must assume some value less than 2^k with some variable $x > 1$. We may assume that when the other variables are fixed then G depends on x and hence has the form

$$H(x) = c_{k-1}x^{k-1} + \ldots + c_1 x + c_0$$

where $c_1, \ldots, c_{k-1}$ are not all zero and $H(2) < 2^k$. But then,

$$2^k \leqslant H(2^k) \leqslant (2^k)^{k-1} H(2) < 2^{k^2} \leqslant 2^n$$

for $n \geqslant k^2$. Hence G assumes some value g with $2^k \leqslant g < 2^n$ while F does not which contradicts the assumption that F and G have the same range. Nevertheless for any n and k, the range of the corresponding F is the non-negative part of the range of a polynomial G in 14 variables with integer coefficients.

We believe that the minimum number of unknowns necessary is less than 13, possibly as small as 3, but new methods will be needed to obtain the optimum result. In a companion paper [12] written with Martin Davis, we give a direct construction of a universal polynomial as well as applications of this paper to the theory of diophantine equations.

N o t a t i o n. Lower-case Latin letters will stand for natural numbers; upper-case Latin letters will stand for integers or particular polynomials with integer coefficients; lower case Greek letters will be used in various ways with their range given at the time or clear from the context. We will use the following notation for familiar relations of number theory: $x \mid y$ for 'x divides y'; $x \perp y$ for 'x is prime to y'; $[\sigma]$ for the greatest integer $\leqslant \sigma$ where σ is real; $\langle\sigma\rangle$ for the nearest integer to σ — we consider $\langle\sigma\rangle$ is undefined if σ is an integer plus a half; $\mathrm{Rem}(x, y)$ for 'the remainder when x is divided by y'; and $x = \square$ for 'x is a square'.

The paper is organized in two parts as follows:

Part I. Definition of the exponential function.

1. The Relation-combining Theorem.
2. Lucas sequences and Pell equations.
3. Definition of the nth solution of a Pell equation.
4. The exponential function.

Part II. Reduction to 13 unknowns.
5. The code.
6. Diophantine equation with coded unknowns.
7. Reduction to the positiveness of a polynomial.
8. Reduction to binomial coefficients.
9. Reduction to a partial binomial expansion.
10. Definition of a partial binomial expansion.
11. Completion of the proof.

Section 4 is not needed for the rest of the paper but contains a diophantine definition with just 5 unknowns of the relation $y = x^n$. We might say that the cost of "polynomializing exponentiation" is not more than 5 unknowns. It also provides a good introduction to the similar but more complicated Section 10. Part I is not used in Part II until Section 9 where the two lemmas of exponential size from Section 2 are needed. Thus, a reader familiar with some account of the negative solution to Hilbert's tenth problem (any of the expository articles by Davis [3], Manin [6], Matijasevič [8], etc. or the original papers by Davis, Putnam, and Robinson [4], Matijasevič [7], and Robinson [10]) may wish to begin with Part II and return to Part I as needed.

Construction of $\bar{P}$. We define a polynomial Q with parameters $a_1, \ldots, a_\mu$ and variables b, c, t such that $P(z_0, \ldots, z_\nu) = 0$ has a solution if and only if there are b and c such that $Q(b, c, t) > 0$ for all t. Q is completely defined by the 14 *-formulas in Sections 5, 6, and 7.

Next we consider an arbitrary polynomial Q in one variable with positive leading coefficient. We find a polynomial $\bar{Q}$ with 11 variables and the same parameters as Q such that $Q(t) > 0$ for all t if and only if $\bar{Q} = 0$ has a solution. We apply this result to $Q(b, c, t)$, treating b and c as parameters, and let $\bar{P}$ be the corresponding $\bar{Q}$. There are 24 **-formulas in Sections 1, 2, 3, 8, 9, and 10 which completely define $\bar{P}$ given Q.

We wish to thank R. M. Robinson for several ideas and suggestions made during the course of this work (some of these have been superseded and others incorporated) and to thank Craig Smorynski who read the manuscript and made helpful comments.

PART I

Definition of the exponential function

1. The Relation-combining Theorem. Certain combinations of relations can be defined more cheaply than by defining each separately by an equation and then combining the equations. For example, let D and E be integers. Then $D > 0$ and $E \neq 0$ both hold if and only if $DE^2 = 1 + x$ for some natural number x. The direct method of combining the equations $D = 1 + x$ and $E^2 = 1 + y$ gives $(D-1-x)^2 + (E^2-1-y)^2 = 0$ which has two unknowns. The Relation-combining Theorem will permit us to save 2 unknowns in the definition of the exponential function and 7 unknowns in the reduction of an arbitrary diophantine equation.

We will need a bound on the absolute values of the complex zeros of a polynomial $F(z) = \varphi_a z^a + \ldots + \varphi_1 z + \varphi_0$ where the φ's are complex numbers and $|\varphi_a| \geqslant 1$. Suppose $F(\tau) = 0$ and $|\tau| \geqslant 1$. Then

$$|\tau^a| \leqslant |\varphi_a \tau^a| \leqslant \sum_{i=0}^{a-1} |\varphi_i| \cdot |\tau|^{a-1}$$

so $|\tau| < 1 + \sum_{i=0}^{a-1} |\varphi_i|$. In particular, if the φ_i are integers then all zeros of $F(z)$ will lie in a circle of radius $1 + \sum_{i=0}^{a-1} \varphi_i^2$.

THEOREM 1. *Suppose $A_1, \ldots, A_q$ are integers. Then $A_1 = \square, \ldots, A_q = \square$ if and only if*

(1) $\qquad J_q(A_1, \ldots, A_q, X) = 0 \quad$ *for some integer X*

where

(2) $\qquad J_q(A_1, \ldots, A_q, X) = \prod (X \pm \sqrt{A_1} \pm \sqrt{A_2}\, W \pm \ldots \pm \sqrt{A_q}\, W^{q-1}),$

(3) $\qquad W = 1 + \sum_{i=1}^{q} A_i^2,$

and the product in (2) *extends over all combinations of signs.*

Proof. Clearly, (1) has a solution if all the A's are squares. On the other hand, we will show by induction on q that (1) implies that all the A's are squares. Obvious for $q = 1$. Suppose it is true for $q = j$. If $A_{j+1} = \square$, then $J_{j+1}(A_1, \ldots, A_{j+1}, X) = 0$ for some integer X if and only if $J_j(A_1, \ldots, A_j, Y) = 0$ for some integer Y. Hence the assertion holds for $q = j+1$. Suppose $A_{j+1} \neq \square$ and some factor ξ of $J_{j+1}(A_1, \ldots, A_{j+1}, X)$ is 0 for X. Let p be a prime which divides A_{j+1} to an odd power and $\bar{\xi}$ be the conjugate of ξ with respect to $\sqrt{p}$. Then

$$\xi - \bar{\xi} = 0 = \pm 2(\sqrt{A_{j+1}}\, W^j \pm \ldots)$$

where ... stands for similar terms of lower degree in W. But this is impossible because $W \geqslant 1 + \sum_{i=1}^{j} |\sqrt{A_i}|$.

Remark. Actually (1) has a solution with X a natural number whenever it has an integer solution. In Theorem 3, we will make use of the fact that in this theorem we ask for integer solutions of (1).

THEOREM 2. *Let $F(X) = F_a X^a + \ldots + F_1 X + F_0$ where $F_a \neq 0$. There is a polynomial H such that if $F_a \perp B$ and $B \neq 0$ then*

$$B \mid C, \quad D > 0, \quad F(X) = 0$$

for some integer X if and only if

$$H(B^2x+C^2) = 0$$

for some natural number x.

Proof. Choose $\bar{F}$ so large that all the integer zeros of F are greater than $-\bar{F}$. Let $G(Y) = F(Y-\bar{F}-C^2)$ then the integer zeros of $G(Y)$ are all greater than C^2. Let $H(Z) = J^a G(Z/J)$. Then the zeros of $H(Z)$ are J times the zeros of $G(Y)$. If $J \perp F_a$ (F_a is the leading coefficient of $G(Y)$) then the integer zeros of $H(Z)$ correspond to integer zeros of $G(Y)$ since the remaining rational zeros of $G(Y)$ have denominators which divide F_a. Also if $J > 0$ the integer zeros of $H(Z)$ will all be positive; otherwise they will all be negative. Now let $J = B^2(2DF_a^2-1)$. Then $J > 0$ if and only if $D > 0$. (Obviously, $J \neq 0$.)

Suppose $H(B^2x+C^2) = 0$. Then $D > 0$ since B^2x+C^2 is positive. Also $B|C$ since B^2x+C^2 is a multiple of J and hence of B^2. By the argument above, $F(X) = 0$ for some integer X. On the other hand, suppose $B|C$, $D > 0$, and $F(X) = 0$. Then $H(B^2x+C^2) = 0$ has a solution if

$$B^2(2DF_a^2-1)(X+\bar{F}+C^2) = B^2x+C^2$$

has. It does, since the left side is a multiple of B^2 and greater than C^2.

Notice that we cannot apply Theorem 2 again since $H(B^2T+C^2) = 0$ will always have an integer solution for T if $B|C$ and $F(X) = 0$.

THEOREM 3. (The Relation-combining Theorem.) *For every q there is a polynomial M_q with integer coefficients such that for all integers $A_1, \ldots, A_q, B, C, D$ with $B \neq 0$ the conditions*

$$A_1 = \square, \quad \ldots, \quad A_q = \square, \quad B|C, \quad D > 0$$

all hold if and only if

$$M_q(A_1, \ldots, A_q, B, C, D, n) = 0$$

for some natural number n.

Proof. Let $F(X) = J_q(A_1, \ldots, A_q, X)$ of Theorem 1. Then $a = 2^q$ and $F_a = 1$. Hence the hypothesis of Theorem 2 is satisfied so we can take for M_q the corresponding polynomial $H(B^2n+C^2)$.

In order to construct M_q explicitly we need to choose $\bar{F}$ and then carry out the steps in the proof of Theorem 2. From the factored form of J_q given by (2), we see that all integer zeros of J_q are greater than $-W^q$. Hence we can put $\bar{F} = W^q$. Also $J = B^2(2D-1)$. Therefore

$$** \quad M_q(A_1, \ldots, A_q, B, C, D, n)$$
$$= \prod \left(B^2n+C^2-B^2(2D-1)(C^2+W^q \pm\sqrt{A_1}\pm\sqrt{A_2}W \pm\ldots\pm\sqrt{A_q}W^{q-1})\right)$$

where the product is over all sets of signs and

$$** \qquad W = 1 + \sum_{i=1}^{q} A_i^2$$

will satisfy Theorem 3. Notice that the parameters of M_q are $A_1, \ldots \ldots, A_q, B, C, D$ and the only variable is n. (W is used only as an abbreviation for the sum.)

2. Lucas sequences and Pell equations. A *Lucas sequence* is a sequence of integers generated by

$$(1) \qquad X_0 = A, \quad X_1 = B, \quad X_{n+2} = CX_{n+1} + DX_n$$

where A, B, C, and D are integers. For example, if $A = 0$, $B = 1$, $C = 3$, $D = -2$ then $X_n = 2^n - 1$.

The ultimate tool in both the definition of exponentiation and the reduction to 13 unknowns is a diophantine definition of the nth term of a special type of Lucas sequences. In this section, we give some elementary facts about Lucas sequences and their relation to Pell equations.

A diophantine equation of the form

$$(2) \qquad x^2 - dy^2 = 1$$

where d is a parameter is called a Pell equation. The basic result about solutions of Pell equations is:

If $d > 0$ and not a square then (2) has infinitely many solutions in natural numbers; if d is a positive square, then the only solution is $x = 1$ and $y = 0$; if $d = 0$, then the solutions are $x = 1$ and $y = 0, 1, \ldots$

We shall be concerned with a special type of Pell equation – those of the form

$$(3) \qquad x^2 - (a^2 - 1)y^2 = 1 \quad \text{for} \quad a > 0.$$

In this case, let the solutions in order of the size of y be

$$x = \chi_a(n), \qquad y = \psi_a(n).$$

In particular, $\chi_1(n) = 1$ and $\psi_1(n) = n$.

There are two convenient ways of generating the solutions of (3): as Lucas sequences and algebraically in terms of the powers of the unit $a + \sqrt{a^2 - 1}$.

Recurrence equations. For $a > 0$,

$$(4) \qquad \chi_a(0) = 1, \quad \chi_a(1) = a, \quad \chi_a(n+2) = 2a \cdot \chi_a(n+1) - \chi_a(n),$$

$$(5) \qquad \psi_a(0) = 0, \quad \psi_a(1) = 1, \quad \psi_a(n+2) = 2a \cdot \psi_a(n+1) - \psi_a(n).$$

Algebraic characterization. For $a > 1$,

$$(6) \qquad (a + \sqrt{a^2 - 1})^n = \chi_a(n) + \psi_a(n)\sqrt{a^2 - 1}.$$

For the proof of these results, see any elementary number theory book which treats the Pell equation. With this background, we can easily derive the properties of χ and ψ which will be needed. The reader may also wish to consult the derivations in expository articles on Hilbert's tenth problem.

For every $a > 0$ and $b > 0$, there are infinitely many n such that $b \mid \psi_a(n)$ since $x^2-(a^2-1)b^2y^2 = 1$ has infinitely many solutions.

Also for $a > 0$, $n \leqslant \psi_a(n)$ by the definition of $\psi_a(n)$ and the fact that (3) has infinitely many solutions.

Congruence rule. For $a > 0$ and $b > 0$, if $a \equiv b \pmod{m}$ then

$$\chi_a(n) \equiv \chi_b(n), \qquad \psi_a(n) \equiv \psi_b(n) \pmod{m}.$$

This is clear since $\chi_a(n)$ and $\psi_a(n)$ for fixed n are polynomials in a by (4) and (5). The special case of the congruence rule which we use over and over is obtained when $b = 1$ and $m = a-1$ so

$$\psi_a(n) \equiv n \pmod{a-1}.$$

Bounds on ψ. For $a > 0$,

$$(2a-1)^n \leqslant \psi_a(n+1) \leqslant (2a)^n.$$

To check this, note that $\psi_a(1) = 1$ and $\psi_a(2) = 2a$. Since $\psi_a(n+1)$ is obtained from $\psi_a(n)$ by multiplying by $2a$ and subtracting something, the upper limit is verified by induction. Similarly, the lower limit can be checked since $\psi_a(n+1)$ is obtained by multiplying by $2a$ and subtracting something less than $\psi_a(n)$, i.e. multiplying by something greater than $2a-1$.

Addition equations. For $a > 0$,

$$\chi_a(n \pm m) = \chi_a(n)\chi_a(m) \pm (a^2-1)\psi_a(n)\psi_a(m), \tag{7}$$

$$\psi_a(n \pm m) = \pm\chi_a(n)\psi_a(m) + \psi_a(n)\chi_a(m). \tag{8}$$

Trivial for $a = 1$. The proof for $a > 1$ is immediate from the algebraic characterization of χ and ψ together with the fact that $(a+\sqrt{a^2-1})^{-1} = a-\sqrt{a^2-1}$. In particular,

$$\chi_a(2n) = 2\chi_a(n)^2-1, \tag{9}$$

$$\psi_a(2n) = 2\chi_a(n)\psi_a(n). \tag{10}$$

In general, we will be interested in the ψ-sequence rather than the χ-sequence. For $a > 0$,

$$(a^2-1)y^2+1 = \square$$

if and only if $y = \psi_a(n)$ for some n.

Next we prove two lemmas which are used in the next section in the definition of $\psi_A(B) = C$.

FIRST STEP-DOWN LEMMA. *For $a > 0$, if $\psi_a(m)^2 \mid \psi_a(n)$ then $\psi_a(m) \mid n$.*

Proof. For convenience, we will drop the subscript a throughout. Trivial for $m = 0$. For $m > 0$ by the addition equation for ψ,

$$\psi(m+k) \equiv \chi(m)\psi(k) \pmod{\psi(m)}.$$

But by (3), $\chi(m) \perp \psi(m)$. Hence $\psi(m) \mid \psi(m+k)$ if and only if $\psi(m) \mid \psi(k)$. If $0 < k < m$, then $0 < \psi(k) < \psi(m)$ and $\psi(m) \nmid \psi(k)$. Hence $\psi(m) \mid \psi(n)$ if and only if $m \mid n$. Now

$$\left(\chi(m) + \psi(m)\sqrt{a^2-1}\right)^k = \chi(km) + \psi(km)\sqrt{a^2-1}.$$

Using the binomial expansion and taking the irrational part, we obtain

$$\psi(km) = \sum_{j \leqslant k, j \text{ odd}} \binom{k}{j} \chi(m)^{k-j} \psi(m)^j (a^2-1)^{(j-1)/2}.$$

Hence

$$\psi(km) \equiv k\chi(m)^{k-1}\psi(m) \pmod{\psi(m)^3}.$$

Since $\chi(m) \perp \psi(m)$, we conclude that if $\psi(m)^2 \mid \psi(km)$ then $\psi(m) \mid k$. The lemma follows.

Remark. This lemma also follows from the classical laws of apparition and repetition of primes in Lucas sequences beginning with $0, 1, \ldots$ We call it a step-down lemma since the conclusion is obtained from the hypothesis by reducing the power of $\psi(m)$ by 1 and replacing $\psi(n)$ by its index n. Thus a particular divisibility condition on $\psi(n)$ implies a corresponding divisibility condition on n.

SECOND STEP-DOWN LEMMA. *For $a > 1$ and $n > 0$, if $\psi_a(j) \equiv \psi_a(i) \pmod{\chi_a(n)}$ then $j \equiv i$ or $j \equiv -i \pmod{2n}$.*

Proof. Here again we omit the subscript a. From the addition equations (8), (9), and (10),

$$\psi(2n \pm m) \equiv \mp \psi(m) \pmod{\chi(n)}.$$

Hence

$$\psi(4n \pm m) \equiv -\psi(2n \pm m) \equiv \pm\psi(m) \pmod{\chi(n)}.$$

Hence $\psi \pmod{\chi(n)}$ has a period of length $4n$. For $m \leqslant n$, we have $\pmod{\chi(n)}$,

$$\begin{aligned} \psi(m) &\equiv \psi(m), \\ \psi(2n-m) &\equiv \psi(m), \\ \psi(2n+m) &\equiv -\psi(m), \\ \psi(4n-m) &\equiv -\psi(m). \end{aligned}$$

In order to prove the lemma, we need to show that

$$(11) \qquad \psi(m) \not\equiv \pm\psi(m') \pmod{\chi(n)} \quad \text{for} \quad m < m' \leqslant n.$$

It would be sufficient to show that

$$(12) \qquad \psi_a(n) < \tfrac{1}{2}\chi_a(n) \quad \text{for} \quad n > 0.$$

This is true for $a > 2$, since

$$4\psi_a(n)^2 < (a^2-1)\psi_a(n)^2+1 = \chi_a(n)^2.$$

Thus the absolutely least residue of $\psi \pmod{\chi(n)}$ is 0 at 0, increases to a maximum of $\psi(n)$ at n, decreases to a minimum of $-\psi(n)$ at $3n$, and increases to 0 at $4n$.

For $a = 2$, (11) holds but not (12). However we can easily check that

$$\tfrac{1}{2}\chi_2(n) \geqslant \chi_2(n)-\psi_2(n) > \psi_2(n-1) \quad \text{for} \quad n > 0,$$

since the recurrence equations yield

$$2\psi_2(n)-\chi_2(n) = \psi_2(n-1) \geqslant 0$$

and

$$\chi_2(n)-\psi_2(n)-\psi_2(n-1) = \chi_2(n-1) > 0.$$

Remark. From the relation in the hypothesis among the terms of Lucas sequences, we obtain almost the same relation among the corresponding indices. Hence we also call this a step-down lemma. It should be compared with Lemma 6 used to define the Fibonacci sequence in [7] as well as similar lemmas used by Kosovskii [5], Čudnovskii [1], and Davis [2] in their definitions of the nth solution of the special Pell equation. All of the definitions used the first step-down lemma.

Next we will prove two lemmas which can be used to force one unknown to be exponentially larger than another. These lemmas will be used only in Section 9.

First lemma of exponential size. *There is a polynomial U with integer coefficients such that*

$$(13) \qquad U(x, y) = \square$$

implies that $y > x^x$ and furthermore, for every x and $m > 0$, there is a y satisfying (13) *such that $m \mid y+1$.*

Proof. By the basic result on Pell equations, for every $d > 1$ and $m > 0$ there are infinitely many z satisfying

$$(d^2-1)(d-1)^2m^2(z+1)^2+1 = \square.$$

We can put $y+1 = m(z+1)$. Hence there are infinitely many solutions for y of

$$(14)\qquad (d^2-1)(d-1)^2(y+1)^2+1 = \square$$

such that $m \mid y+1$. Also $(d-1)(y+1)$ is $\psi_d(w)$ for some w by (14) and $w \equiv (d-1)(y+1) \pmod{d-1}$ by the congruence rule. Since $(d-1)(y+1) \neq 0$, w is a positive multiple of $d-1$ and $\psi_d(w) \geqslant \psi_d(d-1)$. Hence $(d-1)(y+1) \geqslant (2d-1)^{d-2}$ for $d > 2$. If we put $d = x+3$, then

$$(x+2)(y+1) \geqslant (2x+5)^{x+1}$$

so

$$y > x^x.$$

Thus, we can take

$$**(15)\qquad U(x, y) = \big((x+3)^2-1\big)(x+2)^2(y+1)^2+1$$

to satisfy the lemma.

SECOND LEMMA OF EXPONENTIAL SIZE. *There is a polynomial V with integer coefficients such that*

$$(16)\qquad V(x, y, m) = \square$$

implies $y > x^x$ and furthermore, for every x there exists y satisfying (16) *in every residue class* $\pmod{m+1}$.

Proof. Consider the equation

$$(17)\qquad \big((d(m+1)+1)^2-1\big)d^2(y+1)^2+1 = \square$$

where $d > 1$. Now by the congruence rule,

$$\psi_{d(m+1)+1}\big(d(k+1)\big) \equiv d(k+1) \pmod{d(m+1)}$$

so there is a y satisfying (17) and such that

$$d(y+1) \equiv d(k+1) \pmod{d(m+1)},$$

i.e. $y \equiv k \pmod{m+1}$. Also if y satisfies (17), then $d(y+1) = \psi_{d(m+1)+1}(w)$ for some $w \equiv d(y+1) \pmod{d(m+1)}$. Since $d(y+1) > 0$, we must have $w > 0$ so $\psi_{d(m+1)+1}(w) \geqslant \psi_{d(m+1)+1}(d)$. Hence

$$d(y+1) \geqslant \big(2d(m+1)+1\big)^{d-1}$$

and so

$$y \geqslant (2d+1)^{d-2}.$$

If we put $d = x+3$, we will have $y > x^x$ as required. Thus, we can take

$$**(18)\qquad V(x, y, m) = \big(((m+1)(x+3)+1)^2-1\big)(x+3)^2(y+1)^2+1.$$

Remark. Notice that the second lemma does not include the first since V is a function of m while U is not.

3. Definition of the nth solution of a Pell equation. In this section, we will give a diophantine definition of the relation $\psi_A(B) = C$ for $A > 1$, $B > 0$, and $C > 0$. The definition is more economical with respect to the number of unknowns (three) than the previously known definitions given by Kosovskii [5], Čudnovskii [1], and Davis [2].

Let $\mathscr{A}$ be the following system of conditions:

A1 $DFI = \square,\ F \mid H - C,\ B \leqslant C,$
**A2 $D = (A^2-1)C^2+1,$
**A3 $E = 2(i+1)D(e+1)C^2,$
**A4 $F = (A^2-1)E^2+1,$
**A5 $G = A+F(F-A),$
**A6 $H = B+2jC,$
**A7 $I = (G^2-1)H^2+1.$

The parameters of $\mathscr{A}$ are e, A, B, C and the unknowns are i, j, D, E, F, G, H, I. We shall say that $\mathscr{A}$ can be satisfied if there are natural numbers i, j and integers D, E, F, G, H, I such that A1, ..., A7 hold. Of course whether $\mathscr{A}$ can be satisfied depends on the values of the parameters e, A, B, C and hence $\mathscr{A}$ defines a relation among them.

THEOREM 4. *Suppose $A > 1$, $B > 0$, $C > 0$. Then $\psi_A(B) = C$ if and only if $\mathscr{A}$ can be satisfied. [Furthermore,* A3 *and* A4 *imply $e+1 \perp F$.]*

Remarks. We can write $\mathscr{A}$ as a system with only two unknowns i and j since we can eliminate D, E, F, G, H, I in order by means of A2, ..., A7. To see this is possible, notice that the letters on the right side of an equation of $\mathscr{A}$ are either parameters, unknowns defined by earlier equations, i, or j. We have had to put in the extraneous conditions about e in order to apply the theorem later. In Section 11, we will need to combine the condition $F \mid H-C$ of A1 with another divisibility condition with divisor $e+1$ which we can do provided $e+1 \perp F$. If we only want to define $\psi_A(B) = C$, we can put $e = 0$ throughout. The reason we use integer parameters A, B, C and then suppose they are positive is again for applications of the theorem. In general, A, B, C, will be assigned values which are polynomials (with integer coefficients) in parameters and unknowns. We must check in each case that the hypothesis concerning A, B, C is satisfied before substituting $\mathscr{A}$ for $\psi_A(B) = C$ in another system as well as be sure that the set of unknowns of $\mathscr{A}$ is disjoint from the set of unknowns in the other system.

Proof. Suppose $A > 1$, $B > 0$, $C > 0$, and A1, ..., A7 hold. We see immediately that in fact $D, \dots, I$ are all positive. ($F > A$ since $F = (A+1)(A-1)E^2+1$ and $A > 1$.) We will first show that D, F, and I are co-prime and hence each is a square by A1. We obtain in turn

$$E \equiv 0, \quad F \equiv 1, \quad G \equiv 1, \quad I \equiv 1 \pmod{D} \tag{1}$$

by A3, A4, A5, A7 respectively. Next we get

$$G \equiv A, \quad H \equiv C, \quad I \equiv D \pmod{F}, \tag{2}$$

the first two congruences by A5, A1 respectively and then use them to obtain the third. Then both $F \perp D$ and $I \perp D$ by (1), and $I \perp F$ since $I \equiv D \pmod{F}$ and $F \perp D$. Hence D, F, and I are all squares so there are p, q, r such that

$$\begin{aligned} C &= \psi_A(p), & D &= \chi_A(p)^2, \\ E &= \psi_A(q), & F &= \chi_A(q)^2, \\ H &= \psi_G(r), & I &= \chi_G(r)^2. \end{aligned}$$

Now G was chosen in such a way that $\psi_G(r)$ acts as a bridge between r and B on one hand and between r and p on the other, allowing us to conclude that $p = B$ and $C = \psi_A(B)$. Indeed, $G \equiv 1 \pmod{2C}$ by A3, A4, A5 so that $H = \psi_G(r) \equiv r \pmod{2C}$ by the congruence rule and $H \equiv B \pmod{2C}$ by A6 so

$$r \equiv B \pmod{2C}. \tag{3}$$

On the other hand, $G \equiv A \pmod{F}$ by A5 so that $H = \psi_G(r) \equiv \psi_A(r) \pmod{F}$ by the congruence rule and $H \equiv C \pmod{F}$ by A1. Hence

$$\psi_A(r) \equiv \psi_A(p) \big(\operatorname{mod} \chi_A(q)\big) \tag{4}$$

since $C = \psi_A(p)$ and $F = \chi_A(q)^2$. Now $C^2 \mid \psi_A(q)$ by A3 so by the first step-down lemma, $C \mid q$. Hence by (4) and the second step-down lemma,

$$r \equiv \pm p \pmod{2C}. \tag{5}$$

By (3) and (5), we have $B \equiv \pm p \pmod{2C}$. Also $B \leqslant C$ by A1 and $p \leqslant \psi_A(p) = C$ so $B = p$ and $\psi_A(B) = C$ as required.

We still need to show that if $\psi_A(B) = C$ and $A > 1, B > 0, C > 0$ then we can choose natural numbers i, j and integers $D, \ldots, I$ satisfying $\mathscr{A}$. Put $D = \chi_A(B)^2$, then D will satisfy A2. Choose $q > 0$ so that $2D(e+1)C^2 \mid \psi_A(q)$. This is possible since $D > 0$, $C > 0$, and we can always choose a solution ψ of a Pell equation to be a multiple of any positve integer (see Section 2). Put $E = \psi_A(q)$ and choose i satisfying A3. Let F and G be given by A4 and A5. Put $H = \psi_G(B)$. Since A2, …, A5 have already been satisfied, $G \equiv 1 \pmod{2C}$. Hence $\psi_G(B) \equiv B \pmod{2C}$ and $\psi_G(B) \geqslant B$, so we can choose j satisfying A6. Let I be given by A7. Finally, D, F, and I are all squares by the hypothesis and the choice of E and H; $G \equiv A \pmod{F}$ so $H = \psi_G(B) \equiv \psi_A(B) \pmod{F}$; and $B \leqslant C$. Hence A1 is satisfied and the theorem is proved.

COROLLARY. *There is a polynomial Z such that if $A > 1$, $B > 0$, $C > 0$ then $\psi_A(B) = C$ if and only if there are i, j, and k such that $Z(A, B, C, i, j, k) = 0$.*

Proof. Take Z to be the polynomial obtained from

$$M_1(DFI, F, H-C, C-B+1, k)$$

by putting $e=0$ and eliminating $D, \ldots, I$ by A2, ..., A7. Here M_1 is the polynomial satisfying the Relation-combining Theorem for one square condition. Since $F>0$ by A4 and the hypothesis, we can apply the Relation-combining Theorem to see that $Z=0$ has a solution if and only if $\mathscr{A}$ can be satisfied. Hence the corollary follows from the theorem.

4. The exponential function. In this section, we will show that the relation $y=x^n$ can be defined by a diophantine equation in 5 unknowns. We do not use the exponential function in Part II but it seems worthwhile to give the definition — both as an introduction to Section 10 and for its own sake.

Consider the case $x>0$ and $n>0$. The bounds on ψ given in Section 2 yield

$$\psi_A(B) \sim (2A)^{B-1} \quad \text{as} \quad A \to \infty.$$

Hence

$$\frac{\psi_{Mx}(n+1)}{\psi_M(n+1)} \to x^n \quad \text{as} \quad M \to \infty. \tag{1}$$

Let

$$\varrho = \frac{\psi_{Mx}(n+1)}{\psi_M(n+1)}. \tag{2}$$

We wish to find a lower bound M_0 such that for $M>M_0$,

$$\langle \varrho \rangle = x^n. \tag{3}$$

To make the necessary estimates, we will use the elementary inequalities

$$(1-a)^q \geqslant 1-qa > 0 \quad \text{for} \quad 0 \leqslant a < 1/q; \tag{i}$$

$$(1-a)^{-1} \leqslant 1+2a \quad \text{for} \quad 0 \leqslant a \leqslant \tfrac{1}{2}. \tag{ii}$$

Then for $M \geqslant n$,

$$\varrho \leqslant \frac{(2Mx)^n}{(2M-1)^n} = x^n\left(1-\frac{1}{2M}\right)^{-n} \leqslant x^n\left(1-\frac{n}{2M}\right)^{-1} \leqslant x^n\left(1+\frac{n}{M}\right) \tag{4}$$

and

$$\varrho \geqslant \frac{(2Mx-1)^n}{(2M)^n} = x^n\left(1-\frac{1}{2Mx}\right)^n \geqslant x^n\left(1-\frac{n}{2Mx}\right). \tag{5}$$

Hence if M is so large that

$$\frac{nx^n}{M} < \frac{1}{2}, \tag{6}$$

then $|x^n - \varrho| < \frac{1}{2}$ and $\langle\varrho\rangle = x^n$. Also if $M > n$, then

(7) $$\varrho > \tfrac{1}{2}x^n$$

by (5). If $y = \langle\varrho\rangle$, that is

(8) $$(\varrho - y)^2 < \tfrac{1}{4},$$

then $y = x^n$ provided $M > 4n(y+1)$.

We have used two terms of Lucas sequences in obtaining y. This would lead to a definition of exponentiation with nine unknowns. Next we will show that only one term of a Lucas sequence is necessary.

Let $\mathscr{E}$ be the following system:

E0 $C = \psi_A(B)$,

E1 $(M^2-1)L^2+1 = \square$,

E2 $\left(\dfrac{C}{L} - y\right)^2 < \dfrac{1}{4}$, $xyn > 0$,

(or $(L^2 - 4(C-Ly)^2)xyn > 0$ since $L > 0$ by E3, E4),

E3 $M = 4n(y+1)+x+2$,

E4 $L = n+1+l(M-1)$,

E5 $A = Mx$,

E6 $B = n+1$,

E7 $C = k+B$.

The parameters of $\mathscr{E}$ are x, y, n and the variables are k, l, M, L, A, B, C.

LEMMA. *$x > 0$, $n > 0$, and $y = x^n$ if and only if $\mathscr{E}$ can be satisfied.* [E2, E3, E5, E6, *and* E7 *imply* $A > 1$, $B > 0$, $C > 0$.]

Remark. We repeat for emphasis one remark after Theorem 4. Whether $\mathscr{E}$ can be satisfied depends on the values of the parameters x, y, n and our assertion is that $\mathscr{E}$ can be satisfied exactly when $y = x^n$ and x, y, n are all positive. The values of the unknowns k, l must be natural numbers while those of M, L, A, B, C must be integers. Furthermore we could use E3, ..., E7 to eliminate M, L, A, B, C in turn and obtain a system with the same parameters and with only two variables k and l.

Proof. Suppose E0, ..., E7 hold. E2 implies $x > 0$, $y > 0$, and $n > 0$. Also $L > 0$ by E3 and E4 so there is an integer L' such that

$$L = \psi_M\big(n+1+L'(M-1)\big)$$

by E1, E4, and the congruence rule. Also $L' \geqslant 0$ since $M-1 > n+1$ by E3. We will first show that $L' = 0$ by contradiction. For $L' > 0$,

$$\frac{\psi_{Mx}(n+1)}{\psi_M\big(n+1+L'(M-1)\big)} \leqslant \frac{\psi_{Mx}(n+1)}{\psi_M(n+1+M-1)} \leqslant \frac{(2Mx)^n}{(2M-1)^{M+n-1}}$$
$$= \frac{(2M)^n}{\big(4M(M-1)+1\big)^n} \cdot \frac{x^n}{(2M-1)^{M-2n-1}} < \frac{1}{2}$$

provided $2M-1>x$ and $M>2n+1$. Both of these conditions on M follow from E3. Here we used the fact that $n>0$ so $(2M-1)^{2n}>2\cdot(2M)^n$. Thus if $L'>0$ then $y=0$ by the first part of E2 contradicting the second part. Hence $L'=0$ and $L=\psi_M(n+1)$. Then by E2, $y=\langle\varrho\rangle$ which is x^n by the argument above.

On the other hand, suppose $y=x^n$, $x>0$, and $n>0$. Then we will show that $\mathscr{E}$ can be satisfied. Let M, A, B, C be given by E3, E5, E6, E0. Then $B\leqslant C$ (see Section 2) so we can choose k satisfying E7. Let $L=\psi_M(n+1)$ then E1 holds and $L\equiv n+1 \pmod{M-1}$ by the congruence rule. Also $n+1<M-1$ so we can choose l satisfying E4. Finally, E2 holds since $y=x^n=\langle\varrho\rangle$ by the argument above.

Next we eliminate ψ from $\mathscr{E}$. Let $\mathscr{F}$ be the system consisting of $E1,\ldots$ $\ldots$, E7, A1, $\ldots$, A7. The parameters of $\mathscr{F}$ are x, y, n and the variables are $i, j, k, l, A, B, C, D, E, F, G, H, I, L, M$. We have put $e=0$ for this application of Theorem 4.

Lemma. *$x>0$, $n>0$, and $y=x^n$ if and only if $\mathscr{F}$ can be satisfied.*

Proof. Suppose $\mathscr{F}$ is satisfied. Then $A>1$, $B>0$, $C>0$ by the preceding lemma. Hence by Theorem 4, E0 holds and $y=x^n$, $x>0$, and $n>0$. On the other hand, suppose $y=x^n$, $x>0$, and $n>0$. Notice that the sets of unknowns in $\mathscr{A}$ and $\mathscr{E}$ are disjoint. We can choose k, l, M, L, A, B, C so that $\mathscr{E}$ is satisfied. Then $C=\psi_A(B)$ by E0 and $A>1$, $B>0$, $C>0$ by the first lemma. Hence by Theorem 4, we can choose i, j, D, E, F, G, H, I so that $\mathscr{A}$ is satisfied. Therefore $\mathscr{F}$ can be satisfied.

Notice that the condition $B\leqslant C$ in A1 is implied by E7 and hence can be dropped from $\mathscr{F}$. Thus $\mathscr{F}$ consists of two square conditions $DFI=\square$ and $(M^2-1)L^2+1=\square$; one divisibility condition $F\mid H-C$; one inequality $xyn\big(L^2-4(C-Ly)^2\big)>0$; and 11 equations. Hence we are in position to use the Relation-combining Theorem to obtain our final result.

Theorem 5. *There is a polynomial Z with integer coefficients such that $y=x^n$ if and only if $Z(x, y, n, i, j, k, l, m)=0$ for some natural numbers i, j, k, l, m.*

Proof. Let $Z_1(x, y, n, i, j, k, l, m)$ be the polynomial obtained from

$$M_2\big(DFI, (M^2-1)L^2+1, F, H-C, xyn\big(L^2-4(C-Ly)^2\big), m\big)$$

by putting $e=0$ and eliminating $M, L, A, B, C, D, E, F, G, H, I$ in order according to E3, $\ldots$, E7, A2, $\ldots$, A7. Here M_2 is the polynomial constructed after the proof of the Relation-combining Theorem in Section 1. So $Z_1=0$ for some i, j, k, l, m if and only if $\mathscr{F}$ can be satisfied. Therefore $Z_1=0$ can be satisfied if and only if $y=x^n$, $x>0$, and $n>0$. Finally, let

$$Z(x, y, n, i, j, k, l, m)=Z_1\cdot\big(n+(y-1)^2\big)\cdot\big(x+y+(n-1-k)^2\big).$$

Then $Z = 0$ if and only if $x > 0$, $n > 0$, $y = x^n$; or $n = 0$ and $y = 1$; or $x = 0$, $y = 0$, and $n > 0$. Hence Z satisfies the theorem.

PART II

Reduction to 13 unknowns

5. The code. The first step in our reduction of an arbitrary diophantine equation with unknowns $z_0, \ldots, z_\nu$ to one with 13 unknowns is to put a movable ceiling on the size of a solution. Thus, $P(z_0, \ldots, z_\nu) = 0$ has a solution if and only if it has a solution with $z_i \leqslant b$ for some b. This permits us to code $z_0, \ldots, z_\nu$ as digits in the $(b+1)$-ary expansion of a single number. However we will need more flexibility so we let $a_0, \ldots, a_\nu$ be particular positive integers with

$$0 < a_0 < \ldots < a_\nu \tag{1}$$

and choose B so that $B \geqslant b+1$. A number whose B-ary expansion has the form

$$1 + \sum_{i=0}^{\nu} u_i B^{a_i} \tag{2}$$

with $u_i \leqslant b$ will be called a *code with bound b*.

Every integer has a unique *symmetric* expansion in powers of an odd number $B = 2u+1 > 1$ with integer coefficients whose absolute values are $\leqslant u$. Now it turns out that for a suitable choice for the a's there are particular numbers β and λ together with a polynomial A such that if c is the code with bound b corresponding to $u_0, \ldots, u_\nu$ then $\lambda P(u_0, \ldots \ldots, u_\nu)$ is the coefficient of B^β in the symmetric B-ary expansion of $A(B, c)$. Thus, $P(z_0, \ldots, z_\nu) = 0$ has a solution if and only if there is a bound b and a code c with bound b such that the coefficient of B^β in the symmetric expansion of $A(B, c)$ is 0. This means that we can deal with just b and c no matter how many unknowns the original equation has.

The first problem is to express the condition that c is a code with bound b without referring to the digits of c. Let $c = c_0 + c_1 B + \ldots + c_\varrho B^\varrho$ be the usual B-ary expansion of c. Then

$$\mathrm{Rem}(c, B^i) = c_0 + c_1 B + \ldots + c_{i-1} B^{i-1}.$$

It is easy to give conditions on these partial sums for c to be a code with bound b. We give here redundant conditions in order to make the next step of the proof more direct. Namely, let

$$\begin{array}{ll} \mathrm{C}_0 & 0 < \mathrm{Rem}(c, B^{a_0}) < 2, \\ \mathrm{C}_i & 0 < \mathrm{Rem}(c, B^{a_i}) < (b+1) B^{a_{i-1}}, \quad i = 1, \ldots, \nu, \\ \mathrm{C}_{\nu+1} & 0 < c < (b+1) B^{a_\nu}, \end{array}$$

then $\mathrm{C}_0, \ldots, \mathrm{C}_{\nu+1}$ hold if and only if c is a code with bound b. Here C_0 implies $c_0 = 1$ since $a_0 > 0$. For $i \leqslant \nu$, C_i insures that the expansion has the proper gap before B^{a_i} while C_{i+1} insures that the coefficient of B^{a_i} is $\leq b$. Conversely, if c is a code with bound b, then $C_0, \ldots, C_{\nu+1}$ hold.

From $C_0, \ldots, C_{\nu+1}$, we obtain another characterization of codes: c is a code with bound b if and only if there is an integer in each of the open intervals

$$I_0 = \left(\frac{c-2}{B^{a_0}}, \frac{c}{B^{a_0}}\right),$$

$$*(3) \qquad I_i = \left(\frac{c-(b+1)B^{a_{i-1}}}{B^{a_i}}, \frac{c}{B^{a_i}}\right), \qquad i = 1, \ldots, \nu,$$

$$I_{\nu+1} = \left(\frac{c-(b+1)B^{a_\nu}}{(c+1)B^{a_\nu+1}}, \frac{c}{(c+1)B^{a_\nu+1}}\right).$$

The proof of the equivalence with $C_0, \ldots, C_{\nu+1}$ is immediate. In fact, C_i if and only if there is an integer in I_i, is clear for $i = 0, \ldots, \nu$. Also since $I_{\nu+1}$ is included in $(-1, +1)$, the only possible integer in $I_{\nu+1}$ is 0. Hence there is an integer in $I_{\nu+1}$ if and only if $C_{\nu+1}$.

Next we need to express the condition that the coefficient of B^β is 0 in the symmetric expansion of an integer A where $B = 2u+1 > 1$. If the coefficient of B^β is 0, then there is an integer T such that

$$|A - TB^{\beta+1}| \leq u(1+B+\ldots+B^{\beta-1}),$$

i.e.

$$(4) \qquad |A - TB^{\beta+1}| < \tfrac{1}{2}B^\beta.$$

On the other hand, the symmetric expansion of X for $|X| < \frac{1}{2}B^\beta$ contains only powers of B less than B^β. We can see this since there are B^β numbers with such representations and B^β values of X satisfying $|X| < \frac{1}{2}B^\beta$. Hence (4) is also a sufficient condition. Therefore the coefficient of B^β is 0 in the symmetric expansion of A if and only if there is an integer in the open interval

$$*(5) \qquad I_{\nu+2} = \left(\frac{2A-B^\beta}{2B^{\beta+1}}, \frac{2A+B^\beta}{2B^{\beta+1}}\right).$$

6. Diophantine equation with coded unknowns. We now return to the given equation $P(z_0, \ldots, z_\nu) = 0$. We will apply the results of the last section to find conditions on b and c so that c will be the code with bound b of a solution of $P = 0$. Without loss of generality we may assume that P contains some unknowns. Let

$$*(1) \qquad \delta = \text{degree of } P \text{ in } z_0, \ldots, z_\nu.$$

Thus δ is greater than 0. Let

$$*(2) \qquad a_i = (\delta+1)^i, \qquad i = 0, \ldots, \nu.$$

The a's are strictly increasing positive integers as required in the last section. Put

(3) $$C = 1 + \sum_{i=0}^{\nu} z_i B^{(\delta+1)^i}.$$

Then

(4) $$C^\delta = \sum\nolimits^* c_{i_0 \ldots i_\nu} z_0^{i_0} \ldots z_\nu^{i_\nu} B^{i_0 + i_1(\delta+1) + \ldots + i_\nu(\delta+1)^\nu}$$

where the summation extends over all natural numbers $i_0, \ldots, i_\nu$ with $i_0 + \ldots + i_\nu \leqslant \delta$ and

*(5) $$c_{i_0 \ldots i_\nu} = \frac{\delta!}{i_0! i_1! \ldots i_\nu! (\delta - i_0 - i_1 - \ldots - i_\nu)!}.$$

We will continue to use $\sum^*$ to indicate a sum with this range. The exponent of B corresponding to $i_0, \ldots, i_\nu$ is $i_\nu i_{\nu-1} \ldots i_0$ written to the base $\delta+1$. Hence each power of B in the sum comes in exactly once, with the largest exponent being $\delta(\delta+1)^\nu$. Furthermore since all the multinomial coefficients divide $\delta!$, there are $P_{i_0 \ldots i_\nu}$ so that

*(6) $$\delta! P(z_0, \ldots, z_\nu) = \sum\nolimits^* P_{i_0 \ldots i_\nu} c_{i_0 \ldots i_\nu} z_0^{i_0} \ldots z_\nu^{i_\nu}.$$

Here $P_{i_0 \ldots i_\nu}$ are integer multiples of the original coefficients of P.

Now define

(7) $$A(B, C) = C^\delta D(B)$$

where

*(8) $$D(B) = B^{(\delta+1)^\nu + 2} + \sum\nolimits^* P_{i_0 \ldots i_\nu} B^{(\delta+1)^\nu + 1 - i_0 - \ldots - i_\nu(\delta+1)^\nu}.$$

If

(9) $$B \geqslant 1 + \sum\nolimits^* P^2_{i_0 \ldots i_\nu},$$

then the absolute value of $\sum^*$ in (8) is less than $B^{(\delta+1)^\nu + 1 + 1}$ so

$$D(B) > 0.$$

Using (4) to eliminate C from (7), we obtain an expansion of A in powers of B with coefficients which are polynomials in $z_0, \ldots, z_\nu$ and which may be positive or negative. Thus, let

(10) $$A(B, C) = \sum_{j=0}^{(\delta+1)^\nu + 2 + \delta(\delta+1)^\nu} A_j(z_0, \ldots, z_\nu) B^j.$$

We could write out A_j explicitly given P but this will not be necessary. However we will use the fact that

(11) $$A_{(\delta+1)^\nu + 1}(z_0, \ldots, z_\nu) = \delta! P(z_0, \ldots, z_\nu).$$

This follows by direct calculation from (4), …, (10).

We will also need a bound on the absolute value of A_j under the restriction that all the z's are $\leqslant b$. It is easily checked that

$$|A_j(z_0, \ldots, z_\nu)| \leqslant \delta!(1+b^\delta)\left(1+\sum\nolimits^* P^2_{i_0\ldots i_\nu}\right)$$

for $z_0, \ldots, z_\nu \leqslant b$, using (4) and (8).

We are now ready to go from the algebraic identity (10) to the symmetric B-ary expansion of A provided B is sufficiently large and odd. Let

$$*(12) \qquad B = 2\delta!(1+b^\delta)\left(1+\sum\nolimits^* P^2_{i_0\ldots i_\nu}\right)+1$$

and c be a code with bound b. Since (9) holds, $D(B) > 0$. Suppose c is a code for $u_0, \ldots, u_\nu$. Then the digits in the symmetric B-ary expansion of $A(B, c)$ are $A_j(u_0, \ldots, u_\nu)$. Indeed, $A_j(z_0, \ldots, z_\nu)$ is the coefficient of B^j in the algebraic expansion of $A(B, C)$ when C has the algebraic form of a code for $z_0, \ldots, z_\nu$. Hence there is an expansion of $A(B, c)$ into powers of B with $A_j(u_0, \ldots, u_\nu)$ as coefficients. Also B is so large that

$$|A_j(u_0, \ldots, u_\nu)| < B/2$$

and hence the $A_j(u_0, \ldots, u_\nu)$ are the coefficients in the symmetric B-ary expansion of $A(B, c)$. Therefore by (11), we have shown that $P(z_0, \ldots$ $\ldots, z_\nu) = 0$ has a solution if and only if there is a bound b and a code c with bound b so that the coefficient of $B^{(\delta+1)^{\nu+1}}$ in the symmetric B-ary expansion of $A(B, c)$ is 0.

Using the results of the last section, we see that $P(z_0, \ldots, z_\nu) = 0$ has a solution if and only if there is an integer (or natural number) in each of the intervals $I_0, \ldots, I_{\nu+2}$ where

$$*(13) \qquad \beta = (\delta+1)^{\nu+1},$$

$$*(14) \qquad A = A(B, c) = c^\delta D(B),$$

a_i are given by (2), B by (12), and $D(B)$ by (8). (An integer in an open interval of length $\leqslant 1$ with right end-point $\geqslant 0$ is necessarily a natural number.) Note that B and $D(B)$ are polynomials in b with coefficients which are polynomials in the coefficients of $P(z_0, \ldots, z_\nu)$. Also at this point, we are using $\nu+5$ unknowns to express the solvability of $P = 0$. Namely, b and c in the end-points of the intervals together with $t_0, \ldots, t_{\nu+2}$ which are used for the natural number in the corresponding interval, i.e. $t_i \in I_i$.

Now let

$$*(15) \qquad I_i = \left(\frac{E_i(b, c)}{G_i(b, c)}, \frac{F_i(b, c)}{G_i(b, c)}\right) \quad \text{for} \quad i = 0, \ldots, \nu+2;$$

where E_i, F_i, G_i are polynomials determined by I_i according to the definition in Section 5. Thus, $P(z_0, \ldots, z_\nu) = 0$ has a solution if and only if there are $b, c, t_0, \ldots, t_{\nu+2}$ such that

$$E_i(b, c) < t_i G_i(b, c) < F_i(b, c) \quad \text{for} \quad i = 0, \ldots, \nu+2. \tag{16}$$

Remark. Since every I_i is of length $\leqslant 1$, we can transform this system of inequalities into a system of equation in two unknowns by using the greatest integer function. Namely, $P(z_0, \ldots, z_\nu) = 0$ has a solution if and only if there are b and c such that

$$\left[\frac{E_i(b, c)}{G_i(b, c)} + 1\right] = \left[\frac{F_i(b, c) - 1}{G_i(b, c)}\right] \quad \text{for} \quad i = 0, \ldots, \nu+2.$$

7. Reduction to the positiveness of a polynomial. We say a polynomial in one variable t is *positive* if it is positive for all natural numbers t. In this section, we will show that the condition that each of the intervals I_i contains a natural number is equivalent to a particular polynomial Q being positive.

Consider an open interval (σ, τ) where σ and τ are real numbers such that $0 < \tau - \sigma \leqslant 1$. There is an integer in (σ, τ) if and only if there is no integer in the closed $[\tau, \sigma+1]$, (i.e. each integer is on the same side of τ as of $\sigma+1$). Hence there is an integer in (σ, τ) if and only if

$$(T-\tau)(T-\sigma-1) > 0$$

for all integers T. In fact, there can be at most one value of T for which $(T-\tau)(T-\sigma-1) \leqslant 0$. Namely, T equal to the integer in $[\tau, \sigma+1]$ if there is one. Now suppose there is another interval (σ', τ') with $0 < \tau' - \sigma' \leqslant 1$ such that $[\tau, \sigma+1]$ and $[\tau', \sigma'+1]$ are disjoint. The two closed intervals cannot contain the same integer. Hence there are integers in both (σ, τ) and (σ', τ') if and only if

$$(T-\tau)(T-\sigma-1)(T-\tau')(T-\sigma'-1) > 0$$

for all integers T, since $(T-\tau)(T-\sigma-1)$ and $(T-\tau')(T-\sigma'-1)$ are never simultaneously $\leqslant 0$.

Now suppose we have ϱ intervals (σ_i, τ_i) where $0 \leqslant \tau_i$ and $0 < \tau_i - \sigma_i \leqslant 1$ for $i = 0, \ldots, \varrho-1$. We can construct new intervals (σ_i', τ_i') so that the corresponding complementary closed intervals are disjoint by translating the intervals to the right, integral distances. Choose an integer W so that $\sigma_i + 1 < W$ for $i = 0, \ldots, \varrho-2$ and let

$$\sigma_i' = \sigma_i + iW, \quad \tau_i' = \tau_i + iW. \tag{1}$$

Then $[\tau_i', \sigma_i'+1]$ is to the left of $[\tau_{i+1}', \sigma_{i+1}'+1]$ since

$$\sigma_i' + 1 = \sigma_i + iW + 1 < (i+1)W \leqslant \tau_i + (i+1)W = \tau_{i+1}'.$$

Hence there is an integer in each of the intervals (σ_i, τ_i) if and only if

(2) $$\prod_{i=0}^{\varrho-1}(T-\tau_i')(T-\sigma_i'-1) > 0$$

for all integers T. Since $\tau_i \geqslant 0$ and the lengths of all the (open) intervals are $\leqslant 1$, an integer in (σ_i, τ_i) is necessarily $\geqslant 0$. Also (2) holds for all negative integers T since both τ_i' and $\sigma_i'+1$ are $\geqslant 0$. Hence there is a natural number in each of the intervals (σ_i, τ_i) if and only if

(3) $$\prod_{i=0}^{\varrho-1}(t-\tau_i')(t-\sigma_i'-1) > 0$$

for all natural numbers t.

We now return to the intervals $I_0, \ldots, I_{\nu+2}$ defined in Section 6. If we put $I_i = (\sigma_i, \tau_i)$ then $0 \leqslant \tau_i$ and $0 < \tau_i - \sigma_i \leqslant 1$. Hence for W sufficiently large, there is a natural number in each of the intervals I_i if and only if

$$\prod_{i=0}^{\nu+2}\left(t-\frac{F_i}{G_i}-iW\right)\left(t-\frac{E_i}{G_i}-iW-1\right) > 0 \quad \text{for all } t.$$

Now we may take $W = c+1$ since $\sigma_i < \tau_i \leqslant c$ for $i = 0, \ldots, \nu+1$. (No restriction is needed for $i = \nu+2$.) Let

*(4) $$Q(b, c, t) = \prod_{i=0}^{\nu+2}\big((t-ic-i)G_i-F_i\big)\big((t-ic-i-1)G_i-E_i\big).$$

We have shown that $P(z_0, \ldots, z_\nu) = 0$ for some $z_0, \ldots, z_\nu$ if and only if there are b and c such that $Q(b, c, t) > 0$ for all t. The unknowns b and c are the first two of the promised 13. They can be considered as parameters in what follows so we shall write $Q(t)$ instead of $Q(b, c, t)$. It remains to find a polynomial $\bar{Q}$ in 11 variables and the same parameters as Q (that is $a_1, \ldots, a_\mu, b, c$) such that $Q(t) > 0$ for all t if and only if $\bar{Q}(t_1, \ldots$ $\ldots, t_{11}) = 0$ for some $t_1, \ldots, t_{11}$. In constructing $\bar{Q}$ from Q, we shall use the fact that for all b and c, the leading coefficient of $Q(t)$ is greater than 0.

8. Reduction to binomial coefficients. We wish to derive a necessary and sufficient condition that $Q(t) > 0$ for all natural numbers t. Binomial coefficients will come in quite naturally.

Let

**(1) $$Q(t) = Q_0+Q_1t+\ldots+Q_\gamma t^\gamma, \qquad Q_\gamma > 0,$$

(2) $$R = 1+\sum_{i=0}^{\gamma}Q_i^2, \qquad S = R^{\gamma+1}, \qquad T = RS^\gamma.$$

Then the real roots of $Q(t) = 0$ are less than R (see Section 1). If $Q(t) \leqslant 0$ then $t < R$; if $t < R$ then $|Q(t)| < S$; if $t < S$ then $|Q(t)| < T$.

Suppose

$$(3) \qquad d-u \mid Q(d)+v \quad \text{for some } u, v < S.$$

Then

$$(4) \qquad d-u \mid Q(u)+v,$$

since $d-u \mid Q(d)-Q(u)$. Obviously for sufficiently large d, (4) can only hold if $Q(u)+v = 0$. Indeed, if $d > 2S+T$ then $Q(u) = -v$. Conversely, if $Q(u) = -v$ then (3) holds.

Thus for $d > 2S+T$, we have shown that $Q(u) \neq -v$ if and only if $d-u \nmid Q(d)+v$. Now we would like to combine the S^2 conditions

$$(5) \qquad d-u \nmid Q(d)+v \quad \text{for} \quad u, v < S,$$

into one diophantine condition. We are unable to do this directly but the following theorem will be adequate for our purpose.

THEOREM 6. (A). *If* $Q(t) > 0$ *for all* t *and* $S!(S+T)! \mid d+1$, *then*

$$(6) \qquad \binom{d}{S} \perp \binom{Q(d)+S-1}{S}.$$

(B). *If* $d > S!^2+S$ *and*

$$(7) \qquad \binom{d}{S} \Big/ \left(\binom{d}{S}, S!\right) \perp \binom{Q(d)+S-1}{S},$$

then $Q(t) > 0$ *for all* t.

Remark. It will be clear in the next section why we need this unsymmetrical form of the theorem. We can combine the two parts to obtain as a corollary:

A necessary and sufficient condition for Q *to be a positive polynomial is that*

$$\binom{S!(S+T)!-1}{S} \perp \binom{Q\big(S!(S+T)!-1\big)+S-1}{S}$$

where S *and* T *are defined by* (1) *and* (2).

However we will need the theorem itself in order to make further reductions.

LEMMA. *If* $S!(S+T)! \mid d+1$ *and* p *is a prime which divides* $\binom{d}{S}$, *then* $p > S+T$.

Proof. Notice that

$$\binom{d}{S} = \left(\frac{d+1}{1}-1\right)\left(\frac{d+1}{2}-1\right)\cdots\left(\frac{d+1}{S}-1\right)$$

where the factors are all integers. Also

$$(S+T)!\,|\,\frac{d+1}{u+1} \quad \text{for} \quad u < S.$$

Hence a prime divisor of $\binom{d}{S}$ must be greater than $S+T$.

Proof of (A). Suppose Q is a positive polynomial and $S!(S+T)!\,|\,d+1$. If a prime p divides $\binom{d}{S}$ then there is a u less than S such that $p\,|\,d-u$ so $p\,|\,Q(d)-Q(u)$. Now if $p\,|\,Q(d)+v$ for some v less than S then $p\,|\,Q(u)+v$. This is impossible by the lemma since $p > S+T$ and $0 < Q(u)+v < S+T$. Hence $p \nmid \binom{Q(d)+S-1}{S}$.

Proof of (B). We shall suppose that $d > S!^2+S$ and $Q(u) = -v$ for some natural numbers u and v, and show that (7) cannot hold. Then

$$d-u\,|\,Q(d)+v$$

and $u, v < S$ by the choice of S. Hence

$$\text{(8)} \qquad \frac{d-u}{(d-u,\, S!)}\,\bigg|\,\binom{Q(d)+S-1}{S}.$$

Also

$$\text{(9)} \qquad \frac{d-u}{(d-u,\, S!^2)}\,\bigg|\,\frac{\binom{d}{S}}{\left(\binom{d}{S},\, S!\right)}$$

and the left side is greater than 1 by hypothesis. Hence the right sides of (8) and (9) have a common factor greater than 1 which contradicts (7).

In the following section, we will let $Z = Q(d)+S-1$. Note that Z is a polynomial in $a_1, \ldots, a_\mu, b, c$, and d.

9. Reduction to a partial binomial expansion. For $n > 0$, $s > 0$, and $x > n^s$,

$$\text{(1)} \qquad \left[\frac{(x+1)^n}{x^s}\right] = \sum_{i=0}^{n-s} \binom{n}{s+i} x^i$$

and

$$\text{(2)} \qquad \operatorname{Rem}\left(\left[\frac{(x+1)^n}{x^s}\right], x\right) = \binom{n}{s}.$$

To check this, write out the binomial expansion of $(x+1)^n$,

$$1+\ldots+\binom{n}{s-1}x^{s-1}+\binom{n}{s}x^s+\binom{n}{s+1}x^{s+1}+\ldots+x^n.$$

Since $x > n^s$, the first s terms have coefficients $\leqslant x-1$. Hence their sum is less than x^s and (1) holds. Then (2) is an immediate consequence since $x > \binom{n}{s}$. (We call the left side of (1) a *partial binomial expansion.*)

Using (1) and (2), we will see that the conditions given in Theorem 6 can be replaced by a combination of divisibility and square conditions together with one relation of the form

$$Y = \left[\frac{(X+1)^N}{X^S}\right] \tag{3}$$

where S is defined in Section 8.

One tool which we use is the simple observation: If $a \equiv b \pmod{d}$ then

$$\binom{a}{c} \equiv \binom{b}{c} \bmod \left(\frac{d}{(d, c!)}\right). \tag{4}$$

This follows since

$$a(a-1)\ldots(a-c+1) \equiv b(b-1)\ldots(b-c+1) \pmod{d}$$

and we can divide both sides of the congruence by $c!$, provided we divide the modulus by $(d, c!)$.

In Theorem 6, both $\binom{d}{S}$ and $\binom{Z}{S}$ come in. We can avoid defining both binomial coefficients (or the corresponding partial binomial expansions) from scratch, by using (3) where N is sufficiently large and X is divisible by both $N-d$ and $N-Z$. Then by (2) and (4), we have

$$Y \equiv \binom{N}{S} \pmod{X},$$

$$Y \equiv \binom{N}{S} \equiv \binom{d}{S} \bmod \left(\frac{N-d}{(N-d, S!)}\right),$$

$$Y \equiv \binom{N}{S} \equiv \binom{Z}{S} \bmod \left(\frac{N-Z}{(N-Z, S!)}\right).$$

By imposing other conditions on X, Y, and N, we will obtain a definition of $\binom{d}{S}$ and show that

$$Y \equiv \binom{Z}{S} \bmod \left(\frac{\binom{d}{S}}{\left(\binom{d}{S}, S!\right)}\right)$$

and

$$\binom{d}{S} \perp Y$$

which is sufficient to apply Theorem 6.

Let $\mathscr{S}$ be the following system:

S0 $Y = \left[\dfrac{(X+1)^N}{X^S}\right]$,

S1 $U(2S, d) = \Box$,

S2 $V(d(e+1)S, f, d) = \Box$,

S3 $V(N+S, g, e) = \Box$,

S4 $e+1 \mid gY+1$,

**S5 $S = (1+\sum_{i=0}^{\gamma} Q_i^2)^{\gamma+1}$,

**S6 $Z = \sum_{i=0}^{\gamma} Q_i d^i + S - 1$,

**S7 $N = Z + (e+1)f$,

**S8 $X = g(N-d)(N-Z)$,

**S9 $Y = e+1+h(N-d)$.

The parameters of $\mathscr{S}$ are $Q_0, \ldots, Q_\gamma$ and the unknowns are $d, e, f, g, h, S, Z, N, X, Y$. Here U and V are the polynomial functions given in the lemmas on exponential size (see Section 2).

THEOREM 7. *Suppose* $Q(t) = Q_0 + Q_1 t + \ldots + Q_\gamma t^\gamma$, $\gamma > 0$, *and* $Q_\gamma > 0$. *Then* $Q(t) > 0$ *for all* t *if and only if* $\mathscr{S}$ *is satisfied for some* $d, e, f, g, h, S, Z, N, X, Y$. [*Also* S1, ..., S9 *imply that* $Y > 0$, $X > 4N^S$, $N \geqslant S > 0$.]

Remark. We could eliminate the unknowns S, Z, N, X, Y by means of S5, ..., S9. For the purpose of this section, it would be sufficient for $X > N^S$ but later we will need $X > 4N^S$.

Proof. Suppose $\mathscr{S}$ is satisfied. Then we will show that

$$d > S!^2 + S \tag{5}$$

and

$$\frac{\binom{d}{S}}{\left(\binom{d}{S}, S!\right)} \perp \binom{Z}{S}. \tag{6}$$

Hence by part (B) of Theorem 6, $Q(t) > 0$ for all t. Notice that S and Z are defined by S5 and S6 just as in Theorem 6. By S1, S2, S3, and the lemmas on exponential size, we obtain

$$d > (2S)^{2S} \geqslant S^{2S} + S \geqslant S!^2 + S, \tag{7}$$

$$f > \left(d(e+1)S\right)^{d(e+1)S} \geqslant \left(d(e+1)S\right)^{dS}, \tag{8}$$

$$g > (N+S)^{N+S}. \tag{9}$$

Now by (8), $f > 0$ and by (7), d is so large that $Q(d) > 0$ (see the last section). Hence

$$N > Z \geqslant S > 0 \tag{10}$$

by S7, S6, S5 respectively. Also by S7,

$$N > d \geqslant 2 \tag{11}$$

since $f > d$ by (8) and (10). Hence

$$X \geqslant g > N^{N+S} \geqslant 4N^S \tag{12}$$

by S8, (10), and (11). Thus the second part of the theorem holds. Now using the argument at the beginning of this section,

$$Y \equiv \binom{N}{S} \equiv \binom{d}{S} \bmod \left(\frac{N-d}{((N-d),\, S!)}\right) \tag{13}$$

by S0 and the fact that $N-d \mid X$ by S8. Also $Y \equiv e+1 \bmod (N-d)$ by S9, so

$$e+1 \equiv \binom{d}{S} \bmod \left(\frac{N-d}{(N-d,\, S!)}\right). \tag{14}$$

Now $N > d^S + d$ and $N > (e+1)S^S + d$ by (8) since $N > f$, $d \geqslant 2$, and $d^2 > 2d$. Hence

$$\frac{N-d}{S!} > \binom{d}{S}, \qquad \frac{N-d}{S!} > e+1. \tag{15}$$

Therefore $e+1 = \binom{d}{S}$ by (14) and (15). Also $e+1 \mid N-Z$ by S7, so

$$Y \equiv \binom{N}{S} \equiv \binom{Z}{S} \bmod \left(\frac{e+1}{(e+1,\, S!)}\right).$$

Hence by S4,

$$\frac{e+1}{(e+1,\, S!)} \mid g\binom{Z}{S} + 1$$

so (6) holds as required. Hence if $\mathscr{S}$ is satisfied then Q is a positive polynomial.

On the other hand, suppose Q is a positive polynomial. Let S be given by S5 and put

$$T = \left(1 + \sum_{i=0}^{\nu} Q_i^2\right)^{\nu(\nu+1)+1}$$

as in the preceding section. By the first lemma on exponential size, we choose d satisfying both

$$S!(S+T)! \mid d+1 \tag{16}$$

and S1. Let Z be given by S6 and put $e = \binom{d}{S} - 1$ so $e+1 \perp S!$ by the lemma of Section 8 since all primes dividing $\binom{d}{S}$ are greater than $S+T$. Also $e+1 \perp \binom{Z}{S}$ by part (A) of Theorem 6. Choose f satisfying both

$$(17) \qquad Z+(e+1)f-d \equiv 1 \pmod{S!}$$

and S2. This is possible by the second lemma on exponential size and the facts that $S! \mid d+1$ and $e+1 \perp S!$. Let N be given by S7 then $N-d > 0$ and $N-Z > 0$, since $f > 0$ by S2 and $e+1 = \binom{d}{S} \geqslant d$. Also

$$(18) \qquad N-d \perp S!$$

by (17). Choose g satisfying both

$$(19) \qquad e+1 \mid g\binom{Z}{S}+1$$

and S3. This is possible by the second lemma on exponential size since $e+1 \perp \binom{Z}{S}$. Let X be given by S8. Then

$$(20) \qquad X \geqslant g > (N+S)^{N+S} > 4N^S$$

by S3 since $N-d > 0$ and $N-Z > 0$. Let Y be given by S0. Then

$$(21) \qquad Y \equiv \binom{N}{S} \equiv \binom{d}{S} = e+1 \bmod (N-d).$$

The first congruence holds by the argument at the beginning of this section since $X > N^S$ and $N-d \mid X$. The second congruence holds by (18). Also $Y \geqslant \binom{N}{S} > \binom{d}{S} = e+1$ since $N > d$. Hence we can determine h satisfying S9. Furthermore

$$(22) \qquad Y \equiv \binom{N}{S} \equiv \binom{Z}{S} \bmod (e+1)$$

since $e+1 \perp S!$ and $e+1 \mid N-Z$ so $e+1 \mid X$. Hence by (19) and (22), S4 holds and the proof is completed.

Going back to the original equation, $P(z_0, \ldots, z_\nu) = 0$, we have shown that $P = 0$ has a solution if and only if there are b, c, d, e, f, g, h satisfying $\mathscr{S}$ where $Q(t)$ is $Q(b, c, t)$ defined in Section 7. (Here the unknowns S, Z, N, X, Y have been eliminated from $\mathscr{S}$ by means of S5, ..., S9.) S0 is the only condition not obviously diophantine. In the next section, we will give a diophantine definition of that relation.

10. Definition of a partial binomial expansion. The diophantine definition of $Y = [(X+1)^N/X^S]$ will be valid whenever $Y > 0$, $X > 4N^S$, and $N \geqslant S > 0$. The definition is parallel to that of $y = x^n$ in Section 4. Let

$$\sigma = \frac{\psi_{M(X+1)}(N+1)}{\psi_M(N-S+1)\,\psi_{MX}(S+1)}, \tag{1}$$

then by the bounds on ψ from Section 2,

$$\sigma \to \frac{(X+1)^N}{X^S} \quad \text{as} \quad M \to \infty. \tag{2}$$

We wish to find a lower bound M_0 for M so that for $M > M_0$,

$$\langle \sigma \rangle = \left[\frac{(X+1)^N}{X^s}\right] \quad \text{for} \quad X > 4N^S.$$

First we estimate the size of σ for $M \geqslant N$ using the inequalities (i) and (ii) of Section 4.

$$\begin{aligned}\sigma &\leqslant \frac{(2M(X+1))^N}{(2M-1)^{N-S}(2MX-1)^S} \leqslant \frac{(X+1)^N}{X^S}\left(1-\frac{1}{2M}\right)^{-N} \\ &\leqslant \frac{(X+1)^N}{X^S}\left(1-\frac{N}{2M}\right)^{-1} \leqslant \frac{(X+1)^N}{X^S}\left(1+\frac{N}{M}\right)\end{aligned}$$

and

$$\begin{aligned}\sigma &\geqslant \frac{(2M(X+1)-1)^N}{(2M)^{N-S}(2MX)^S} = \frac{(X+1)^N}{X^S}\left(1-\frac{1}{2M(X+1)}\right)^N \\ &\geqslant \frac{(X+1)^N}{X^S}\left(1-\frac{N}{2M(X+1)}\right).\end{aligned}$$

Now if

$$|\varrho - \sigma| < \tfrac{1}{4} \quad \text{and} \quad \varrho - [\varrho] < \tfrac{1}{4}$$

then $\langle \sigma \rangle = [\varrho]$. For $M \geqslant N$,

$$\sigma \geqslant \frac{1}{2}\frac{(X+1)^N}{X^S}.$$

Let $Y = \langle \sigma \rangle$ so $Y+1 > \dfrac{1}{2}\dfrac{(X+1)^N}{X^S}$. Therefore, if $M > 8N(Y+1)$ then

$$\frac{(X+1)^N}{X^S} - \frac{1}{4} < \sigma < \frac{(X+1)^N}{X^S} + \frac{1}{4}.$$

Also

$$\frac{(X+1)^N}{X^S} - \left[\frac{(X+1)^N}{X^S}\right] = \sum_{i=0}^{S-1} \binom{N}{i} X^{i-S} \leqslant \frac{SN^{S-1}}{X} \leqslant \frac{N^S}{X} < \frac{1}{4}$$

provided $X > 4N^S$. Hence $Y = [(X+1)^N/X^S]$.

In defining σ, we used three terms of Lucas sequences $\psi_{M(X+1)}(N+1)$, $\psi_M(N-S+1)$, and $\psi_{MX}(S+1)$. Just as in Section 4, we can define the last two in terms of the first.

Let $\mathcal{T}$ be the following system:

T0 $C = \psi_A(B)$,

T1 $(M^2-1)K^2+1 = \square$,

T2 $(M^2X^2-1)L^2+1 = \square$,

T3 $\left(\frac{C}{KL} - Y\right)^2 < \frac{1}{4}$ (or $K^2L^2 - 4(C-KLY)^2 > 0$),

**T4 $M = 8N(X+Y)+2$,

**T5 $K = N-S+1+k(M-1)$,

**T6 $L = S+1+l(MX-1)$,

**T7 $A = M(X+1)$,

**T8 $B = N+1$,

**T9 $C = m+B$.

The parameters of $\mathcal{T}$ are X, Y, N, S and the unknowns are $k, l, m, A, B, C, K, L, M$.

THEOREM 8. *Suppose* $X > 4N^S$, $Y > 0$, *and* $N \geqslant S > 0$. *Then* $Y = [(X+1)^N/X^S]$ *if and only if* $\mathcal{T}$ *is satisfied for some* $k, l, m, A, B, C, K, L, M$. [*Also the hypothesis on* X, Y, N, S *together with* T4, T7, T8, T9 *implies that* $A > 1$, $B > 0$, $C > 0$.]

Remark. We can eliminate the integer unknowns M, K, L, A, B, C in turn from $\mathcal{T}$ by T4, T5, T6, T7, T8, T9 so that $\mathcal{T}$ becomes a system with unknowns k, l, m. The equivalence of the two forms of T3 is clear since $K > 0$, $L > 0$ by T4, T5, T6, and the hypothesis.

Proof. The second part of the theorem is trivial. Suppose that T0, ..., T9 hold. We will show first that

$$K = \psi_M(N-S+1), \tag{3}$$

$$L = \psi_{MX}(S+1). \tag{4}$$

Notice that by T4,

$$M > 8NX+2 \tag{5}$$

since $Y > 0$. Also K and L are positive, $K \equiv N-S+1 \bmod (M-1)$, and $L \equiv S+1 \bmod (MX-1)$ by T5 and T6. Hence by T1, T2, and the congruence rule,

$$K = \psi_M\big(N-S+1+K'(M-1)\big), \qquad L = \psi_{MX}\big(S+1+L'(MX-1)\big)$$

for some integers K' and L'. Since $N-S+1 < M-1$ and $S+1 < MX-1$ by (5) both K' and L' are non-negative. We need to show that $K' = L' = 0$. Now

$$\frac{(2M(X+1))^N}{(2M-1)^{M-1}} = \frac{(2M)^N}{(4M(M-1)+1)^N} \cdot \frac{(X+1)^N}{(2M-1)^{M-2N-1}} < \frac{1}{2}$$

since $2M-1 \geqslant X+1$ and $M-2N-1 \geqslant N$. Hence both

$$\frac{\psi_{M(X+1)}(N+1)}{\psi_M(N-S+1+M-1)} < \frac{1}{2}, \qquad \frac{\psi_{M(X+1)}(N+1)}{\psi_{MX}(S+1+MX-1)} < \frac{1}{2}$$

by the bounds on ψ in Section 2. Therefore, if either K' or L' were $\geqslant 1$ then T3, T0, T7, T8 would imply $Y = 0$ contrary to hypothesis. Hence (3) and (4) have been proved so T3 implies $Y = \langle\sigma\rangle$. Then by T4, $M > 8N(Y+1)$ so

$$Y = \left[\frac{(X+1)^N}{X^S}\right]$$

by the argument at the beginning of this section.

On the other hand, suppose $Y = [(X+1)^N/X^S]$, $X > 4N^S$, $N \geqslant S > 0$. We must show that there are natural numbers k, l, m and integers M, K, L, A, B, C so that T0, ..., T9 hold. Let M, A, B, C be given by T4, T7, T8, T0. Put $K = \psi_M(N-S+1)$ and $L = \psi_{MX}(S+1)$. Then T1 and T2 hold. Also there are k and l satisfying T5 and T6 by the congruence rule and the facts that $M-1 > N-S+1$ and $MX-1 > S+1$ which follow from the hypothesis and T4. Next choose m so that T9 holds which is possible since $\psi_A(B) \geqslant B$ for $A > 0$. Finally, T3 holds since

$$0 < \frac{(X+1)^N}{X^S} - Y < \frac{1}{4}, \qquad \left|\frac{(X+1)^N}{X^S} - \frac{C}{KL}\right| < \frac{1}{4}.$$

11. Completion of the proof. Let $\mathscr{U}$ be the system consisting of A1, ..., A7, S1, ..., S9, T1, ..., T9. The parameters of $\mathscr{U}$ are $Q_0, \ldots, Q_\gamma$ and the unknowns are $d, e, f, g, h, i, j, k, l, m, A, B, C, D, E, F, G, H, I, K, L, M, N, S, X, Y, Z$.

THEOREM 9. *Let $Q(t) = Q_0+Q_1t+\ldots+Q_\gamma t^\gamma$ and $Q_\gamma > 0$. Then $Q(t) > 0$ for all t if and only if $\mathscr{U}$ can be satisfied.*

Proof. Suppose $\mathscr{U}$ is satisfied. Then $Y > 0$, $X > 4N^S$, $N \geqslant S > 0$ by Theorem 7. Hence $A > 1$, $B > 0$, $C > 0$ by Theorem 8. Hence by Theorem 4, T0 holds so by Theorem 8, S0 holds. Therefore Q is a positive polynomial by Theorem 7.

On the other hand, suppose Q is a positive polynomial. Notice that the sets of unknowns in the three systems $\mathscr{A}$, $\mathscr{S}$, and $\mathscr{T}$ are disjoint.

We can choose $d, e, f, g, h, S, Z, N, X, Y$ so that $\mathscr{S}$ is satisfied by Theorem 7. Hence $Y = [(X+1)^N/X^S]$ and $Y > 0$, $X > 4N^S$, $N \geqslant S > 0$. Then by Theorem 8, we can choose $k, l, m, A, B, C, K, L, M$ so that $\mathscr{T}$ is satisfied and $A > 1$, $B > 0$, $C > 0$. Therefore $C = \psi_A(B)$ so by Theorem 4 we can find i, j, D, E, F, G, H, I satisfying $\mathscr{A}$.

We are now ready to construct $\bar{Q}$ the polynomial described at the end of Section 7.

THEOREM 10. *There is a polynomial $\bar{Q}$ with* 11 *variables $d, e, f, g, h, i, j, k, l, m, n$ and the same parameters as Q such that $Q(t) > 0$ for all t if and only if $\bar{Q} = 0$ for some $d, e, f, g, h, i, j, k, l, m, n$.*

Proof. First we modify $\mathscr{U}$ by replacing S4 by

S4′ $\quad (e+1)F \mid (e+1)(H-C)+F(gY+1)$,

and omitting the second and third conditions in A1. This system has exactly the same solutions as $\mathscr{U}$. Indeed, S4′ implies both S4 and $F \mid H-C$ since $e+1 \perp F$ by A3 and A4. Conversely, S4 and $F \mid H-C$ imply S4′. Also $B \leqslant C$ by T9. Now we have a system consisting of six square conditions A1, S1, S2, S3, T1, and T2; one divisibility condition S4′; one inequality; and 17 equations. Let M_6 be the polynomial constructed after the proof of the Relation-combining Theorem in Section 1 corresponding to 6 square conditions. Let

$$** \quad \bar{Q}(d, e, f, g, h, i, j, k, l, m, n)$$
$$= M_6\big(DFI,\ U(2S, d),\ V\big(d(e+1)S, f, d\big),\ V(N+S, g, e),$$
$$(M^2-1)K^2+1,\ (M^2X^2-1)L^2+1,\ (e+1)F,$$
$$(e+1)(H-C)+F(gY+1),\ K^2L^2-4(C-KLY)^2,\ n\big),$$

that is, $\bar{Q}$ is the polynomial obtained from the right side by substituting the coefficients of Q for $Q_0, \ldots, Q_\gamma$ and eliminating all the integer unknowns by means of the 17 equations. To see that this is possible, we carry out the elimination in the following order $S, Z, N, X, Y, M, K, L, A, B, C, D, E, F, G, H, I$ using S5, …, S9, T4, …, T9, A2, …, A7. Here each unknown is expressed in terms of earlier ones and $Q_0, \ldots, Q_\gamma$. By the Relation-combining Theorem, $\bar{Q} = 0$ has a solution if and only if $\mathscr{U}$ can be satisfied. Hence Q is a positive polynomial if and only if $\bar{Q} = 0$ has a solution by Theorem 9.

In Section 7, we showed that the given diophantine equation $P(z_0, \ldots, z_\nu) = 0$ has a solution if and only if there are b and c such that $Q(b, c, t) > 0$ for all t. Here Q is the particular polynomial defined by the *-formulas in Sections 5, 6, and 7. Finally, let

$$** \qquad \bar{P}(a_1, \ldots, a_\mu, b, c, d, e, f, g, h, i, j, k, l, m, n) = \bar{Q}$$

where $\bar{Q}$ corresponds to $Q(b, c, t)$. We thus obtain our final result:

THEOREM 11. *Given a polynomial P there is a polynomial $\overline{P}$ such that* $P(a_1, \ldots, a_\mu, z_0, \ldots, z_\nu) = 0$ *for some* $z_0, \ldots, z_\nu$ *if and only if*

$$\overline{P}(a_1, \ldots, a_\mu, b, c, d, e, f, g, h, i, j, k, l, m, n) = 0$$

for some $b, c, d, e, f, g, h, i, j, k, l, m, n$.

References

[1] Г. Чудновский, *Диофантовы предикаты*, Успехи мат. наук 25: 4 (1970), pp. 185–186.

[2] Martin Davis, *An explicit diophantine definitions of the exponential function*, Comm. Pure Appl. Math. 24 (1971), pp. 137–145.

[3] — *Hilbert's tenth problem is unsolvable*, Amer. Math. Monthly 80 (1973), pp. 233–269.

[4] Martin Davis, Hilary Putnam, and Julia Robinson, *The decision problem for exponential diophantine equations*, Ann. of Math. 74 (1961), pp. 425–436 = Математика 8:5 (1964), pp. 69–79.

[5] Н. Косовский, *О диофантовых представлениях последовательности решений уравнения Пелля*, Зап. научн. семинаров Ленингр. отд. Матем. ин-та им. В. А. Стеклова 20 (1971), pp. 49–59. J. of Soviet Math. 1:1 (1973), pp. 28–35.

[6] Ю. Манин, *Десятая проблема Гильберта*, Современные проблемы математики 1 (1973), pp. 5–37.

[7] Ю. Матиясевич, *Диофантовость перечислимых множеств*, Докл. АН СССР 191 (1970), pp. 279–282 = Soviet Math. Doklady 11 (1970), pp. 354–357.

[8] — *Диофантовы множества*, Успехи мат. наук 27:5 (1972), pp. 185–222 = Russian Math. Surveys 27:5 (1972), pp. 124–164.

[9] Hilary Putnam, *An unsolvable problem in number theory*, J. Symb. Logic 25 (1960), pp. 220–232 = Математика 8 : 5 (1964), pp. 55–67.

[10] Julia Robinson, *Existential definability in arithmetic*, Trans. Amer. Math. Soc. 72 (1952), pp. 437–449 = Математика 8: 5 (1964), pp. 3–14.

[11] Thoralf Skolem, *Diophantische Gleichungen*, Berlin 1938.

[12] Martin Davis, Yuri Matijasevič, and Julia Robinson, *Hilberts tenth problem. Diophantine equations: Positive aspects of a negative solution*, Proceedings of the Symposium on Hilbert's problems, Amer. Math. Soc. 1975.

STEKLOV MATHEMATICAL INSTITUTE
OF ACADEMY OF SCIENCES OF USSR
Leningrad

UNIVERSITY OF CALIFORNIA
Berkeley, California

Received on 28. 2. 1974 (538)

Proceedings of Symposia in Pure Mathematics
Volume 28, 1976

HILBERT'S TENTH PROBLEM. DIOPHANTINE EQUATIONS: POSITIVE ASPECTS OF A NEGATIVE SOLUTION

Martin Davis[1], Yuri Matijasevič, and Julia Robinson

ABSTRACT

Applications (including the negative solution of Hilbert's tenth problem) and extensions are surveyed of the Main Theorem on Diophantine sets: Every listable (recursively enumerable) set is Diophantine. Key steps in the proof of the Main Theorem are outlined and applied to obtain prime representing polynomials, a universal Diophantine equation, and a sharp form of Gödel's incompleteness theorem. Many famous problems are reduced to the solvability of Diophantine equations. The number, size and effectiveness of solutions are discussed. Relationships are explored with the theory of algorithms (recursion theory), model theory, and algebraic number theory.

INTRODUCTION

Hilbert's tenth problem asked for an algorithm to test polynomial Diophantine equations for solvability in integers (or equivalently in the natural numbers $N = \{0,1,2,3,\ldots\}$). The development of the theory of algorithms (also called recursion theory or computability theory) since the 1930's suggested that the tenth problem might be solved negatively, i.e. that it might be shown that there is no algorithm of the kind sought. Work in this direction had its fruition in 1970 when Matijasevič [34] gave just such a negative solution. Full accounts of this work are available in the expository articles in various languages: Azra [4], Davis [13], Manin [32], and Matijasevič [36]. Other expository articles on the subject are: Fenstad [16], Havel [21], Hermes [22], Ruohnen [52].

The methods used to achieve the negative solution of Hilbert's tenth problem have their positive side as well, and it is this positive side that we propose to stress in this article. (The key negative results are summarized in this introductory section.) From the perspective suggested by these methods, a Diophantine equation is viewed primarily as defining a set or relation on the natural numbers, in a manner which is explained below. This leads to considering the class of Diophantine sets, consisting of all

AMS (MOS) subject classifications (1970). Primary 02F25, 02F50, 10B99, 10N05; Secondary 10B05, 10B10, 12L05.

[1]Supported by the National Science Foundation under Grant DCR71-02039A 03.

sets which can be so defined. The Main Theorem is that the Diophantine sets are precisely those sets which can be *listed* (again this is explained more precisely below) by means of some definite algorithms. The equivalence of these two notions, one arising from the theory of numbers, the other in the theory of algorithms (also known as recursive function theory and as computability theory) permits a quite surprising interplay between the two fields, and leads to an entirely novel collection of problems and results in the theory of Diophantine equations.

In this paper we survey some of the problems and results which arise from this viewpoint. Although familiarity with one of the expository papers on Hilbert's tenth problem cited above will be helpful, we have not in fact assumed such knowledge on the reader's part. Nor does this paper presuppose any knowledge of the theory of algorithms.

If $P(a_1,\ldots,a_\mu,z_1,\ldots,z_\nu)$ is a polynomial with integer coefficients, in the variables $a_1,\ldots,a_\mu,z_1,\ldots,z_\nu$ we may think of the Diophantine equation

$$P(a_1,\ldots,a_\mu,z_1,\ldots,z_\nu) = 0 \tag{1}$$

as *defining* the set of μ-tuples $\langle a_1,\ldots,a_\mu\rangle$ of natural numbers for which (1) has a solution in natural numbers in the unknowns $z_1,\ldots,z_\nu$. (Here $a_1,\ldots a_\mu$ are the *parameters* of the equation.) Any set which can be defined in this way is called *Diophantine*.

We shall also have occasion to deal with *exponential Diophantine equations*, that is equations of the form (1) in which P is the sum of terms $c\alpha_1^{\beta_1}\alpha_2^{\beta_2}\cdots\alpha_k^{\beta_k}$ where c is an integer and where each α_i and each β_i is either a positive integer or one of the variables $a_1,\ldots,a_\mu,z_1,\ldots,z_\nu$. In such a case we speak of an *exponential Diophantine definition* of an *exponential Diophantine set*.

Now if S is the set of μ-tuples of natural numbers defined by (1), we can set up the following procedure for making a *list* of the members of S: Fix some definite enumeration of all $(\mu+\nu)$-tuples of natural numbers. (One such enumeration is defined later in this introduction.) Proceeding in order through the $(\mu+\nu)$-tuples, compute the value of P for each $(\mu+\nu)$-tuple; and whenever $P = 0$ for $\langle a_1,\ldots,a_\mu,z_1,\ldots,z_\nu\rangle$, place the μ-tuple $\langle a_1,\ldots,a_\mu\rangle$ on the list. Thus the members of S and only the members of S will be placed on the list (perhaps more than once).

A set S of μ-tuples of natural numbers is called *listable* (or *recursively enumerable)* if there is a well defined algorithm for making a list of precisely the members of S. Of course only finitely many members of S can be placed on the list in any finite number of steps. However what is crucial is that each member will *eventually* be placed on the list and that no nonmember will ever be placed on the list. So we have at once:

Every Diophantine set is listable.

Now while we will all agree that the method given above of listing a Diophantine set is an algorithm, we need a definition of algorithm to make our definition of listable set precise. Suitable definitions can be found in books on the theory of algorithms (e.g. Davis [9], Malcev [31], Markov [33], Rogers [51]). However readers of this paper who wish to learn the nature and scope of the results presented will manage perfectly well if they rely on the heuristic account just given.

The main result in the theory of Diophantine sets (hereafter called the Main Theorem) is (cf. Matijasevič [34]):

THEOREM. *Every listable set is Diophantine.*

Thus every set of μ-tuples of natural numbers which can be listed by means of any algorithm whatever can, in fact, be listed in the manner described above where P is a particular polynomial with μ parameters.

Actually the proof of this theorem is constructive and we will use this fact in what follows. Hence we restate the theorem indicating this.

MAIN THEOREM. *There is a procedure which can be applied to an arbitrary algorithm for lising a set S of μ-tuples of natural numbers to obtain a polynomial P with integer coefficients such that* $P(a_1,\ldots,a_\mu,x_1,\ldots,x_\nu) = 0$ *has a solution in natural numbers if and only if* $\langle a_1,\ldots,a_\mu\rangle \in S$.

A set $S \subseteq N$ is *computable* if there is an algorithm to determine in a finite number of steps whether an arbitrary natural number belongs to S. It is easy to see that if S is computable then both S and N-S are listable. Conversely, if both S and N-S are listable, then each number will occur on one or the other list eventually. So we can combine the algorithms for listing S and N-S into an algorithm for computing S, i.e. for telling whether an arbitrary natural number belongs to S.

A basic result in the theory of algorithms (see any of the books referred to above) is:

THEOREM. *There is a listable set* $K \subset N$ *which is not computable.*

It is this theorem which has made it possible to prove that various mathematical problems are algorithmically unsolvable. In §5, we will sketch the construction of such a K. Since K is listable, K is Diophantine. Let

$$P(a,z_1,\ldots,z_\nu) = 0 \tag{2}$$

be a *Diophantine definition* of K. That is, (2) has a solution in natural numbers $z_1,\ldots,z_\nu$ for a given $a \in N$ if and only if $a \in K$. Then we have

COROLLARY. *There is no algorithm for telling whether* (2) *has a solution in* $z_1,\ldots,z_\nu$ *for given values of the parameter* a.

This corollary gives a negative solution of Hilbert's tenth problem in a strong form: Not only is there no algorithm for testing Diophantine equations in general for solvability, but there is not even such an algorithm for a particular one parameter family of Diophantine equations.

Notice that this result has little practical effect on the problem of finding which Diophantine equations have solutions. This is because we can only hope to find the answer for sufficiently simple equations in general. For example, a is composite if and only if

$$(x + 2)(y + 2) = a \qquad (3)$$

has a solution. Now there is a trivial algorithm to determine whether (3) has a solution, since it is sufficient to try all values of x less than $\sqrt{a}$. But it is clear that this method cannot in fact be used to show that a is prime if a has say 50 digits. However the largest known prime $2^{19937} - 1$ is much larger than that and was found using a more sophisticated algorithm that also has its limitations [63]. Thus a given algorithm can be carried out only for a finite set of values of the parameters. Also we may be able to devise ad hoc methods for particular equations without a general algorithm. Thus the negative solution to Hilbert's tenth problem does not preclude our eventually being able to determine exactly which of the equations that can be written with one line of type have solutions.

By using a simple device due to Hilary Putnam [40], Diophantine sets of natural numbers can be defined in a particularly striking manner. Thus let S be the set of natural numbers a for which the equation $P(a,z_1,\ldots,z_\nu) = 0$, has a solution. Let

$$Q(x,z_1,\ldots,z_\nu) = (x+1)\{1 - P^2(x,z_1,\ldots,z_\nu)\} - 1 .$$

Then, S *consists precisely of the nonnegative values assumed by* Q *as* $x,z_1,\ldots,z_\nu$ *assume all values in* N. (For, $P(a,z_1,\ldots,z_\nu) = 0$ implies $Q(a,z_1,\ldots,z_\nu) = a$; moreover $Q(x,z_1,\ldots,z_\nu) = a \geq 0$ implies $1-P^2(x,z_1,\ldots z_\nu) > 0$ which implies $P(x,z_1,\ldots,z_\nu) = 0$ and $a = x$.) Since all the familiar sets of natural numbers which occur in the theory of numbers (such as the primes, the square-free numbers, or the perfect numbers) are listable and hence Diophantine, they can each be represented as the set of nonnegative values assumed by a polynomial.

It is important to remark that a Diophantine set can be defined by a simultaneous system of Diophantine equations just as well as by a single equation. This is because the system $P_1 = 0, P_2 = 0, \ldots, P_n = 0$ is equivalent to the single equation $P_1^2 + P_2^2 + \ldots + P_n^2 = 0$. (And, of course, the same holds for exponential Diophantine equations.) From the point of view of logic, this remark means that Diophantine definitions can be combined using the operation "&". The same is true for "v". Namely, the condition

$$P_1 = 0 \vee P_2 = 0 \vee \ldots \vee P_n = 0$$

is equivalent to the single equation

$$P_1 P_2 \cdots P_n = 0 .$$

We will have occasion to make use of Cantor's function:

$$J(x,y) = (1 + 2 + \ldots + (x+y)) + x$$
$$= ((x+y)^2 + 3x + y)/2 .$$

The equation $J(x,y) = a$ obviously has a unique solution $\langle x,y\rangle$ for each natural number a: x is determined as the excess of a over the largest "triangular" number which is $\leq$ a. Thus, the function $J(x,y)$ provides an enumeration of all ordered pairs ($J(x,y)$ is the number of the pair $\langle x,y\rangle$ in the enumeration). The unique solution of $J(x,y) = a$ is written $x = K(a)$,

$y = L(a)$, and the functions J, K, L are called *Cantor's pairing functions*.

Proceeding inductively, we define:

$$J_1(z_1) = z_1$$

$$J_{n+1}(z_1,\ldots,z_{n+1}) = J(J_n(z_1,\ldots,z_n),z_{n+1}) \ .$$

Then $J_2(z_1,z_2) = J(z_1,z_2)$, and generally the function J_n provides a one-one correspondence between the natural numbers and the n-tuples of natural numbers. It is important to notice that for each n, $J_n(z_1,\ldots,z_n)$ is a *polynomial with rational coefficients*. Hence if $J_n(z_1,\ldots,z_n)$ is substituted for a variable in a Diophantine equation, the result will also be a Diophantine equation.

The notions of computability and listability apply not only to μ-tuples of natural numbers, but also to any collection of objects to which algorithmic procedures can apply. What this turns out to mean is that we can speak of computability and listability in connection with objects which have natural representations as *expressions on an alphabet*, i.e. as finite sequences of a finite or countably infinite set of objects, called *symbols*. For example we can speak of a computable or listable set of strings on the symbols *a*, *b*, *c* (e.g. the set of such strings in which no *b* is immediately followed by *c* is computable and hence listable).

An instance (which is in some sense prototypical) is provided by *formalized mathematical theories*. Here the symbols are those needed for logical deductions ($\sim$ & $\vee$ $\rightarrow$ $\leftrightarrow$ $\forall$ $\exists$ =, parentheses, and a supply of variables) and those needed to express the basic concepts of the part of mathematics being "formalized" [in the case of number theory, these symbols might be e.g. 0, S (for the successor function), +, $\cdot$, <]. Among all of the expressions which can be written using these symbols a certain computable subset is singled out as the *formulas* (or w.f.f.'s) of the theory [e.g. $(\forall x)(x\cdot x = x \rightarrow (x = 0 \vee x = S(0)))$]. And once again a particular computable set of these formulas is singled out as the *mathematical axioms* of the theory (e.g. Peano's axioms). It is clear of course that in order to use a set of axioms we must be able to check that an alleged axiom really is one -- hence the requirement that the axioms form a computable set is essential. Finally the *theorems* of the theory are the formulas obtained by beginning with the axioms and applying rules of logical deduction. Now what is important is that *the set of theorems of a formalized mathematical theory is listable* (though in general it may not be computable).

To see this it is necessary to consider one of the various (equivalent) formulations of the rules of logic. For our purposes it is sufficient to remark that there exist formulations of logic (cf. Rosser [51a]) which begin by specifying a computable set of formulas called the *logical axioms*. One then employs the rule of *detachment* (also called modus ponens) which proceeds from A and (A $\rightarrow$ B) to obtain B, where A and B are each formulas of the theory. Thus in this formulation the *theorems* of the theory are those formulas obtained from the (logical and mathematical) axioms by iteratively employing the rule of detachment.

Now it is easy to describe an algorithm for making a list of the theorems of a theory: We begin with an algorithm for listing the axioms (thus we only use the listability of the axioms although we have assumed that they are computable). The algorithm we are constructing operates alternatively in two phases. In Phase I it employs the axiom listing algorithm to place one more axiom on our list of theorems. In Phase II it searches through the list of theorems so far obtained for pairs to which detachment can be applied; finding such a pair it applies detachment and adjoins the result to the list. It is clear that all theorems and only theorems will eventually be placed on the list; hence the theorems are listable.

1. DIOPHANTINE REPRESENTATION OF THE SET OF PRIMES.

As we mentioned in the introduction, one consequence of the Main Theorem is that there exists a polynomial P with integer coefficients such that the nonnegative range of P over the natural numbers is the set of primes. This corollary was deduced by Putnam in 1960 from the then conjectured Main Theorem and it was considered by some to be an argument against its plausibility. The set of primes is probably the most popular set in number theory and the quest for prime representing functions is an old one. It is remarkable that until 1970 it was not known that there is a representation of such a simple form. To construct a Diophantine definition of the set of primes, we need only a part of the machinery necessary for an arbitrary listable set. (In particular we do not need the Bounded Quantifier Theorem of §4.) In fact the whole construction is purely number-theoretical, which suggests that it may be more informative than Diophantine definitions of other sets. However to avoid disappointment, we should warn the reader that no magic is involved. The definition is really a direct translation of a familiar fact! However now that we know there is a Diophantine equation, which furnishes a Diophantine definition of the primes, we can look forward to successively simpler equations being found.

Our first aim is to find an exponential Diophantine definition of primes. We give here in an improved form the original definition by Robinson [46]. Clearly, p is prime if and only if

$$p = s + 1 = r + 2 \tag{1}$$

$$q = s! \tag{2}$$

$$ap - bq = 1 \tag{3}$$

has a solution for a, b, s, r, q in N (we could use Wilson's theorem and replace bq by q). Hence we need only transform (2) into an exponential Diophantine equation.

Now

$$s! = \frac{t(t-1)\ldots(t-s+1)}{\binom{t}{s}} \quad \text{for} \quad t \geq s .$$

Replacing the numerator by the exponential function t^s, we have

$$s! \leq s!(1 + \frac{1}{t-1}) \ldots (1 + \frac{s-1}{t-s+1}) = \frac{t^s}{\binom{t}{s}} \to s!$$

as $t \to \infty$. It is easy to verify that

$$s! = \left[\frac{t^s}{\binom{t}{s}} \right]$$

provided $t \geq 2\, s^{s+2}$. (As usual $[\alpha]$ is the integer n such that $n \leq \alpha < n+1$.) Hence we can replace (2) by

$$t = 2s^{s+2} \tag{4}$$

$$t^s = qu + w \tag{5}$$

$$u = \binom{t}{s} \tag{6}$$

$$u = w + x + 1 \tag{7}$$

since (5) and (7) have a solution in x, w if and only if $q = [t^s/u]$. It remains to transform (6) into an exponential Diophantine equation.

The equation

$$(y + 1)^t = \sum_{i=0}^{t} \binom{t}{i} y^i \tag{8}$$

may be thought of as defining the binomial coefficients when it is considered an identity in y. Also if y is sufficiently large (greater than 2^t) then the binomial coefficients are just the digits in the y-ary expansion of $(y + 1)^t$. Thus the numbers $\binom{t}{i}$ are uniquely defined by saying that they are nonnegative integers satisfying (8) for at least one value of $y > 2^t$. So we can replace (6) by

$$y = 2^t + 1 \tag{9}$$

$$z = y + 1 \tag{10}$$

$$z^t = \ell y^{s+1} + uy^s + m \tag{11}$$

$$u+v = 2^t \tag{12}$$

$$m+n+1 = y^s . \tag{13}$$

Thus, the system (1), (3)-(5), (7), (9)-(13) is an exponential Diophantine definition of the set of primes. (We can combine the system of equations into one equation as indicated in the Introduction.)

Notice that we have incidentally given an exponential Diophantine definition of the binomial coefficient. Indeed, for t and s not both 0,

$u = \binom{t}{s}$ if and only if (9)-(13) can be satisfied for $y, \ell, m, v, z, n \in N$.

Also we have an exponential Diophantine definition of $q = s!$. Namely,

$q = s!$ if and only if (4), (5), (7); (9)-(13) can be satisfied for $u, t, w, x, y, \ell, m, v, z, n \in N$.

To obtain a Diophantine definition of the set of primes and the relations $u = \binom{t}{s}$ and $q = s!$, it suffices now to obtain a Diophantine definition of the exponential function $a = b^c$. That is we need to find a polynomial A in three parameters and say ν unknowns such that

$$A(a,b,c,x_1,\ldots,x_\nu) = 0$$

has a solution if and only if $a = b^c$. It was shown in Robinson [46] that such an A exists provided that there is some Diophantine relation of exponential growth. The proof involved used elementary properties of the equation (with one parameter a):

$$x^2 + (a^2-1)y^2 = 1 \quad \text{for} \quad a > 0 \qquad (*)$$

As a specially simple Pell equation, much is known about the infinite set of its solutions. We will number the solutions in order of size $x = \chi_a(0)$, $y = \psi_a(0)$; $x = \chi_a(1)$, $y = \psi_a(1)$; ... . Matijasevič [34], [74], showed that the sequence of even Fibonacci numbers (which obviously grows exponentially) formed a Diophantine relation. However shortly afterwards several mathematicians [7], [11], [26] showed that the relation given by $y = \psi_a(k)$ is Diophantine. From this, the Diophantine definability of exponentiation follows directly. Here we will give a definition of $y = \psi_a(k)$ from Matijasevič and Robinson [39] which can be expressed with just 3 unknowns. Here and below we write e.g. $x = \square$, to mean: x is a perfect square.

THEOREM. $c = \psi_a(b)$ *if and only if the system*

$$DFI = \square \ , \quad F \mid H-c \ , \quad b \le c \qquad (a > 1)$$

has a solution in the unknowns i *and* j *where* D, F, H, I *are defined as follows:*

$$\begin{aligned} D &= (a^2-1)c^2 + 1 \\ E &= 2(i+1)Dc^2 \\ F &= (a^2-1)E^2 + 1 \\ G &= a + F(F-a) \\ H &= b + 2jG \\ I &= (G^2-1)H^2 + 1 . \end{aligned}$$

Finally, the easily obtained estimate for the nth solution of (*),

$$(2a-1)^n \le \psi_a(n+1) \le (2a)^n$$

enables us to define the exponential function quite simply in terms of ψ. Clearly, for $x > 0$,

$$\frac{\psi_{Mx}(n+1)}{\psi_M(n+1)} \to x^n \quad \text{as} \quad M \to \infty ,$$

and hence for sufficiently large M the equation $y = x^n$ is equivalent to

$$\left(y - \frac{\psi_{Mx}(n+1)}{\psi_M(n+1)}\right)^2 < \frac{1}{4} . \qquad (**)$$

The explicit construction of a Diophantine definition of $y = x^n$ with just 5 unknowns is described at length in [39].

In fact, if we are interested only in a Diophantine definition of primes then all we have to do is to find a good approximation to $t^s/\binom{t}{s}$. This can be done directly in terms of ψ and leads to a polynomial in 12 variables whose positive range is the set of primes. In [35], the first such polynomial published had 24 unknowns and was of the 37th degree; this was improved in the English translation to 21 variables and the 21st degree. H. Wada independently found a polynomial with 12 variables. J. P. Jones (cf. [25]) has been trying to minimize the number of symbols necessary to write down such a polynomial. His best result so far is 325 symbols in the following

THEOREM (Jones). *The set of primes is identical with the set of positive values assumed by the polynomial*

$$\begin{aligned}&(k+2)\{1-([wz+h+j-q]^2+[(gk+2g+k+1)\cdot(h+j)+h-z]^2+[16(k+1)^3\cdot(k+2)(n+1)^2\\&+1-f^2]^2+[2n+p+q+z-e]^2+[e^3\cdot(e+2)\cdot(a+1)^2+1-o^2]^2+[(a^2-1)y^2+1-x^2]^2\\&+[16r^2y^4\cdot(a^2-1)+1-u^2]^2+[((a+u^2\cdot(u^2-a))^2-1)\cdot(n+4dy)^2+1-(x+cu)^2]^2\\&+[(a^2-1)\ell^2+1-m^2]^2+[ai+k+1-\ell-i]^2+[n+\ell+v-y]^2+[p+\ell(a-n-1)\\&+b(2an+2a-n^2-2n-2)-m]^2+[q+y(a-p-1)+s\cdot(2ap+2a-p^2-2p-2)-x]^2\\&+[z+p\ell(a-p)+t(2ap-p^2-1)-pm]^2)\}\end{aligned}$$

as the variables range over natural numbers.

Can we learn anything about primes from the fact that they are Diophantine? Definitely yes see §6.

2. FAMOUS PROBLEMS

In this section we shall see that the Main Theorem not only implies that there is no algorithm for solving Hilbert's tenth problem, but also shows that the tenth problem is far more all-embracing than one would have suspected. That is, many famous problems, even some explicitly mentioned in Hilbert's address, can be transformed, by using the Main Theorem, into assertions to the effect that some particular Diophantine equation has no solution.

A simple case is Fermat's conjecture that the equation

$$x^n + y^n = z^n \tag{1}$$

has no positive integer solution with $n > 2$. Of course a positive solution of the tenth problem would have enabled us to settle each instance of (1) with $n = 3,4,\ldots$. But (cf. §1) we can give explicitly a Diophantine definition of the exponential function, that is we can exhibit a polynomial $A(a,b,c,w_1,\ldots,w_k)$ such that the Diophantine equation

$$A(a,b,c,w_1,\ldots,w_k) = 0$$

is solvable in unknowns $w_1,\ldots,w_k$ if and only if the parameters a, b, c satisfy the condition $a = b^c$. Thus equation (1) is solvable for the same values of n for which there are solutions of equation

$$A^2(p,x,n,w_1,\ldots,w_k) + A^2(q,y,n,v_1,\ldots,v_k) + A^2(p+q,z,n,u_1,\ldots,u_k)=0 \tag{2}$$

Hence, an equivalent formulation of Fermat's conjecture is the assertion that the Diophantine equation

$$A^2(p,x+1,n+3,w_1,\dots,w_k) + A^2(q,y+1,n+3,v_1,\dots,v_k) + A^2(p+q,z,n+3,u_1,\dots u_k) = 0$$

has no solution in nonnegative integers $n, p, q, u_1,\dots,u_k, v_1,\dots,v_k, w_1,\dots,w_k, x, y, z$. Thus the validity not only of each special case, but of the entire conjecture could be established by a supposed algorithm for testing an arbitrary Diophantine equation.

We would easily write out (2) explicitly. In fact, Keijo Ruohnen [52] did this, based on the construction of A described in Matijasevič [34]. Ruohnen's equation (more precisely, system of three equations which could be easily combined into one equation) has 72 unknowns and occupies less than a page. Today implementing the ideas of Matijasevič and Robinson [39] where a simpler construction of A is given, one can construct (2) with fewer than a dozen unknowns.

More interesting and less evident is the situation with Goldbach's conjecture mentioned in the eighth problem. The conjecture states that every even integer greater than 2 is the sum of two prime numbers. Now we can construct a Diophantine equation

$$B(p,w_1,\dots,w_k) = 0 \tag{3}$$

which is solvable in unknowns $w_1,\dots,w_k$ if and only if p is prime. It is easy to check that the equation

$$(u+1)(1-B^2(p_1,w_1,\dots,w_k) - B^2(p_2,w'_1,\dots,w'_k) - (2u+4-p_1-p_2)^2 - t) = a \tag{4}$$

is solvable for all nonpositive a (with $u = 0$, $t = -a$, $p_1 = p_2 = 2$, values of $w_1,\dots,w_k$, $w'_1,\dots,w'_k$ determined by the values of a solution of (3) with $p = 2$) and for those and only those positive a for which $2a+2$ is the sum of two primes (with $u = a-1$, $t = 0$, $p_1+p_2 = 2a+2$, values of $w_1,\dots,w_k$, $w'_1,\dots,w'_k$ determined by the solutions of (3) with $p = p_1$ and $p = p_2$ correspondingly). Thus we can reformulate Goldbach's conjecture as the assertion that the polynomial on the left side of (4) represents every integer when the variables range over the nonnegative integers. Replacing each variable by the sum of squares of four new variables we obtain a new polynomial which represents every integer when the variables range over the integers if and only if the Goldbach conjecture is valid.

So far we have reduced Goldbach's conjecture to Diophantine equations, but the conjecture is still outside of the scope of the tenth problem, since Hilbert asked for treatment of single equations, not infinite systems of them. Nevertheless we can reduce Goldbach's conjecture to the case of a single Diophantine equation. Before dealing with the conjecture itself, we consider this possibility in a general setting.

If we have an algorithm to tell in a finite number of steps whether an arbitrary natural number has a particular property P (written $P(n)$), then we can reduce the assertion that every natural number has the property P to the unsolvability of a particular Diophantine equation and give instructions for writing it down. Namely, from the given algorithms, the Main Theorem, and

Putnam's device mentioned in the Introduction, we can find polynomials R and S such that

$$P(n) \leftrightarrow (\exists w_1 \dots w_k)[R(w_1,\dots,w_k) = n]; \sim P(n) \leftrightarrow (\exists z_1 \dots z_m)[S(z_1,\dots z_m)=n].$$

Then the problem under discussion can be reformulated as the assertion that the Diophantine equation

$$R(w_1,\dots,w_k) = n$$

is solvable for each natural number n or equivalently as the assertion that the Diophantine equation

$$S(z_1,\dots,z_m) = z_0$$

has no solution at all.

Now returning to the Goldbach conjecture, we can easily give an algorithm for testing whether 2n+4 is the sum of two primes and hence Goldbach's problem can be formulated not only in the form (4) but also as the assertion that some single Diophantine equation has no solution. Thus a positive solution of the tenth problem would have enabled us (in principle) to settle the Goldbach conjecture.

Many other famous problems even some remote from number theory can be formulated in the form $(\forall n)P(n)$ with a suitable decidable P. For example, the classical four color conjecture in the theory of graphs states that every map on a plane or sphere can be colored in 4 colors in such a manner that no two adjacent countries would be colored the same. (For exact definitions and numerous equivalent formulations, cf. Saaty [53].) It is well known that the four color conjecture is equivalent to the statement that every positive integer n has the property: "every map with fewer than n countries can be properly colored in four colors." Since this property is obviously decidable, we see that there is some particular Diophantine equation whose unsolvability is equivalent to the four color conjecture. In §4 we show how such an equation can be explicitly obtained.

Another example is the problem of the consistency of a given axiomatic system mentioned in general form in Hilbert's second problem. A formalized mathematical theory (cf. the Introduction) is called *consistent* if there is no formula R such that R and $\sim$R are both theorems. We have already seen that the set of theorems of a formalized mathematical theory is a listable set. Now for a given formalized mathematical theory consider the following property P of a natural number n: no contradiction arises when n steps of some definite theorem-listing algorithm (e.g. that outlined in the Introduction) for the theory are carried out. Clearly, we have an algorithm for P: we need only scan a finite list for a pair of formulas, R, $\sim$R. Since the assertion that the theory is consistent is just the statement that every positive integer has property P, the consistency of the given formalized mathematical theory is also equivalent to the assertion that some definite Diophantine equation has no solutions.

Hilbert's first problem is Cantor's continuum hypothesis. The difficulties of settling this problem led to consideration of the questions: is the continuum hypothesis consistent with the traditional Zermelo-Fraenkel axioms of set theory? is its negation? By our remarks, each of these questions (in fact both were answered affirmatively by Gödel [19] and Cohen [6] respectively) is equivalent to the unsolvability of a specific Diophantine equation.

We mention in passing that various so called large cardinal axioms studied by contemporary set theorists imply the consistency of certain sets of axioms. Thus from the existence of a *strongly inaccessible cardinal* it is known that the consistency of Zermelo-Fraenkel set theory follows. Hence there is a Diophantine equation such that the existence of a strongly inaccessible cardinal implies that the equation has no solution, whereas it cannot be proved from the Zermelo-Fraenkel axioms alone that the equation has no solution.

Now we proceed to consider in some detail another famous problem, namely Riemann's conjecture about the ζ function mentioned in Hilbert's eighth problem. It is not as evident as before that the conjecture can be formulated in the desired form $\forall n\ P(n)$. Nevertheless it has long been realized that this is the case. (Kreisel outlined a proof in [28] based on his earlier work on computable estimates for analytic functions.)

Riemann's function $\zeta(s)$ is defined by

$$\zeta(s) = \sum_{n=1}^{\infty} n^{-s} .$$

The series converges for all s with $\mathrm{Re}(s) > 1$ to an analytic function which can be extended to the entire complex plane except for $s = 1$ where ζ has its unique pole. Points $-2,-4,-6,\ldots$ are known to be zeros of ζ. All other zeros are called *nontrivial*. They are known to lie inside the so-called critical strip $0 < \mathrm{Re}(s) < 1$ symmetrically about the line $\mathrm{Re}(s) = 1/2$. The Riemann hypothesis states that all the nontrivial zeros in fact lie on the line $\mathrm{Re}(s) = 1/2$. By symmetry, the Riemann hypothesis is equivalent to the statement that $\zeta(s)$ has no zeros for $\frac{1}{2} < \mathrm{Re}(s) < 1$.

To see heuristically that the Riemann hypothesis has the desired form, note that we can express the region $\frac{1}{2} < \mathrm{Re}(s) < 1$ as the union of countably many bounded regions each with rectilinear boundary. The number of zeros of $\zeta(s)$ in each of the regions is given by the integral of $\zeta'(s)/2\pi i\zeta(s)$ taken about its boundary (if there is a zero on one of the boundaries, we can replace the region by a neighboring one). So the Riemann hypothesis is equivalent to the statement that for each such bounded region the corresponding integral is less than (say) $\frac{1}{2}$. To see whether the n-th region has this property is basically an elementary exercise in numerical analysis, finding appropriate rational approximations for $\zeta'(s)/2\pi i\zeta(s)$ and for its integral. However the details are cumbersome and we propose to use a more indirect method which relies on the classical theory of the distribution of primes.

We introduce the computable function, defined for positive integers:

$$\delta(x) = \prod_{n<x} \prod_{j\le n} \eta(j)$$

where $\eta(j) = 1$ unless j is a prime power, and $\eta(p^k) = p$. Then, we have:

THEOREM. *The Riemann hypothesis is equivalent to the assertion that*

$$\left(\sum_{k\le\delta(n)} \frac{1}{k} - \frac{n^2}{2}\right)^2 < 36\, n^3 \quad \textit{for} \quad n = 1,2,3,\ldots . \tag{5}$$

(As the reader will readily verify, the constant 36 can easily be improved.)

Obviously the condition (5) is a decidable property of n. And, it would be quite a simple matter, using the methods of proof of the Main Theorem, to obtain explicitly a Diophantine equation which has no solutions, just in case (5) always holds.

In proving the theorem,[2] we make use of the following facts.

(a) $\frac{\Gamma'(3/2)}{\Gamma(3/2)} = 2 - \gamma - \log 4$, where $\gamma = .577\ldots$ is Euler's constant. (Cf. Magnus, Oberhettingher, & Soni [30], p. 15.)

(b) Letting R be the set of nontrivial zeros of $\zeta(s)$,

$$\frac{\zeta'(s)}{\zeta(s)} = \log 2\pi - 1 - \frac{\gamma}{2} - \frac{1}{s-1} - \frac{1}{2}\,\frac{\Gamma'(\frac{s}{2}+1)}{\Gamma(\frac{s}{2}+1)} + \sum_{\rho\in R}\left(\frac{1}{s-\rho} + \frac{1}{\rho}\right) .$$

$$\lim_{s\to 1}\left(\frac{\zeta'(s)}{\zeta(s)} + \frac{1}{s-1}\right) = \gamma. \quad \text{(Cf. Landau [29], pp. 165, 316-317.)}$$

(c) For positive real x, we write

$$\psi(x) = \sum_{p^k\le x} \log p = \sum_{j\le x} \log \eta(j)$$

$$\psi_1(x) = \int_1^x \psi(u)\, du ,$$

so that for x integral, $\psi(x) \le \sum_{j\le x} \log x = x \log x \le x^{3/2}$ and $\psi_1(x) = \log \delta(x)$. Then for $x\ge 1$,

$$\psi_1(x) = \frac{x^2}{2} - \sum_{\rho\in R} \frac{x^{\rho+1}}{\rho(\rho+1)} - x\,\frac{\zeta'(0)}{\zeta(0)} + \frac{\zeta'(-1)}{\zeta(-1)} - \sum_{r=1}^{\infty} \frac{x^{1-2r}}{2r(r-1)} .$$

(Cf. Ingham [24], p. 73.)

(d) $\zeta'(0)/\zeta(0) = \log 2\pi$, $\zeta'(-1)/\zeta(-1) = 1.985\ldots$. (Cf. Walther [65], p. 400.)

(e) For n a positive integer,

$$\left(\sum_{k\le n} \frac{1}{k}\right) - 1 < \log n < \sum_{k\le n} \frac{1}{k} ,$$

as is readily seen on estimating $\log n = \int_1^n du/u$. We have:

---

[2] We are grateful to Harold N. Shapiro who suggested this approach and provided most of the technical details.

LEMMA.

$$\sum_{\rho\in R} \frac{1}{\rho(1-\rho)} = \gamma + 2 - \log 4\pi .$$

Proof: Using (a) and (b),

$$\sum_{\rho\in R} \frac{1}{\rho(1-\rho)} = \sum_{\rho\in R} \left(\frac{1}{1-\rho} + \frac{1}{\rho}\right)$$

$$= \lim_{s\to 1} \left(\frac{\zeta'(s)}{\zeta(s)} + \frac{1}{s-1}\right) - \log 2\pi + 1 + \frac{\gamma}{2} + \frac{1}{2}\,\frac{\Gamma'(3/2)}{\Gamma(3/2)}$$

$$= \gamma + 2 - \log 4\pi .$$

Now, *assume the Riemann hypothesis*; i.e. $\rho \in R$ implies $\mathrm{Re}(\rho) = 1/2$. Then for $\rho \in R$, $\bar{\rho} = 1-\rho$. Hence, using the lemma,

$$\sum_{\rho\in R} \frac{1}{|\rho|\cdot|\rho+1|} \le \sum_{\rho\in R} \frac{1}{|\rho|^2} = \sum_{\rho\in R} \frac{1}{\rho(1-\rho)} = \gamma + 2 - \log 4\pi .$$

We can use this estimate in the expression for $\psi_1(x)$ given in (c), noting that the Riemann hypothesis implies that for $x \ge 1$, $\rho \in R$, we have $|x^{\rho+1}| = x^{3/2}$, so that

$$|\psi_1(x) - \frac{x^2}{2}| \le x^{3/2}\left[\sum_{\rho\in R} \frac{1}{|\rho|\cdot|\rho+1|} + \left|\frac{\zeta'(0)}{\zeta(0)}\right| + \left|\frac{\zeta'(-1)}{\zeta(-1)}\right| + \sum_{r=1}^{\infty} \frac{1}{2r(2r-1)}\right]$$

Here,

$$\sum_{r=1}^{\infty} \frac{1}{2r(2r-1)} = \sum_{r=1}^{\infty} \left(\frac{1}{2r-1} - \frac{1}{2r}\right) = \sum_{n=1}^{\infty} \frac{(-1)^{n+1}}{n} = \log 2 ,$$

so using (d),

$$|\psi_1(x) - \frac{x^2}{2}| \le x^{3/2}\left[\gamma + 2 - \log 4\pi + \log 2\pi + 2 + \log 2\right]$$

$$= x^{3/2}[4 + \gamma]$$

$$< 5x^{3/2}$$

But by (e)

$$\left|\sum_{k\le\delta(x)} \frac{1}{k} - \psi_1(x)\right| < 1 .$$

Hence

$$\left|\sum_{k\le\delta(x)} \frac{1}{k} - \frac{x^2}{2}\right| < 1 + 5x^{3/2} \le 6x^{3/2} .$$

Squaring now gives (5).

Conversely, suppose that (5) holds for all positive integers. Then by (e),

$$|\psi_1(x) - \frac{x^2}{2}| < 1 + 6x^{3/2}$$

for $x = 1,2,3,\ldots$. Thus, for any *real* $x \geq 1$,

$$|\psi_1(x) - \frac{x^2}{2}| \leq \int_{[x]}^{x} \psi(u)\,du + |\psi_1([x]) - \frac{[x]^2}{2}| + |\frac{[x]^2-x^2}{2}|$$

$$< \psi([x]) + 1 + 6x^{3/2} + x \leq 9x^{3/2} ,$$

using the inequality in (c). Now (cf. Ingham [24], p. 18) for $\mathrm{Re}(s) > 1$,

$$-\frac{\zeta'(s)}{\zeta(s)} = s\int_1^\infty \frac{\psi(x)}{x^{s+1}}\,dx$$

$$= s\int_1^\infty \frac{d\psi_1(x)}{x^{s+1}}$$

$$= s(s+1)\int_1^\infty \frac{\psi_1(x)}{x^{s+2}}\,dx .$$

Hence,

$$-\frac{\zeta'(s)}{\zeta(s)} - \frac{s(s+1)}{2(s-1)} = s(s+1)\int_1^\infty \frac{\psi_1(x) - \frac{x^2}{2}}{x^{s+2}}\,dx \qquad (6)$$

So, assuming our inequality,

$$\left|\frac{\psi_1(x) - \frac{x^2}{2}}{x^{s+2}}\right| \leq \frac{9}{\left|x^{s+\frac{1}{2}}\right|}$$

Hence, the integral in (6) converges uniformly on the closed region $\mathrm{Re}(s) \geq \sigma_0$ for any $\sigma_0 > 1/2$, (6) must be valid for $1/2 < \mathrm{Re}(s)$, and the function on the left is analytic in this domain. But this implies that $\zeta(s)$ does not vanish for $1/2 < \mathrm{Re}(s) < 1$, i.e. that the Riemann hypothesis holds.

Of course the Main Theorem does not permit us to reduce every famous problem to the unsolvability of a Diophantine equation. For example, let us consider the twin prime conjecture:

$$(\forall n)\{(\exists p)[p>n \ \&\ p \text{ prime} \ \&\ p+2 \text{ prime}]\} . \qquad (7)$$

Note that the set of n's satisfying the condition in braces is certainly a computable set. This set is either the set of all natural numbers or it is finite. Nevertheless we cannot reduce the twin prime conjecture to the unsolvability of a particular Diophantine equation because we don't know which of these alternatives holds and therefore don't know a specific algorithm for deciding whether n satisfies the condition. However if we strengthen the conjecture:

$$(\forall n)\{(\exists p)[n+4<p<2^{n+4} \ \&\ p \text{ prime} \ \&\ p+2 \text{ prime}]\} , \qquad (8)$$

then we will have an algorithm to determine whether n satisfies the condition in braces. Therefore this strong twin prime conjecture can be transformed into the unsolvability of a particular Diophantine equation. Number theorists certainly believe that both the twin prime conjecture and this stronger version are true.

The most intriguing question concerning such transformations of problems into the unsolvability of particular Diophantine equations is the one of usefulness: Is there any mathematical advantage to be gained by such a translation or does it represent a mere curiosity? One of the charms of mathematics is the constant discovery of unexpected almost unbelievable connections. Whatever is logically possible may be true! Diophantine sets as a class have been studied for only 25 years. Their richness has surprised experts and perhaps their usefulness will also.

Of course not much is to be expected from the routine uniform translation of a famous problem into its equivalent Diophantine equation. An approach which may be more promising is to look for decidable conditions sufficient to imply that a given Diophantine equation has no solution and then try to verify (possibly with the aid of a computer) the validity of one of these conditions for the equations under consideration. The simplest condition of this kind is the unsolvability of the equation modulo p. However our present method of obtaining a Diophantine definition of a listable set uses as an intermediate step inequalities $P \leq Q$ which are then replaced by $P+x = Q$ where x is a new unknown. But the corresponding congruences $P+x \equiv Q \pmod p$ always have a solution for x. Thus passing to the equation modulo p eliminates the condition $P \leq Q$ and hence could not lead to a proof of unsolvability of an equation corresponding to a serious problem. This suggests that we should look for new proofs of the Main Theorem in the hope of obtaining equations which would not automatically have solutions modulo p for every p. See [47].

Another possibility is that eventually Diophantine equations and definitions may be classified and studied so that the reduction of a problem to an equation of a particular type may in fact yield additional information. For example sets defined with polynomial test functions (§9) may have certain combinatorial or density properties which then would follow for any set so defined. Of course this can only be expected to work if the translation is simpler than it would be in the general case.

Finally, the translation of a theorem of the appropriate form in some part of mathematics shows that the corresponding Diophantine equation has no solution. Hence whatever methods went into proving the theorem can in fact be used to show that a particular Diophantine equation has no solution. It is possible that the same methods can be used to show that a class of equations including perhaps an equation of interest in itself are unsolvable. Such an example providing a new tool for solving Diophantine equations would be a considerable breakthrough. In any case, any mathematical method that has been used to prove a theorem of the appropriate form has in fact been used to show that a particular Diophantine equation has no solution. Thus all mathematical methods can be tools in the theory of Diophantine equations and perhaps we should consciously attempt to exploit them.

3. EFFECTIVENESS AND EQUATIONS WITH FINITELY MANY SOLUTIONS

The negative solution of Hilbert's tenth problem has given impetus to the treatment of other arithmetical problems from the algorithmic point of view. In this section, we deal with the problem of effectiveness in solving equations in integers.

Suppose that we have somehow proved that the Diophantine equation

$$P(x_1,\dots,x_n) = 0$$

has at most finitely many solutions. Depending on the nature of our proof, we can have very different knowledge about the actual solutions. In the luckiest case we have the list of all the solutions; but if reductio ad absurdum has been used then often we have nothing more than a deduction of the contradiction.

As a classical example of the latter situation we can point to the famous theorem by Thue according to which a Diophantine equation of the form

$$F(x,y) = a \tag{1}$$

has at most finitely many solutions where F is an irreducible form of degree at least 3. The original proof by Thue gives no way to find all the solutions and it is only recently, exploiting new ideas, that Baker [5] has given another proof enabling us to calculate (in principle) all the solutions of (1) for a given a.

The methods of the theory of algorithms enable us to construct very easily a computable function ϕ such that for every value of the parameter a,

$$\phi(a,x_1,\dots,x_n) = 0 \tag{2}$$

has at most a finite number of solutions, but there exists no computable function τ such that

$$\phi(a,x_1,\dots,x_n) = 0 \Rightarrow x_1 + \dots + x_n \le \tau(a) \ . \tag{3}$$

Thus there is no uniform procedure which can be used given a, to calculate the list of all the corresponding solutions of (2).

The straightforward method for constructing such a ϕ leads to a very complicated function. The question arises: How simple can a function ϕ with the above properties be? For example, could equation (2) be a Diophantine equation? Baker's result shows that ϕ cannot be of the form $F(x_1,x_2) - a$ with F as above. On the other hand it is shown in Matijasevič [37] that (2) can actually be taken to be an exponential Diophantine equation.

More precisely, it was shown that we can effectively find a polynomial A with integer coefficients such that

(i) for every natural number value of the parameter a the equation

$$A(a,x_1,\dots,x_n) = y + 4^y \tag{4}$$

has at most one solution in natural numbers $x_1,\ldots,x_n,y$; hence for every a there is an integer t_a such that equation (4) implies:

$$x_1 + \ldots + x_n + y < t_a ,$$

(ii) for each computable function τ there are natural numbers $a,x_1,\ldots,x_n,y$ such that (4) holds but $x_1 + \ldots + x_n + y > \tau(a)$.

In a sense, the negative answer to Hilbert's tenth problem justifies the past treatment of individual Diophantine equations by ad hoc methods (instead of developing a general theory which is impossible). In a similar sense the above result justifies our proving the finiteness of the number of solutions of an exponential Diophantine equation by reductio ad absurdum and other noneffective methods; these methods may well be the best available to us because universal methods for giving all the solutions do not exist.

Let us outline the main ideas in constructing the polynomial A. Suppose (as can readily be arranged) that (2) has at most one solution for every a, but still there is no computable τ satisfying (3). Then there is no procedure for determining given a whether (2) is solvable. In fact, otherwise we could define τ in the following way: apply the procedure to find if there is a solution; if there isn't, then put $\tau(a) = 0$, if there is, then look over in succession all the n-tuples until finding one satisfying (2); then put $\tau(a) = x_1 + \ldots + x_n$.

On the other hand, if (2) has for every a at most one solution and there is no procedure to determine whether it is solvable then there is no computable τ satisfying (3). This is clear since there are only finitely many n-tuples satisfying $x_1 + \ldots + x_n \leq \tau(a)$ and only such an n-tuple could be the solution.

The situation here is rather like the situation in Thue's proof. His method enables us to find a bound for all the solutions of (3) as soon as we find one solution which is sufficiently (in an appropriate sense) large. But until Baker's result, no method was known for determining whether such a solution exists.

Now our purpose is to find an algorithmically unsolvable one-parameter equation which never has more than a single solution. We start from a one-parameter Diophantine equation

$$P(a,z_1,\ldots,z_k) = 0 \tag{5}$$

for which there is no procedure for determining given a whether there is a solution in natural numbers $z_1,\ldots,z_k$ (cf. the Introduction). We shall construct an exponential Diophantine equation among whose unknowns will be $z_1,\ldots,z_k$. While both equations will have solutions for the same values of a, the new equation will have at most one solution for given a, and indeed one in which $z_1,\ldots,z_k$ will form the least solution of (5) while the other unknowns can easily be expressed in terms of $z_1,\ldots,z_k$. By "the least solution" we mean the least one in Cantor's enumeration of all the k-tuples, i.e. the one for which $J_k(z_1,\ldots,z_k)$ is least (cf. the Introduction).

The k-tuple $\langle z_1,\ldots,z_k\rangle$ is the least solution if and only if it actually is a solution and for each value of w less than $J_k(z_1,\ldots,z_k)$ the following equation has a solution.

$$(J_k(v_1,\ldots,v_k)-w)^2 + (P^2(a,v_1,\ldots,v_k)-u-1)^2 = 0 .$$

Note that $v_1,\ldots,v_k,u$ are uniquely determined by w.

A method for reducing the problem of the solvability of a Diophantine equation for all values of one of the variables (in this case w) not exceeding some bound to the problem of solving an exponential Diophantine equation has been known since 1961 (cf. Davis, Putnam, Robinson [14]): several improvements of this original method are known today (cf. §4), but they all lead to exponential Diophantine equations which have either infinitely many solutions or no solutions. Nevertheless, the original method can easily be modified to give an exponential Diophantine equation having at most one solution provided that the original Diophantine equation has at most one solution for each value of the variable w up to the bound imposed on it.

Obtaining the desired equation of the special form (4) is now a matter of easy transformations involving the introduction of new unknowns. For details see Matijasevič [37].

The above result suggests that it is worthwhile to introduce the notion of *singlefold* (Russian: odnokratno) *representation*. Namely, we say that an existential representation

$$(\exists x_1 \ldots x_n)\ R(a_1,\ldots,a_m,x_1,\ldots,x_n)$$

for some predicate $P(a_1,\ldots,a_m)$ is a singlefold representation if not only

$$P(a_1,\ldots,a_m) \leftrightarrow (\exists x_1 \ldots x_n)\ R(a_1,\ldots,a_m,x_1,\ldots,x_n) \tag{6}$$

but also

$$P(a_1,\ldots,a_m) \leftrightarrow (\exists !\ x_1 \ldots x_n)\ R(a_1,\ldots,a_m,x_1,\ldots,x_n) \tag{7}$$

where $(\exists !\ x_1\ldots x_n)$ means as usual "there is a unique n-tuple $x_1,\ldots,x_n$ such that". Note that (7) does not imply (6) and that $(\exists x_1 x_2)\ R(a,x_1,x_2)$ being singlefold implies $(\exists x_1)[(\exists x_2)\ R(a,x_1,x_2)]$ being singlefold but not conversely.

In this terminology the main result of Matijasevič [37] is that every listable predicate has a singlefold exponential Diophantine representation. It would be very interesting to extend this result to singlefold Diophantine representations and thus to show that for (2) one could take a Diophantine equation. To do this it would suffice to show that the relation $a = b^c$ (even with fixed $b \geq 2$) has a singlefold Diophantine representation and then apply the main result of Matijasevič [37]. Unfortunately, all of the known Diophantine definitions of $a = b^c$ are infinitefold, and there is no obvious way to modify any of them into a singlefold or even into a finitefold representation.

OPEN PROBLEM. *Is there a finitefold (or better a singlefold) Diophantine definition of* $a = b^c$?

An application of such a definition is mentioned in §13.

In this connection we wish to draw attention to a result which preceded Matijasevič's discovery of the existence of Diophantine definitions of $a = b^c$. It was shown in Davis [10] that a Diophantine representation for $a = 2^c$ could be constructed if the equation

$$9(u^2+7v^2)^2 - 7(r^2+7s^2)^2 = 2 \tag{8}$$

has no solution except for the trivial solution $u = r = 1$, $v = s = 0$. The conjecture about the uniqueness of the solution of (8) was refuted by Herrmann in [23]. However this does not spoil the approach entirely: indeed it can be shown that a *singlefold* Diophantine representation of $a = 2^c$ can be constructed if the equation (8) has only a *finite number* of solutions.

We conclude this section by suggesting that a particular arithmetical problem be considered from the algorithmic point of view. Thue's theorem mentioned above was obtained by him as a consequence of the following theorem on Diophantine approximation: for every algebraic number θ of degree n the inequality

$$|\theta - \frac{p}{q}| < \frac{1}{q^\nu} . \tag{9}$$

has at most finitely many solutions in integers p,q provided $\nu > n/2 + 1$. This was a sharpening (for the case $n > 2$) of a similar result by Liouville with $\nu > n$. Thue's result was in turn sharpened by various authors. The latest and the best possible result (due to Roth) is for the case $\nu > 2$. Thue's proof and the proofs of all the improvements suffer from noneffectiveness -- they give no way to find all the solutions. It may actually be the case that there is no algorithm given θ and $\nu > 2$ for finding all the solutions of (9). If so, it seems very difficult to prove the fact.

4. THE BOUNDED QUANTIFIER THEOREM

In 1931 Gödel [18] revolutionized mathematical logic when he showed that no system of axioms is sufficient to decide all statements of number theory correctly (see §5). In the course of the proof, he needed an arithmetically definable way of representing arbitrary finite sequences of natural numbers. Gödel's elementary solution of this problem using the Chinese remainder theorem is a cornerstone of the negative solution of Hilbert's tenth problem. We shall state the result here in an explicit form needed later.

GÖDEL'S LEMMA. *For every* a *and sequence* $a_1,\ldots,a_n$ *with each* $a_i < a$ *there is a uniquely determined* b *such that*

$$b < \prod_{i=1}^{n} (1 + n!ai)$$

and

$$a_i = \mathrm{Rem}(b,\ 1+n!ai) \quad \textit{for} \quad i = 1,\dots,n.$$

(Here $\mathrm{Rem}(b,c)$ is the least nonnegative remainder when b is divided by c.)

Now suppose $P(a,z_1,\dots,z_\nu) = 0$ defines a Diophantine set S whose complement $N-S$ is not Diophantine. Then

$$a \in N-S \leftrightarrow (\forall k)(\exists z_0 \dots z_\nu)[k = J_\nu(z_1,\dots,z_\nu) \ \&\ P(a,z_1,\dots,z_\nu)^2 = 1+z_0].$$

Here J_ν is the ν-tuple enumerating function defined in the Introduction. Thus, there is a polynomial Q such that the set T defined by

$$a \in T \leftrightarrow (\forall k)(\exists z_0 \dots z_\nu)[Q(a,k,z_0,\dots,z_\nu) = 0]$$

is not Diophantine i.e. not listable. However we shall show that if we put a bound x on k, we get a Diophantine relation $a \in T_x$ with

$$a \in T_x \leftrightarrow (\forall k)_{\le x}(\exists z_0 \dots z_\nu)[Q(a,k,z_0,\dots,z_\nu) = 0] \tag{1}$$

whatever polynomial Q we start with. Note that a bound on k implies a corresponding bound on $z_0,\dots,z_\nu$. Thus,

$$a \in T_x \leftrightarrow (\exists y)(\forall k)_{\le x}(\exists z_0 \dots z_\nu)_{\le y}[Q(a,k,z_0,\dots,z_\nu)=0] \quad ,$$

since if (1) is satisfied for a given x there are only a finite number of values of the z's needed to verify it.

We shall prove a slightly more general theorem in which the polynomial may depend on the bounds (see Davis, Putnam, and Robinson [14] for an earlier and less satisfactory version of this theorem).

BOUNDED QUANTIFIER THEOREM. *Let* $P(x,y,k,z_1,\dots,z_\nu)$ *be a polynomial with* x, y, *and* k *among its parameters and* $z_1,\dots,z_\nu$ *its variables. Then*

$$(\forall k)_{\le x}(\exists z_1 \dots z_\nu)_{\le y}\ [P(x,y,k,z_1,\dots,z_\nu) = 0]$$

$$\leftrightarrow (\exists b_1 \dots b_\nu)\left[\binom{b_1}{y+1} \equiv \dots \equiv \binom{b_\nu}{y+1} \equiv P(x,y,Q!-1,b_1,\dots,b_\nu) \equiv 0 \left(\mathrm{mod}\binom{Q!-1}{x+1}\right)\right] \tag{2}$$

where Q *is a polynomial such that*

$$Q = Q(x,y) > |P(x,y,k,z_1,\dots,z_\nu)| + 2x + y + 1 \tag{3}$$

for all $k \le x$, $z_1 \le y$, $\dots$, $z_\nu \le y$. *Also* $b_1,\dots,b_\nu$ *may be chosen all* $< \binom{Q!-1}{x+1}$.

Note. If P has additional parameters not listed then they will also occur in Q.

Remark. Since $x = y!$ and $x = \binom{y}{z}$ are Diophantine as we saw in §1, the relation defined by (2) is also Diophantine. Hence working from the inside out, we can eliminate the bounded universal quantifiers from an arithmetical definition in which all the universal quantifiers are bounded and obtain an equivalent Diophantine definition. This theorem is one of the steps in the proof of the Main Theorem.

PROOF. Looking first at the consequences of (3), we see that

$$\binom{Q!-1}{x+1} = (Q!-1)\left(\frac{Q!}{2}-1\right)\cdots\left(\frac{Q!}{x+1}-1\right) \tag{4}$$

where the factors on the right are all integers since $Q > x+1$. All primes $\leq Q$ divide $Q!/(k+1)$ for $k \leq x$ since $Q \geq 2x+2$. Hence a prime which divides one of the factors is $> Q$. If a prime divides both $(Q!/(i+1))-1$ and $(Q!/(j+1))-1$ then it divides $|i-j|$ which is $\leq x < Q$. Hence $i = j$; thus the factors in (4) are pairwise relatively prime. Finally, let p_k be a prime which divides $(Q!/(k+1))-1$. Then

$$Q!-1 \equiv k \pmod{p_k} \tag{5}$$

so

$$P(x,y,Q!-1,b_1,\ldots,b_\nu) \equiv P(x,y,k,\mathrm{Rem}(b_1,p_k),\ldots,\mathrm{Rem}(b_\nu,p_k)) \pmod{p_k}\ ,\ k \leq x. \tag{6}$$

(This part of the proof parallels the proof of Gödel's Lemma.)

Suppose the right side of (2) holds. Then $p_k | b(b-1)\cdots(b-y)$ so $\mathrm{Rem}(b,p_k) \leq y$ for $b = b_1,\ldots,b_\nu$. Hence by (3) the absolute value of the right side of (6) is less than Q and hence less than p_k. Also the left side of (6) is $\equiv 0 \pmod{p_k}$ by hypothesis. Hence

$$P(x,y,k,\mathrm{Rem}(b_1,p_k),\ldots,\mathrm{Rem}(b_\nu,p_k)) = 0$$

and the first half of the theorem holds.

Conversely, suppose there are $z_{1k} \leq y, \ldots, z_{\nu k} \leq y$ such that

$$P(x,y,k,z_{1k},\ldots,z_{\nu k}) = 0 \quad \text{for} \quad k \leq x\ . \tag{7}$$

Then we can find $b_i < \binom{Q!-1}{x+1}$ for $i = 1,\ldots,\nu$ satisfying the system of congruences

$$b_i \equiv z_{ik} \pmod{\frac{Q!}{k+1}-1} \quad \text{for} \quad k \leq x\ , \tag{8}$$

by the Chinese remainder theorem since the moduli are pairwise relatively prime. Then

$$\frac{Q!}{k+1}-1 \;\Big|\; b_i(b_i-1)\cdots(b_i-y) \quad \text{for} \quad i = 1,\ldots,\nu\ , \tag{9}$$

since $z_{ik} \leq y$. Furthermore since the divisors in (9) are pairwise relatively prime, their product $\binom{Q!-1}{x+1}$ divides the right side. Also since all the primes dividing the left side of (9) are greater than Q which is greater than $y+1$, we obtain

$$\binom{Q!-1}{x+1} \;\Big|\; \binom{b_i}{y+1} \quad \text{for} \quad i = 1,\ldots,\nu. \tag{10}$$

Finally, by (7) and (8),

$$P(x,y,Q!-1,b_1,\ldots,b_\nu) \equiv P(x,y,k,z_{1k},\ldots,z_{\nu k}) \pmod{\frac{Q!}{k+1}-1}. \tag{11}$$

Since the right side of (11) is zero and the moduli are pairwise relatively prime, we have

$$\binom{Q!-1}{x+1} \,\Big|\, P(x,y,Q!-1,b_1,\ldots,b_\nu)$$

and the second half of the theorem is proved.

Example. To show the usefulness of the bounded quantifier theorem, we will construct a Diophantine definition of the set S of superpowers, i.e.

$$\{1,2^2,3^{3^3},\ldots\}.$$

Notice that $m \in S$ if and only if there is a sequence $t_0,t_1,\ldots,t_n$ such that $t_0 = 1$, $t_{k+1} = n^{t_k}$ for $k \le n$, and $t_n = m$. Hence by Gödel's lemma,

$$\begin{aligned} m \in S \leftrightarrow (\exists\, n,b,d)\ (\forall k)_{\le n}\ &[\mathrm{Rem}(b,\ 1+d) = 1 \\ &\ \&\ \mathrm{Rem}(b,\ 1+(k+2)d) = n^{\mathrm{Rem}(b,1+(k+1)d)} \\ &\ \&\ \mathrm{Rem}(b,\ 1+(n+1)d) = m]\ . \end{aligned} \tag{12}$$

(Here we could take $d = (1+n^m)(n+1)!$ and $b < \prod_{i=1}^{n+2}(1+id)$ by the explicit form of Gödel's lemma stated above.) Since the relations given by $x^y = z$ (cf. §1) and $\mathrm{Rem}(x,y) = z$ are Diophantine, we can transform (12) into the form

$$m \in S \leftrightarrow (\exists\, n,a,d)\ (\forall k)_{\le n}\ (\exists z_1 \ldots z_\nu)\ [P(m,n,a,d,k,z_1,\ldots,z_\nu) = 0]. \tag{13}$$

Finally by the bounded quantifier theorem, the right side of (13) can be shown to be Diophantine and the set of superpowers is Diophantine.

As another application of the bounded quantifier theorem and Gödel's lemma, we will show explicitly how to reduce the four color hypothesis (already briefly discussed in §2) to the unsolvability of a particular Diophantine equation. In fact we shall construct a Diophantine equation $F = 0$ with one parameter a which has a solution for given $a > 0$ just in case there exists a plane map which can*not* be colored in a colors.

To begin with, it is sufficient to consider maps contained in a bounded portion of the plane which has been marked off into a lattice of unit squares and to assume that the regions consist of connected sets of these squares. We can then further reduce the problem by coloring just the centers of the squares which we take to be points with integer coordinates. Note that two of these points in the same region can be joined by a lattice path (with adjacent points one unit apart) lying within the region. The conditions on a coloring then become:

(a) If $\langle x,y\rangle$ and $\langle u,v\rangle$ are in the same region, they must have the same color;

(b) If $\langle x,y\rangle$ and $\langle x+1,y\rangle$ are in different regions, they must have different colors;

(c) If $\langle x,y\rangle$ and $\langle x,y+1\rangle$ are in different regions, they must have different colors.

In particular, it is enough to consider maps contained in $T_0, T_1, \ldots$ where T_n is the triangle given by $0 \leq x+y \leq n$. Note that T_n consists of all points x,y such that $J(x,y) \leq Q$ where

$$Q = J(n,0) = (n^2+3n)/2 \ . \tag{14}$$

Now any sequence $t_0,\ldots,t_Q$ of natural numbers can be considered a coloring of T_n by associating different colors with different natural numbers where t_k is the number of the color of the point x,y with $J(x,y) = k$. It is important to note that such a coloring determines the corresponding map since two points are in the same region if and only if there is a lattice-path between them with all of its points the same color. Hence two sequences $t_0,\ldots,t_Q$ and $u_0,\ldots,u_Q$ are *colorings of the same map* exactly if adjacent points in one sequence have the same color if and only if they do in the other. That is, if and only if,

$$\begin{aligned} t_{J(x,y)} = t_{J(x+1,y)} &\leftrightarrow u_{J(x,y)} = u_{J(x+1,y)} \\ t_{J(x,y)} = t_{J(x,y+1)} &\leftrightarrow u_{J(x,y)} = u_{J(x,y+1)} \end{aligned} \tag{*}$$

for all x,y with $x+y < n$.

Let $t_0,\ldots,t_Q$ be a sequence corresponding to a coloring of some map in T_n. Now suppose that for every sequence $u_0,\ldots,u_Q$ with $u_i < a$, the conditions (*) are not all satisfied. This would mean that the map of which $t_0,\ldots,t_Q$ is a coloring cannot be colored with a colors. It is this proposition that we will translate into a Diophantine equation with one parameter a.

By Gödel's lemma, given a coloring of T_n, $t_0,\ldots,t_Q$, there are s and t such that

$$t_i = \mathrm{Rem}(t, 1+s(i+1)) \quad \text{for} \quad i = 0,\ldots,Q \ . \tag{15}$$

Furthermore every sequence $u_0,\ldots,u_Q$ with each $u_i < a$ can be represented by some $u \leq R$ such that

$$u_i = \mathrm{Rem}(u, 1+(Q+1)!\,a(i+1)) \quad \text{for} \quad i = 0,\ldots,Q \ . \tag{16}$$

where

$$R = (1+(Q+2)!\,a)^{Q+1}. \tag{17}$$

Hence there is a map which cannot be colored in a colors if and only if

$$\begin{aligned}(\exists n,t,s)\ (\forall u \le R)\ (\exists x,y)\ [(x+y \le n\ \&\ u_{J(x,y)} \ge a)\ \vee \\ \{x+y\ n\ \&\ [(t_{J(x,y)} = t_{J(x+1,y)}\ \&\ u_{J(x,y)} \ne u_{J(x+1,y)}) \\ \vee\ (t_{J(x,y)} \ne t_{J(x+1,y)}\ \&\ u_{J(x,y)} = u_{J(x+1,y)}) \\ \vee\ (t_{J(x,y)} = t_{J(x,y+1)}\ \&\ u_{J(x,y)} \ne u_{J(x,y+1)}) \\ \vee\ (t_{J(x,y)} \ne t_{J(x,y+1)}\ \&\ u_{J(x,y)} = u_{J(x,y+1)})]\}].\end{aligned}$$

Hence using (14)-(17), Diophantine definitions of exponentiation and factorial, and the bounded quantifier theorem, we can obtain a Diophantine equation

$$F(a,n,t,s,\ldots) = 0 \tag{18}$$

which has a solution only if there is a map in T_n which cannot be colored with a colors.

Thus, the situation for (18) is as follows: for $a \ge 5$ (18) has no solution (since every map can be colored using 5 colors); for $0 < a \le 3$ (18) is solvable (since there are maps which require at least 4 colors); *for* $a = 4$ *the four color hypothesis is equivalent to* (18) *having no solutions*. Note that, e.g. for $a = 5$, the fact that the Diophantine equation (18) has no solution was proved using, in effect, topological methods.

5. THE UNIVERSAL EQUATION

In 1936, Turing [64] proved a fundamental result in the theory of algorithms: there exists an all encompassing algorithm. In Turing's development this meant a universal Turing machine, that is a single Turing machine that could be programmed to carry out any calculation that any Turing machine could do. In our case, we obtain:

UNIVERSAL EQUATION THEOREM. *There is a Diophantine equation* $U(a,n,x_1,\ldots,x_\nu) = 0$ *such that to every Diophantine set* $\mathcal{D}$ *there is an* n *with:*

$$a \in \mathcal{D} \leftrightarrow (\exists x_1 \ldots x_\nu)\ [U(a,n,x_1,\ldots,x_\nu) = 0]\ .$$

(Below we write $P_n(a,x_1,\ldots,x_\nu)$ for $U(a,n,x_1,\ldots,x_\nu)$.)

Although this theorem is an immediate consequence of the Main Theorem in the context of the theory of algorithms, we will give a direct proof of it without relying on the theory of algorithms. In fact we will sketch the construction of the polynomial U in such a way that given a Diophantine definition of $\mathcal{D}$, we can calculate a corresonding n.

We have treated Diophantine equations as equations of the form P = 0 where P is a polynomial with integer coefficients. Another way of looking at Diophantine equations is to consder them to be equations between terms built up from $0,1,x_0,x_1,\ldots$ by repeated additions and multiplications. Clearly, every Diophantine equation without parameters is equivalent to such an equation. There is a convenient way of numbering equations in this form. We first number the terms as follows:

$$\begin{aligned}
&\tau_0 = 0\,, \\
&\tau_1 = 1\,, \\
&\tau_{3k+2} = x_k\,, \\
&\tau_{3k+3} = \tau_{K(k)} + \tau_{L(k)}\,, \\
&\tau_{3k+4} = \tau_{K(k)} \cdot \tau_{L(k)}\,, \quad \text{for } k = 0,1,\ldots,
\end{aligned} \tag{1}$$

where K and L are the Cantor pairing functions defined in the Introduction. Let the nth Diophantine equation $\mathcal{E}_n$ be $\tau_{K(n)} = \tau_{L(n)}$. In order to number the Diophantine sets, we will consider x_0 to be a parameter and define $\mathcal{D}_n$ to be the set of values of x_0 for which $\mathcal{D}_n$ has a solution. Thus $m \in \mathcal{D}_n$ if and only if there is a sequence $t_0, t_1, \ldots, t_n$ of natural numbers such that

$$\begin{aligned}
&t_0 = 0,\ t_1 = 1,\ t_2 = m,\ t_{3k+3} = t_{K(k)} + t_{L(k)}\,, \\
&t_{3k+4} = t_{K(k)} \cdot t_{L(k)}\,, \quad t_{K(n)} = t_{L(n)} \quad \text{for } k \le n\,.
\end{aligned} \tag{2}$$

In fact these conditions yield

$$t_{K(n)}(m, t_5, t_8, \ldots) = t_{L(n)}(m, t_5, t_8, \ldots)$$

where no restriction is put on $t_5, t_8, \ldots$. Hence any solution of

$$\tau_{K(n)}(m, x_1, \ldots) = \tau_{L(n)}(m, x_1, \ldots)$$

is represented by some sequence satisfying the conditions above.

Now we use Gödel's lemma to represent the sequence $t_0, \ldots, t_n$. That is, there are a and d so that

$$t_\ell = \operatorname{Rem}(a,\ 1+(\ell+1)d)\ .$$

Hence $m \in \mathcal{D}_n$ if and only if there are a, d, q, r so that for all $k \le n$ the system

$$\begin{aligned}
&k = J(i,j), \\
&n = J(q,r), \\
&\operatorname{Rem}(a,1+d) = 0, \\
&\operatorname{Rem}(a,1+2d) = 1, \\
&\operatorname{Rem}(a,1+3d) = m, \\
&\operatorname{Rem}(a,1+(3k+4)d) = \operatorname{Rem}(a,1+(i+1)d) + \operatorname{Rem}(a,1+(j+1)d), \\
&\operatorname{Rem}(a,1+(3k+5)d) = \operatorname{Rem}(a,1+(i+1)d) \cdot \operatorname{Rem}(a,1+(j+1)d), \\
&\operatorname{Rem}(a,1+(q+1)d) = \operatorname{Rem}(a,1+(r+1)d),
\end{aligned} \tag{3}$$

can be satisfied. Finally since (3) is Diophantine we can replace it by a single equation, say

$$P(m,n,a,d,q,r,k,i,j,z_1,\ldots,z_\nu) = 0\ . \tag{4}$$

Since the unknowns used to define $\operatorname{Rem}(a,b) = c$ are $\le a$, we can apply the bounded quantifier theorem with $x = n$ and $y = n + a$ together with the definitions of the binomial coefficient and factorial from §1 to obtain a universal equation $U(m,n,x_1,\ldots,x_\mu) = 0$. The proof sketched here uses methods developed to prove the Main Theorem and of course was not known until after the Main Theorem was proved. Given an algorithm for listing a set M, we can calculate a particular natural number n_0 so that

$$m \in M \leftrightarrow U(m,n_0,x_1,\ldots,x_\mu) = 0 .$$

For example, M could be the set of primes, the set of superpowers, the set of Gödel numbers of the theorems of set theory, the set of numbers of solvable Diophantine equations, etc. (This last can also be defined by considering m as an unknown in the universal equation.)

Let K be the Diophantine set $\{n\colon U(n,n,x_1,\ldots,x_\mu) = 0 \text{ has a solution}\}$. Then N-K is not Diophantine. The proof is by contradiction. Suppose $N - K = \{a\colon U(a,k,x_1,\ldots,x_\mu) = 0 \text{ has a solution}\}$ for some k. Then

$$k \in N - K \leftrightarrow U(k,k,x_1,\ldots,x_\mu) = 0 \text{ has a solution} \leftrightarrow k \in K ,$$

which is a contradiction. Thus, *there is a Diophantine set whose complement is not Diophantine.* Also, in view of the Main Theorem K is a listable set which is not computable. Another example of a listable set which is not computable is the set of numbers of all solvable Diophantine equations. (If this last set were computable, Hilbert's tenth problem would be decidable.)

We will now derive a variant of Gödel's incompleteness theorem. In the first place, we have already sketched a proof (cf. the Introduction) that the set of theorems of a formalized mathematical theory is listable. Moreover as we have just observed the set of unsolvable Diophantine equations is not. Thus there is no formalized mathematical theory which can determine correctly the solvability of Diophantine equations. Roughly, Gödel started with an axiomatization of number theory and then exhibited a statement of number theory which was true by construction but not provable from the given axioms. We can now take as the unprovable proposition the unsolvability of a Diophantine equation which is constructed in such a way as to be unsolvable in natural numbers. (Cf. also the remarks on notation in formalized mathematical theories in the Introduction.)

GÖDEL'S INCOMPLETENESS THEOREM. *Let A be a system of axioms in a language including the mathematical symbols* 0 , S, +, $\cdot$, <, *and satisfying*

(a) *A is consistent* (*i.e. it is not possible to prove both ϕ and $\sim\phi$ for any sentence ϕ*);

(b) *A is listable;*

(c) *A is sufficiently strong to prove every true statement of the forms*

$$\alpha + \beta = \gamma$$
$$\alpha \cdot \beta = \gamma$$
$$\alpha < \beta$$

where α, β, γ *are among* 0, S0, SS0, ... *and* S *is the successor function.*

Then we can construct a Diophantine equation $F(x_1,\ldots,x_\nu) = 0$ *corresponding to* A *such that* $F = 0$ *has no solutions in natural numbers but we cannot derive*

$$\sim (\exists x_1 \ldots x_\nu)\ [F(x_1,\ldots,x_\nu) = 0] \tag{5}$$

from A.

REMARK. Any system of axioms satisfying (c) is sufficient to prove that a Diophantine equation solvable in natural numbers has a solution. For example, the system of identities

$$\begin{aligned} x + 0 &= x \\ x + S(y) &= S(x + y) \\ x \cdot 0 &= 0 \\ x \cdot S(y) &= x \cdot y + x \end{aligned} \tag{$\mathcal{D}$}$$

is sufficient to prove the addition and multiplication tables for 0, S0, SS0, SSS0, ...

Note also that if any sentence of the form (5) is even *consistent* with all the statements in (c) (or with $\mathcal{D}$) then it is true.

PROOF: Let $A = \{a: \sim(\exists x_1 \ldots)\ [P_a(a,\ldots) = 0]$ is provable from the axioms $A\}$. Then, since the theorems are listable, A is listable and hence A is $\mathcal{D}_k$ for some k. Hence

$k \in \mathcal{D}_k \leftrightarrow$ The equation $P_k(k,x_1,\ldots) = 0$ has a solution,

$k \in \mathcal{D}_k \leftrightarrow$ The formula which expresses $\sim(\exists x_1 \ldots)[P_k(k,x_1,\ldots)=0]$ is provable from the axioms A

Now if $k \in \mathcal{D}_k$ we have a contradiction derivable from A since if $P_k(k,x_1,\ldots) = 0$ has a solution in N by (c) we can prove in A that it has a solution. But A is consistent by (a) so we must have $k \notin \mathcal{D}_k$. This means that $P_k(k,x_1,\ldots) = 0$ has no solution in N but we cannot prove from the axioms A that it has no solution.

REMARK. Notice that if the axioms A are true statements of number theory then A is incomplete. However, if A is an axiomatization of the field of real numbers, then A is complete but gives the wrong answer for $P_k(k,x_1,\ldots) = 0$. If we replace (c) by the stronger condition that A is an extension of Peano's axioms, then A would be incomplete with respect to unsolvability of Diophantine equations.

6. REDUCTIONS

We say a Diophantine equation or definition can be *reduced* to another of a simpler form in some sense if there is an equivalent equation in the simpler form, that is an equation which has a solution for exactly the same values of the parameters. If α is the degree of a universal equation $U(m,n,z_1,\ldots,z_\nu) = 0$ in $m,z_1,\ldots,z_\nu$ then every Diophantine set can be defined by an equation of degree α with ν unknowns. It was this surprising consequence of the Main Theorem that led many mathematicians to believe prior to 1970 that not all listable sets were Diophantine. In §5, we have given a proof of the universal equation theorem independent of the Main

Theorem, but using methods of proof used in proving the Main Theorem.

REDUCTION OF DEGREE (Skolem [58]). *Every Diophantine equation can be reduced to an equation of degree* ≤ 4.

PROOF: Skolem observed that a Diophantine equation can be reduced to a system of equations of the form

$$\begin{aligned} \alpha + \beta &= \gamma \\ \alpha \cdot \beta &= \gamma \end{aligned} \tag{1}$$

by first writing the equation in the form A = B where A and B are polynomials with positive integer coefficients and then introducing new unknowns for the partial terms needed to build up A and B. Since the equations (1) are of at most degree 2, they can be combined into one equation of degree ≤ 4 by replacing

$$A_i = B_i \ , \qquad i = 1, \ldots, k$$

by

$$\sum_{i=1}^{k} (A_i - B_i)^2 = 0 \ .$$

In particular, there is a universal equation of degree 4.

REDUCTION OF THE NUMBER OF UNKNOWNS. (Matijasevič and Robinson [39]). *Given a Diophantine equation* $P(a_1, \ldots, a_\mu, z_1, \ldots, z_\nu) = 0$ *there is a Diophantine equation* $Q(a_1, \ldots, a_\mu, x_1, \ldots, x_{13}) = 0$ *which has a solution for exactly the same* μ*-tuples* $a_1, \ldots, a_\mu$.

COROLLARY. *There is a universal Diophantine equation* $U(a, n, x_1, \ldots, x_{13}) = 0$.

The corollary follows because we already know there is some universal equation. J. P. Jones (cf. [25]) is currently studying pairs (δ, ν) such that there is a universal equation of degree δ in ν unknowns, and has given the following examples: (4,153); (6,129); (8,108); (10,107); (20,86); (44,83); (1952,80). These results show how with present methods we can reduce the degree at the expense of the number of unknowns and conversely. We can think of the pair (δ, ν) as a measure of the complexity of a given Diophantine equation.

The equation $U(J_\mu(a_1, \ldots, a_\mu), x_1, \ldots, x_{13}) = 0$ (where J_μ is as defined in the Introduction) is clearly universal for Diophantine sets of μ-tuples. Note that its degree in the parameters increases to ∞ as $\mu \to \infty$.

OPEN PROBLEM. *Are there absolute constants* δ, ν *such that for every* μ *there is a universal Diophantine equation for Diophantine sets of* μ*-tuples of total (parameters and variables) degree* δ *and in* ν *unknowns?*

Most likely, the answer is negative.

R. M. Robinson [50] proposed considering systems of Diophantine equations of the following sort. Let $W_{j,k}$ be the problem of telling whether a system of equations in j+k variables $u_1, \ldots, u_j, v_1, \ldots, v_k$ is solvable in natural numbers, with the agreement that $u_1, \ldots, u_j$ must have the same values in all the equations, but $v_1,\ldots,v_k$ may have different values in different equations. Viewed as an ordinary system of Diophantine equations, $W_{j,k}$ would have $j+\lambda k$ unknowns where λ is the number of equations in the system. He pointed out that the decision problem for both $W_{n,0}$ and $W_{0,n}$ reduced to that of a single Diophantine equation in n unknowns. This is true for $W_{n,0}$ since a system of equations can be reduced to a single equation. In the case of $W_{0,n}$, a decision method for Diophantine equations could be applied separately to each equation of the system. He proved that there was no decision method for $W_{2,3}$ or $W_{1,4}$. Later Matijasevič [38] improved this to $W_{2,2}$ and $W_{1,3}$. These results were all obtained by showing that an arbitrary Diophantine equation could be reduced to the appropriate $W_{i,j}$.

Another measure of the complexity of a Diophantine equation is the total number of additions and multiplications necessary to verify a solution. We may get a smaller number if we write the equation as an equivalent system of equations so we will allow this. Now what we are really interested in is the complexity of a Diophantine definition of a particular set so we can search for the simplest (or at least simpler) definitions. Here there are many options. For example, we may treat constants, parameters, and unknowns alike. Thus, the operations $x+1$ and $x \cdot y$ would each contribute 1 to the total. Another possibility is to count only operations on two parameters or unknowns since these will generally be large compared to the coefficients of the equation or system. For expediency we might consider that multiplication is so time consuming compared to addition that we count only multiplications of parameters and unknowns. The choice of measure and the resulting minimal definitions is part of the theory of the complexity of computation.

A particularly interesting set in the theory of numbers is the set of prime numbers. If a number c is composite there exists a one-step proof of the fact consisting of multiplying two factors together and obtaining c. The work of searching for the factors is considered scratch work and is thrown away. It is not part of the proof. Now the number of steps (divisions) in the trivial algorithm for proving a number p is prime is $[\sqrt{p}]$. Until the set of primes was shown to be Diophantine, it was not known that *there is a proof that* p *is prime consisting of a bounded number of additions and multiplications*. Given some Diophantine definition of the set of primes, the proof that a prime p is a prime consists in substituting suitable values of the unknowns in the defining equation and verifying that it is a solution. Finding the solution is again considered scratch work to be discarded.

By using a universal equation to define an arbitrary listable set, we obtain a proof that a given number belongs to the set in which the number of steps is independent of the size of the number and of the particular set given.

For example, suppose we number the statements in the language of set theory and let T be the set of numbers of the theorems of some axiom system.

Then as we have seen T is a listable set. Thus, we can find a Diophantine definition of T and calculate its number t_0. Then a statement ϕ with number f is a theorem of set theory if and only if $U(f,t_0,x_1,\ldots,x_{13}) = 0$ has a solution in N. Hence if ϕ is in fact a theorem, then there is a metamathematical proof of ϕ whose length is a constant plus the length of the calculation of f. Hence the length of this proof depends only on the length of ϕ. [The usual way of numbering statements is to number each symbol in the language. Then the number of the statement is some code for the finite sequence of numbers corresponding to the sequence of symbols in the statement.]

Another example is obtained by letting $\mathcal{D}_{p_0}$ be the Diophantine set of numbers of theorems in the theory of Peano's axioms. Then if f is the number of a statement of number theory ϕ we see that ϕ is a consequence of Peano's axioms if and only if

$$(\exists x_1 \ldots x_{13})\ [U(f,p_0,x_1,\ldots,x_{13}) = 0]$$

is a consequence of the axioms $\mathcal{D}$ in §5. So that even though it is a theorem of Ryll-Nardzewski that Peano arithmetic cannot be finitely axiomatized, there is a complete image of it contained in the fragment derivable from the usual recursive definitions of + and $\cdot$. There are finite axiomatizations of stronger theories than Peano's arithmetic (see for example Robinson [48]).

Another kind of reduction is that carried out by Adler [3]:

REDUCTION OF HOMOGENEOUS EQUATIONS. *Every homogeneous Diophantine equation* $P(x_1,\ldots,x_\nu) = 0$ *can be reduced effectively to a system of homogeneous quadratic equations* $Q_1(y_1,\ldots,y_\lambda) = 0, \ldots, Q_k(y_1,\ldots,y_\lambda) = 0$.

Hence every homogeneous Diophantine equation is equivalent to a single homogeneous quartic equation.

These are especially interesting because an outstanding:

OPEN PROBLEM. *Is there an algorithm to tell whether an arbitrary Diophantine equation has a solution in rational numbers?*

It can be shown that this problem is equivalent to the decision problem for nontrivial solutions in integers of homogeneous Diophantine equations.

Another interesting result of Adler is that an arbitrary Diophantine equation can be reduced to an equation

$$F = 1$$

where F is a homogeneous polynomial of degree 4.

7. THE BOUNDARY BETWEEN THE DECIDABLE AND THE UNDECIDABLE

From what we have stated in the previous section, we see that no algorithm can exist for testing even Diophantine equations in 13 unknowns for solvability in natural numbers. On the other hand it is obvious that an algorithm exists for 1 unknown. For 2 unknowns, a decision method has been given by Baker and his collaborators for a wide class of equations but not yet for the general case and for 3 unknowns, almost nothing is known. Thus the problem of locating the precise boundary between the decidable and the undecidable, with respect to the number of unknowns, remains very much an open question.

From the point of view of the total degree of the Diophantine equation, we are somewhat better off. An algorithm is known (cf. Siegel [57]) for Diophantine equations of degree 2 in arbitrarily many unknowns.

There are two directions in which Siegel's result could be generalized. First, to single equations of a fixed degree. The unsolvability of Hilbert's tenth problem together with Skolem's reduction (see §6) shows that there is no algorithm for testing the solvability of equations of degree 4.

OPEN QUESTION. *Is Hilbert's tenth problem decidable for equations of degree 3?*

Second, the generalization could extend Siegel's result to systems of equations of degree 2. But since every Diophantine equation can be put in the form of a system of equations of degree 2 (cf. §6), we see that there is an N such that there is no decision method for systems of N equations. Today the gap between Siegel's result and the least known value of N is very great.

For yet another form of the problem, we can obtain a sharp answer: It is shown in Matijasevič and Robinson [39] that every Diophantine equation is equivalent to a system of equations of the form

$$[\sigma_1(x,y)] = [\sigma_2(x,y)]$$

where the σ's are rational functions with rational coefficients of x and y. Hence there is no algorithm for telling whether such a system of equations has a solution in N. On the other hand, Straus [61] has given a method for telling whether an arbitrary system of equations of the form

$$[\tau_1(x)] = [\tau_2(x)]$$

where the τ's are rational functions with rational coefficients is solvable in N. Namely, consider first one such equation. Then,

$$\tau_i(x) = P_i(x) + \gamma_i + \delta_i(x) , \qquad i = 1,2$$

where P_i is a polynomial with rational coefficients and without a constant term, γ_i is a rational number, and $\delta_i(x)$ is a rational function of negative degree. Then

$$|P_1(x)-P_2(x)| \to \infty \text{ as } x \to \infty$$

$$a + b = c \leftrightarrow (\exists w)((w \dot{-} a) \dot{-} b = w \dot{-} c \ \&\ 1 \dot{-} (w \dot{-} c) = 0)$$

$$a \cdot b = c \leftrightarrow c + c + a^2 + b^2 = (a + b)^2 .$$

Also a system of equations $\alpha_i = \beta_i$ $(i = 1,\dots,\nu)$ is equivalent to a single equation

$$(\dots((1 \dot{-} (\alpha_1 \dot{-} \beta_1)) \dot{-} (\beta_1 \dot{-} \alpha_1)) \dot{-} \dots) \dot{-} (\alpha_\nu \dot{-} \beta_\nu)) \dot{-} (\beta_\nu \dot{-} \alpha_\nu) = 1$$

Hence we obtain:

There is no algorithm for listing identities (over N) of the form

$$F(x_1,\dots,x_n) = 0$$

where F *is obtained by composition from* $\dot{-}$ *and the operation of squaring applied iteratively to particular natural numbers and* $x_1,\dots,x_n$.

Also, *there is no algorithm for listing identities in one variable of the form*

$$F(x) = 0$$

where F *is obtained by composition from the functions* $\dot{-}$, $[\sqrt{x}]$, Rem *(the remainder function).* For these results see Matijasevič [38].

Another example comes from the fact that + can be existentially defined in terms of S (successor and $\cdot$ since

$$z = x + y \leftrightarrow S(Sx\cdot Sz)\cdot S(y\cdot Sz) = S(Sz\cdot Sz\cdot S(Sx\cdot y))$$

(see Robinson [45]). Hence

Every listable set is existentially definable in terms of S *and* $\cdot$.

Thus a Diophantine equation can be reduced to a system of equations of the forms $\alpha = S\beta$ and $\alpha\cdot\beta = \gamma$ where α, β, γ are natural numbers, parameters, or unknowns. Furthermore, such a system is equivalent to a single equation since $a_i = b_i$ $(i = 0,\dots,\nu)$ if and only if

$$\prod_i S(a_i M^{2^i}) = \prod_i S(b_i M^{2^i}) \tag{1}$$

where $M = \prod_i (Sa_i \cdot Sb_i)$. To see this, note that M is so large that (1) can hold only if the corresponding coefficients of the powers of M are equal and that the coefficients of M^{2^i} are a_i and b_i respectively.

It is easy to see that S and $\cdot$ are existentially definable in terms of $z = [x/y]$ where $[x/0] = 0$. In fact, we have

$$y = 2x \ \&\ z = 2x+1 \leftrightarrow (x = 0 \ \&\ y = 0 \ \&\ z = 1) \vee ([\tfrac{y}{2}] = [\tfrac{z}{2}] = x \ \&\ [\tfrac{y}{z}] = 0)$$

$$y = x+1 \leftrightarrow (x = 0 \ \&\ y = 1) \vee ([\tfrac{x}{x}] = 1 \ \&\ (\exists z)\{[\tfrac{z}{2x+1}] = 2x+1 \ \&\ [\tfrac{z}{2(2x)}] = y\})$$

$$z = xy \leftrightarrow (z=0 \ \&\ x=0) \vee (z=0 \ \&\ y=0) \vee (\exists t)\{[\tfrac{z}{y}] = x \ \&\ z=t+1 \ \&\ [\tfrac{t}{y}]+1=x\}.$$

Thus, any Diophantine equation can be reduced to a disjunction of systems of equations of the form $[\alpha/\beta] = \gamma$ where α,β,γ are natural numbers, parameters or unknowns. In particular, *there is no algorithm for testing whether a system of such equations has a solution.*

Kosovskii [27] showed that $\cdot$ is existentially definable in terms of $+$, $|$, and any relation S such that

(a) $$S(x, x^2)$$
and
(b) $$S(x,y) \rightarrow y \leq x^2 .$$

It is an immediate consequence that $\cdot$ is existentially definable in terms of $+$, $|$, and $\square$. Indeed, we can define the required relation S satisfying (1) and (2) by

$$S(x,y) \leftrightarrow y = \square \ \& \ y + 2x + 1 = \square .$$

This result suggests the:

OPEN PROBLEM. *Is there an algorithm to test whether a system of conditions of the forms*

$$\alpha < \beta, \quad \alpha | \beta, \quad \alpha = \square$$

(where α, β *stand for natural numbers or unknowns) has a solution?*

Recently Edward Schwartz [54] has found various systems of functions and relations in terms of which $+$ and $\cdot$ are existentially definable. Thus he showed that $+$ is existentially definable in terms of $\cdot$, x^2+y^2, and $<$. For,

$$a = p + q \leftrightarrow (\exists b,x,y)\{[p = q \ \& \ a = 2\cdot p] \vee [p=0 \ \& \ a=q] \vee [q=0 \ \& \ a=p]$$
$$\vee \ [(p^2+q^2)^2 + 0^2 = x^2 + (2\cdot(p\cdot q))^2 \ \& \ x = a\cdot b$$
$$\& \ (a^2+b^2)^2 + 0^2 = y^2 + (2\cdot(a\cdot b))^2 \ \& \ y = 4\cdot(p\cdot q) \ \& \ b < a]\} .$$

Similar methods were used by Schwartz to show that $+$ and $\cdot$ are existentially definable in each of the following systems of functions and relations:

(1) $+$ and x^2+y^2

(2) x^2+y^2, $2x$, and $<$

(3) $x\cdot y$ and x^2-y^2

(4) $x+y$ and x^2-y^2

(5) x^2-y^2 and x^2+y^2

(6) $x\cdot y$, $<$, and x^2-dy^2, d fixed

(7) $x\cdot y$, $<$, and x^2+dy^2, d fixed.

9. SIZE OF SOLUTIONS

A function ϕ is a *test function* for a Diophantine equation $P(a_1,\ldots,a_\mu,x_1,\ldots,x_\nu) = 0$ if whenever $P(a_1,\ldots,a_\mu,x_1,\ldots,x_\nu) = 0$ has a solution for given values of the parameters $a_1,\ldots,a_\mu$ it has a solution in which each unknown is less than $\phi(a_1,\ldots,a_\mu)$. Clearly, a Diophantine equation defines a computable relation if and only if it has a computable test function. Since there are Diophantine sets which are not computable, unless $P_1 = P_2$ identically. Also $\delta_i(x) \to 0$ as $x \to \infty$. Hence either there is an easily calculated bound on possible solutions or $P_1 = P_2$. Throwing out the trivial cases, we can without loss of generality take

$$\tau_1(x) = \frac{P(x)}{n} + \delta_1(x)$$
$$\tau_2(x) = \frac{P(x)}{n} + \frac{m}{n} + \delta_2(x) \qquad (1)$$

where P is a polynomial with integer coefficients and m,n are particular natural numbers. There is an easily computed x_0 so that for $x > x_0$, $\delta_1(x)$ and $\delta_2(x)$ have a constant sign and have absolute value less than 1/n. Then any solution of (1) is either less than x_0 or lies in one of an easily determined set of residue classes (mod n). Also all numbers greater than x_0 in these residue classes will be solutions of (1). Since it is easy to determine whether a finite set of such conditions can be satisfied simultaneously, we can determine whether a system of such equations has a solution. Thus for this problem, we have a sharp result:

For systems of equations in 2 or more unknowns there is no algorithm while for systems of such equations in one variable there is an algorithm.

8. EXISTENTIAL DEFINABILITY AND UNIVERSAL STATEMENTS

We consider formulas which can be written in a language in which there are variables ranging over the natural numbers, symbols for =, &, ∨, and $(\exists x)$, and for each fixed natural number, and symbols for certain given functions and/or relations on N. Such a formula defines a relation on N in its free variables (parameters). And any relation which is so definable is called *existentially definable* in terms of the given function and relations. Obviously a Diophantine relation is simply one which is existentially definable in terms of + and ·. Now it is clear that if a relation R is existentially definable in terms of given functions and relations and if each of the given functions and relations is in turn existentially definable in terms of a second set of functions and relations then R must be existentially definable in terms of this second set. In particular, any class C of functions and relations in terms of which + and · are existentially definable has the property that any Diophantine relation (and hence any listable relation) is existentially definable in terms of C. Also there can be no algorithm to tell whether such a system of equations and relations can be satisfied in N.

Before turning to examples, we want to remark that the negative solution of Hilbert's tenth problem can also be used to show that there is no algorithm for listing certain types of systems of conditions which hold for all natural number values. For example:

There is no algorithm for listing all polynomials P and Q with positive integer coefficients such that

$$P(x_1,\ldots,x_n) < Q(x_1,\ldots,x_n) \qquad (1)$$

for all natural number values of $x_1,\ldots,x_n$.

In fact, by the reduction theorem we can take n = 13. This is clear since there is no algorithm for listing unsolvable Diophantine equations

$$F(x_1,\ldots,x_{13}) = G(x_1,\ldots,x_{13}) .$$

But F = G has no solution if and only if

$$(F - G)^2 > 0 \quad \text{for all} \quad x_1,\ldots,x_{13} .$$

REMARKS. This result contrasts with two earlier related results. First, the corresponding problem for real values of any number of variables $x_1,\ldots,x_n$ where (1) is equivalent to a Boolean combination of inequalities among polynomials in the coefficients of P and Q by Tarski's elimination of quantifiers theorem. Hence:

There is an algorithm to determine whether

$$P(x_1,\ldots,x_n) < Q(x_1,\ldots,x_n)$$

for all real $x_1,\ldots,x_n$.

Second, consider the problem of recognizing identities (over N) of terms built up from natural numbers and x by additions, multiplications, and exponentiations. Richardson [43] obtained an algorithm to solve this problem. See also Goodstein and Lee [20].

For undecidability results in real analysis see Richardson [42] and Adler [2].

In general if there is no algorithm for telling which of a class of existential formulas can be satisfied then there is no algorithm for listing those of the *dual formulas* (& replaced by v, v by &, = by ≠, each relation by its complement) which hold for all natural numbers.

For example,

$$p \neq q$$

is equivalent to the equation

$$(1 \dot{-} (p \dot{-} q)) \dot{-} (q \dot{-} p) = 0$$

where $\dot{-}$ is defined as usual by

$$x \dot{-} y = \begin{cases} x - y & \text{if } x \geq y \\ 0 & \text{otherwise} \end{cases} .$$

Hence

There is no algorithm for listing identities (over N) of the form

$$F(x_1,\ldots,x_n) = 0$$

where F *is obtained by composition from* $+$, $\cdot$, $\dot{-}$ *applied to natural numbers and* $x_1,\ldots,x_n$.

This brings up the problem of finding minimal subclasses of identities which cannot be listed. For example we can replace $+$, $\cdot$, $\dot{-}$ in the result above by $\dot{-}$, x^2. To see this note that $+$, $\cdot$ can be existentially defined in terms of $\dot{-}$ and x^2 by there are Diophantine equations with one parameter which do not have computable test functions.

Given a Diophantine equation, we know we cannot in general find a computable test function for it. However there are two natural types of questions about test functions which we can hope to answer.

I. *Given some familiar set* S *of natural numbers such as the set of primes, can we find a Diophantine definition of* S *with a polynomial test function? an iterated exponential test function? etc.*

Here a function is called *iterated exponential* if it can be built up from its arguments and particular natural numbers by addition, multiplication, and exponentiation. Thus a^{a^a} is an iterated exponential function but

$$\phi(a) = a^{a^{\cdot^{\cdot^{\cdot^{a}}}}} \Big\} \text{ a levels}$$

is not.

II. *What can be said about sets which can (or cannot) be defined by Diophantine equations with polynomial test functions? with iterated exponential test functions? etc.*

By examination of the proofs of the Diophantine definitions of $z = x^y$, $z = \binom{x}{y}$, $z = x!$, and the set of primes sketched in §1 and given in detail in the expository articles referred to in the Introduction, we see that all of these relations have definitions with iterated exponential test functions. Also if we look at the definition of the set of superpowers in §4, we see that $t_1,\ldots,t_{n+1} \le n^m$, $d \le a$, $n \le m$, and a is less than an exponential function of m by Gödel's lemma. Hence by the bounded quantifier theorem, the set of superpowers can also be defined by an equation with an iterated exponential test function.

An example of a set definable by an equation with a polynomial test function is the set of composite numbers. More generally, the same can be said for a set defined by an equation in which one side is a polynomial with positive coefficients not containing the parameter and the other side is a polynomial in the parameter not containing unknowns, or for a set defined by an equation with one unknown.

Another interesting case of sets with polynomial test functions can be obtained using the Cantor pairing function $J(x,y) = ((x+y)^2+3x+y)/2$ (cf. Introduction). We first note that the condition

$$2J(x,y) + 1 = \square$$

is equivalent to $x = y$. (Obviously $2J(x,x)+1 = (2x+1)^2$; conversely if $2J(x,y)+1 = (2a+1)^2$, then $x = y = a$ is a solution, and by the one-one property of the pairing function, the unique solution.) Now we consider an arbitrary Diophantine equation $A(x_1,\ldots,x_\nu) = B(x_1,\ldots,x_\nu)$ where A and B have positive coefficients and no parameters. Then the equation

$$2J(A(x_1,\dots,x_\nu),B(x_1,\dots,x_\nu)) + 1 = a^2 \tag{1}$$

is equivalent to $A = B$ if a is regarded as an unknown. Hence we have:

THEOREM. *There is no algorithm for telling whether an arbitrary polynomial with positive integer coefficients assumes a square value.*

If however a is regarded as a parameter, then (1) has a polynomial test function. In fact, all the x's must be less than a. Furthermore the set defined by (1) is infinite if and only if $A = B$ has infinitely many solutions. This proves:

THEOREM. *There is no algorithm for telling whether an arbitrary polynomial with positive integer coefficients represents infinitely many squares.*

In these examples of equations with polynomial test functions, the test functions are in fact bounds for all of the solutions of the equation. This of course is not the case in general.

If $P(a,x_1,\dots,x_\nu) = 0$ has a polynomial test function say a^n then we can introduce new unknowns and obtain an equivalent equation which has the identity function for a test function. Namely, let $x_i = u_0^{(i)} + \dots u_{n-1}^{(i)}a^{n-1}$ then it will be sufficient to consider u's less than a. Also if a set S can be defined by an equation with a polynomial test function, then it has a definition of the form

$$a \in S \leftrightarrow (\exists x_1 \dots x_\nu\ y_1 \dots y_\nu)[P(a,x_1,\dots,x_\nu)=0\ \&\ x_1+y_1=a\ \&\dots\&\ x_\nu+y_\nu=a]. \tag{2}$$

The conditions in brackets can be combined in the usual way into a single equation which for each a has a finite number of solutions. Thus,

THEOREM. *If a set S can be defined by an equation with a polynomial test function, then it has a finitefold Diophantine definition.*

REMARK. It is possible that the converse of this theorem is true. This would eliminate the possibility of singlefold definitions for all Diophantine sets raised in §3. It would also mean that the iterated exponential bounds that Baker obtained for a class of equations in 2 unknowns could actually be replaced by polynomial bounds.

We can list all definitions of the form of (2) and hence all sets which can be defined with polynomial test functions. Let (cf. p. 25) $C_n = \{a\colon (\exists x_1 \dots)_{\leq a}[P_n(a,x_1,\dots) = 0]\}$. Then the complement of the diagonal set i.e. $\{n\colon n \notin C_n\}$ cannot be defined by an equation with a polynomial test function although it is clearly computable.

Similarly, we can list the sets definable by equations with iterated exponential test functions. First number the iterated exponential functions of one variable: $E_1, E_2, \dots$. Then consider the sets

$$S_{ij} = \{m\colon (\exists x_1 \dots)_{\leq E_j(m)}\ [P_i(m,x_1,\dots) = 0]\}.$$

All sets which can be defined by equations with iterated exponential test functions are among the S_{ij}. Also since exponentiation can be defined by an equation with an iterated exponential test function, so also can S_{ij}. Again the set $\{J(i,j): J(i,j) \notin S_{ij}\}$ cannot be defined by an equation with an iterated exponential test function although it is clearly computable.

Obviously the union and intersection of two Diophantine sets can each be defined by an equation with a test function equal to the maximum of given test functions for the two sets. However, we have:

OPEN QUESTION. *Is the complement of a set defined by an equation with a polynomial test function always definable by an equation with a polynomial test function?*

Although this question remains open, we can see that the complement of a set defined by an equation with an iterated exponential test function can be defined by an equation with an iterated exponential test function. Indeed, suppose S is defined by $P(a,x_1,\dots,x_\nu) = 0$ and $E(a)$ is a test function for $P = 0$. Then $a \notin S$ if and only if there is a sequence $t_0,\dots,t_n$ such that

$$n = J_\nu(E(a),\dots,E(a)) \;\&\; (\forall k)_{\leq n} \; (\exists x_1 \dots x_\nu y)$$

$$[k = J_\nu(x_1,\dots,x\ ) \;\&\; t_k = P^2(a,x_1,\dots,x_\nu) \;\&\; t_k = 1+y] \;. \qquad (3)$$

Here we can put bounds on the x's, y's, and n which are polynomial in E(a) and hence iterated exponential functions of a. Finally using Gödel's lemma and the bounded quantifier theorem, we can transform (3) into a Diophantine definition with an iterated exponential test function.

10. APPLICATIONS OF THE THEORY OF ALGORITHMS

Since the listable sets are identical with the Diophantine sets, all of the theorems about listable sets in the theory of algorithms can be translated into theorems about Diophantine equations. Of course in most cases, the resulting statements seem quite unnatural, and answer questions about Diophantine equations that no one would have been much inclined to ask. In this section we discuss a few examples that we believe are of some interest.

The most spectacular example has already been discussed in §5: the existence of a universal Diophantine equation.

We begin with an application of the universal equation.

THEOREM. *There is a Diophantine equation*

$$W(a,t,x_1,\dots,x_\nu) = 0 \qquad (1)$$

such that if C_t *is the set of* a *for which* (1) *has a solution, then* $C_0, C_1, \dots$ *is an enumeration of all computable sets.*

PROOF. Let $\mathcal{D}_0, \mathcal{D}_1, \dots$ be the sets listed by the universal Diophantine equation $U(a,n,z_1,\dots,z_\nu) = 0$ as $n = 0,1,\dots$. We will construct $C_{J(m,n)}$ in such a way that if $\mathcal{D}_m$ and $\mathcal{D}_n$ are complementary sets then $C_{J(m,n)} = \mathcal{D}_m$ and otherwise $C_{J(m,n)}$ is at least computable.

Let $b \in C_{J(m,n)}$ if the following conditions hold:

I. $b \in \mathcal{D}_m$ is verified with the unknowns in the universal equation equal to $y_1,\ldots,y_\nu$, where no "earlier" values $z_1,\ldots,z_\nu$ of the unknowns (earlier in the sense that $J_\nu(z_1,\ldots,z_\nu) < J_\nu(y_1,\ldots,y_\nu)$) serve to verify that $b \in \mathcal{D}_n$.

II. For each $a < b$, either $a \in \mathcal{D}_m$ or $a \in \mathcal{D}_n$.

Now if $\mathcal{D}_m = N - \mathcal{D}_n$ then $C_{J(m,n)} = \mathcal{D}_m$. Hence every computable set is some C_t. Also if there is a number k which does not belong to either $\mathcal{D}_m$ or $\mathcal{D}_n$ then no number $\geq k$ will be in $C_{J(m,n)}$. Hence it is finite and certainly computable. Finally, if $C_{J(m,n)}$ is infinite we can check whether $b \in \mathcal{D}_n$ is verified before $b \in \mathcal{D}_m$ is and hence tell whether b belongs to $C_{J(m,n)}$. The conditions I and II can easily be written out using bounded universal quantifiers. Namely,

$$\begin{aligned} b\in C_t \leftrightarrow (\exists m,n)\{t=J(m,n) \;&\&\; (\exists y y_1\ldots y_\nu)[y=J_\nu(y_1,\ldots,y_\nu) \,\&\, U(b,m,y_1,\ldots,y_\nu)=0 \\ &\&\; (\forall z)_{<y} \; (\exists z_1\ldots z_\nu)[z = J_\nu(z_1,\ldots,z_\nu) \;\&\; U(b,n,z_1,\ldots,z_\nu) \neq 0]] \\ &\&\; (\forall a)_{<b} \; (\exists z_1\ldots z_\nu)[U(a,m,z_1\ldots z_\nu)\cdot U(a,n,z_1,\ldots,z_\nu) = 0]\}. \end{aligned}$$

We can now use the bounded quantifier theorem to obtain the required polynomial W.

This theorem is essentially that of Suzuki [62], and could have been obtained immediately by combining Suzuki's result with the Main Theorem.

Notice that we escape from a diagonal counterexample because there is no method of listing the set $\{n: n \notin C_n\}$. On the other hand, there can be no Diophantine equation which lists all and only the graphs of computable functions of one variable because otherwise we could compute the diagonal function, add 1 to it, and have a computable function not on the list.

Now we do the best we can. Namely, we start with Peano's axioms and make a list of all Diophantine equations $P(a,b,x_1,\ldots) = 0$ such that we can prove

$$(\forall a)(\exists! b)(\exists x_1,\ldots)[P(a,b,x_1,\ldots) = 0] .$$

For such a P, we call the function F defined by $P(a,F(a),x_1,\ldots) = 0$ for some $x_1,\ldots$ a *provably computable function*. Using our list of equations, we can define a computable function G which grows faster than any provably computable function. Notice that to construct G, we have to step outside Peano's axioms; otherwise G would be a provably computable function which is impossible.

It has been suggested by Kreisel [28] that possibly the reason that a proof of the infinitude of certain sets has not been found is that the function H which enumerates the set in question in order of size grows faster than any provably computable function, just as our diagonal function G does.

There are proof methods in the theory of algorithms which would be very difficult to translate into intuitive number theoretical arguments. For example, Friedberg using his "priority method" showed that there is a listable set M with an infinite complement such that every listable set which includes M either differs from M or from N by a finite number of elements. Cf. [51], [55]. Hence the corresponding theorem about Diophantine sets is obtained.

THEOREM. *There is a Diophantine set M with an infinite complement such that every Diophantine set which includes M either differs from M or from N by a finite set.*

COROLLARY. *Every infinite Diophantine set contains infinitely many elements of M.*

11. RINGS OF ALGEBRAIC INTEGERS

In this section we discuss Diophantine definability in rings of algebraic integers and the related question of the existence of an algorithm to tell whether an arbitrary Diophantine equation has a solution in a given ring. There are two natural generalizations of the notion of Diophantine set: one in which the coefficients of the defining polynomial equation remain rational integers while the parameters and unknowns can be assigned values in the given ring, and one in which the coefficients are chosen from the ring as well. We shall see that the two generalizations are very closely related.

Let I be the ring of integers in a subfield F of the field A of all algebraic numbers. Let G be the group of automorphisms of F (or equivalently, I). The degree of F may be finite or infinite over the rational field Q. A relation $T \subseteq I^n$ is called *Diophantine over* I if there is a polynomial equation with coefficients in I

$$P(a_1,\ldots,a_n,y_1,\ldots,y_m) = 0 \tag{1}$$

such that $T(a_1,\ldots,a_n)$ holds if and only if (1) has a solution in I for the unknowns $y_1,\ldots,y_m$. The equation $P = 0$ is then called a *Diophantine definition* of the relation T. Furthermore, T is called a *pure Diophantine* relation if it has a Diophantine definition with coefficients in Z (the rational integers). (Note that we are writing, e.g. $T(a_1,\ldots,a_n)$ to indicate that $\langle a_1,\ldots,a_n\rangle \in T$.)

Except possibly for the ring of all algebraic integers, it is immaterial whether we consider sets defined by single equations or by Boolean combinations of equations. In particular, let

$$A(x) = \sum_{i=0}^{n} a_i x^i = 0, \quad a_n \neq 0, \quad a_i \in I .$$

have no solution in F. Let

$$B(x,y) = \sum_{i=0}^{n} a_i x^i y^{n-i}.$$

Then in I, $u = 0$ & $v = 0$ if and only if $B(u,v) = 0$. Clearly, $B(0,0) = 0$. Conversely, suppose $v \neq 0$. Then $B(u,v) = 0$ implies $A(u/v) = 0$ where $u/v \in F$ which contradicts the hypothesis on A. So $v = 0$ and hence $u = 0$. Since an arbitrary finite algebraic extension of $K \subset A$ can always be made by adjoining a root of a single equation to K, the only field F for which there is no equation without a solution is the field A of all algebraic numbers. Finally, for any I, we have:

$$u=0 \vee v=0 \leftrightarrow uv=0$$

and

$$u\neq 0 \leftrightarrow (\exists xyz)\ [ux = (2y+1)(3z+1)]\ .$$

This last equivalence holds because when $u \neq 0$, x can be found so that ux is a positive rational integer which is the product of a power of 4 and an odd number.

The general question which we wish to raise here for each ring I of algebraic integers is:

What sets and relations of I *are Diophantine over* I*?*

The related question corresponding to Hilbert's tenth problem is:

Is there an algorithm to tell whether a Diophantine equation has a solution in I *?*

(Both these questions can also be asked with "pure Diophantine" in place of "Diophantine".)

Now the Main Theorem yields the answer to both questions for I = Z. In the first real advance since 1970, Denef [15] has completely answered both questions for the case where I is the ring of integers in a quadratic extension of Q. Before discussing his results, we need to define the notion of a *listable relation on* I. This can be done in any number of equivalent ways. For example, if I is of finite dimension k as a vector space over Z, an n-ary relation on I can be regarded simply as a kn-ary relation on Z. In general, if the elements of I can be numbered systematically, then a relation $T \subseteq I^n$ is *listable* if the corresponding relation among the numbers of the elements of I is a listable relation of natural numbers. Clearly, as in the case of *N*, any Diophantine relation over I is listable provided I itself is. We will also say that a relation T is *selfconjugate* if $T(x_1,\ldots,x_n)$ implies that $T(x_1^\sigma,\ldots,x_n^\sigma)$ for all automorphisms σ of I. Note that if a relation is pure Diophantine it must be selfconjugate.

THEOREM (Denef). *A relation is Diophantine over the ring* I *of integers of a quadratic extension of* Q *if and only if it is listable. A relation is pure Diophantine if and only if it is listable and selfconjugate over* I.

COROLLARY (Denef). *There is no algorithm to tell whether an arbitrary pure Diophantine equation has a solution in* I *(the ring of integers of a quadratic extension of* Q*).*

R. M. Robinson [49] settled for all I the question:

Which elements of I *have pure Diophantine definitions?*

THEOREM (R. M. Robinson). *Let* I *be the ring of integers in a field of algebraic numbers and suppose an element* α *of* I *is fixed under all automorphisms of* I *(i.e.* α *is selfconjugate). Then there are polynomials* P *and* Q *with rational integer coefficients such that*

$$x = \alpha \leftrightarrow (\exists y)[P(y) = 0 \ \& \ x = Q(y)] . \qquad (2)$$

Note that we have changed Robinson's theorem to conform to our statement of the problem -- the original version applied to arithmetical definitions of elements of an algebraic field. The two equations in (2) can be combined into one except in the case of the ring of all algebraic integers. In that case the rational integers are the only selfconjugate numbers and they can be defined directly.

Some negative results follow from considerations of selfconjugacy. For example,

THEOREM. *The set of real algebraic integers is not Diophantine over the ring of all algebraic integers.*

PROOF: Since there are automorphisms of A which take a real number into a non-real number, the set of real algebraic integers is not pure Diophantine. Suppose $P(a,y_1,\dots,y_m) = 0$ is a Diophantine definition over the ring of all algebraic integers of the ring of real algebraic integers and the coefficients of P lie in $I = Z(\alpha_1,\dots,\alpha_q)$. We can find an automorphism σ of A which leaves I fixed but still takes some real algebraic integer b into a nonreal integer. Hence $P(b,x_1,\dots,x_m) = 0$ implies $P(b^\sigma,x_1^\sigma,\dots,x_m^\sigma) = 0$. But this is impossible since b^σ is not real yet lies in the set defined by $P = 0$.

The next theorem gives the exact connection between pure Diophantine and Diophantine relations.

THEOREM. *Let* I *be the ring of integers of a field* F *of algebraic numbers. A relation* $A \subseteq I^n$ *is pure Diophantine if and only if* A *is Diophantine and selfconjugate.*

PROOF: Suppose A has the Diophantine definition

$$P(a_1,\dots,a_n,y_1,\dots,y_m) = 0$$

with coefficients in Z. Then A is of course Diophantine. If $A(x_1,\dots,x_n)$ holds, so that for suitable $y_1,\dots,y_m \in I$,

$$P(x_1,\dots,x_n,y_1,\dots,y_m) = 0$$

then (since σ is an automorphism leaving Z fixed)

$$P(x_1^\sigma,\dots,x_n^\sigma,y_1^\sigma,\dots,y_m^\sigma) = 0$$

so that $A(x_1^\sigma,\dots,x_n^\sigma)$. Hence A is selfconjugate.

Conversely, let A be Diophantine and selfconjugate, and let

$$P(a_1,\dots,a_n,y_1,\dots,y_m) = 0$$

be a Diophantine definition of A with coefficients in I. Let $\sigma \in G$ and let P^σ be obtained from P by replacing each coefficient by its image under σ. There are only a finite number k of different polynomials P^σ since each coefficient goes into some root of its irreducible equation. Let $\sigma_1,\dots,\sigma_k$ be the corresponding automorphisms which are the identity function on numbers not in the ring generated by the coefficients of P.

Let

$$Q(x_1,\dots,x_n,y_1,\dots,y_m) = \prod_{j=1}^{k} P^{\sigma_j}(x_1,\dots,x_n,y_1,\dots,y_m) .$$

since the different σ's enter symmetrically into the calculation of the coefficients of Q, these coefficients are fixed under G. Hence each coefficient has a pure Diophantine definition of the form (2). Thus, we can obtain a pure Diophantine definition of the same relation that is defined by $Q = 0$. We shall show that this relation is in fact A, thus completing the proof.

Suppose $A(x_1,\dots,x_n)$. Then $P(x_1,\dots,x_n,y_1,\dots,y_m) = 0$ for some $y_1,\dots,y_m \in I$. Hence (since the identity automorphism belongs to G), $Q(x_1,\dots,x_n,y_1,\dots,y_m) = 0$.

Finally, let $Q(x_1,\dots,x_n,y_1,\dots,y_m) = 0$. Then for some j

$$P^{\sigma_j}(x_1,\dots,x_n,y_1,\dots,y_m) = 0$$

Let τ be the inverse of σ_j. Then,

$$P(x_1^\tau,\dots,x_n^\tau,y_1^\tau,\dots,y_m^\tau) = 0 ,$$

so that $A(x_1^\tau,\dots,x_n^\tau)$ holds. Since A is selfconjugate and $x_i = (x_i^\tau)^{\sigma_j}$, we have $A(x_1,\dots,x_n)$.

THEOREM. *Let* F *be a finite extension of* Q *and* I *the ring of integers of* F. *Every listable relation on* I *is Diophantine over* I *if and only if the set* Z *is Diophantine over* I.

PROOF: Since Z is a listable subset of I, we need only show that if Z is Diophantine over I then every listable relation is Diophantine over I.

Let $A \subseteq I^n$ be listable and $\{\eta_1,\dots,\eta_k\}$ be an integral basis for I over Z. Let $R \subseteq Z^{nk}$ be defined by letting $R(u_1^{(1)},\dots,u_k^{(1)},\dots,u_1^{(n)},\dots,u_k^{(n)})$ hold just in case $A(u_1^{(1)}\eta_1+\dots+u_k^{(1)}\eta_k,\dots,u_1^{(n)}\eta_1+\dots+u_k^{(n)}\eta_k)$ holds. Thus R is a listable relation over Z and hence is Diophantine. Let $P(a_1^{(1)},\dots,a_k^{(1)},\dots,a_1^{(n)},\dots,a_k^{(n)},y_1,\dots,y_m) = 0$ be a Diophantine definition of R over Z. Then, we have (*with all variables ranging over* I):

$$A(x_1,\dots,x_n) \leftrightarrow (\exists u_1^{(1)},\dots,u_k^{(1)},\dots,u_1^{(n)},\dots,u_k^{(n)},y_1,\dots,y_m)$$

$$[u_1^{(1)} \in Z \ \& \ \ldots \ \& \ u_k^{(1)} \in Z \ \& \ \ldots \ \& \ u_1^{(n)} \in Z \ \& \ \ldots \ \& \ u_k^{(n)} \in Z \ \&$$
$$y_1 \in Z \ \& \ \ldots \ \& \ y_m \in Z \ \&$$
$$x_1 = u_1^{(1)}\eta_1 + \ldots + u_k^{(1)}\eta_k \ \&$$
$$\vdots$$
$$x_n = u_1^{(n)}\eta_1 + \ldots + u_k^{(n)}\eta_k \ \&$$
$$P(u_1^{(1)},\ldots,u_k^{(1)},\ldots,u_1^{(n)},\ldots,u_k^{(n)},y_1,\ldots,y_m) = 0] \ .$$

By hypothesis, there is a Diophantine definition for a condition of the form $\alpha \in Z$. Hence since Diophantine relations over I are closed under '&', we are done.

REMARK. In order to obtain the above theorem for an infinite extension F of Q, we would need some way of representing an arbitrary algebraic integer α of F as a listable relation over Z such that we can identify α as the number represented by means of a Diophantine relation over I. For example in Z, we can represent any algebraic number as a pair of Dedekind cuts. But in the ring of all algebraic integers, as we have seen, there is no way to define the set of real algebraic integers and the same argument also shows that it could not be defined in terms of Z.

Returning to the question of an algorithm to tell whether a Diophantine equation has a solution in I, we have already noted that if Z is definable there can be no such algorithm. It seems likely that for rings of integers of finite extensions of Q it will be possible to define Z and hence that there will be no algorithm in these cases. Thus, the most promising place to look for an algorithm is for the ring of all algebraic integers.

12. THE NUMBER OF SOLUTIONS OF A DIOPHANTINE EQUATION

In a Diophantine equation D with coefficients in Z, the following are all natural questions:

(1) Does D have a solution?

(2) Does D have infinitely many solutions?

(3) Does D have fewer than 100 solutions?

(4) Does D have an even (or a prime, etc.) number of solutions?

Here we will assume that solutions are being sought in some fixed domain R which may be the natural numbers *N*, the positive integers *P*, the rational integers Z, or the ring I of a proper subfield of all algebraic numbers. And we will assume that the analogue of Hilbert's tenth problem for R (i.e. question (1) in the above list) is known to be unsolvable. (So on the basis of current knowledge, R could be *N*, *P*, Z, or (by Denef's work [15], mentioned in §11) the ring I of integers of a *quadratic* extension of the rationals.) We shall show that all the questions (1)-(4) (and any others like them) are unsolvable. This was originally shown for P in Davis [12]; the present discussion follows the somewhat simplified treatment of Smorynski [60].

Let $a_1,\ldots,a_n$ be n given rational integers. Then we can make up a Diophantine equation with rational integer coefficients and unknowns $x_1,\ldots,x_n$ which has the unique solution in R: $x_1 = a_1, \ldots, x_n = a_n$. To see this we just combine these n equations into one equation using repeatedly the method of combining two equations into one given in §11.

Now, let $P(x_1,\ldots,x_n) = 0$ be an arbitrary Diophantine equation, and let $\langle a_1,\ldots,a_n\rangle$ be the first (in some suitable ordering e.g. using J_n) n-tuple of positive integers for which $P(a_1,\ldots,a_n) \neq 0$. And let $Q(x_1,\ldots,x_n) = 0$ be the equation constructed above whose unique solution in R is $x_1 = a_1, \ldots, x_n = a_n$. Then, we set:

$$T^+(P) = P\cdot Q .$$

We will write $\#(F)$ for the cardinal number of the set of solutions in R of the equation $F = 0$. (Thus for any F, $0 \leq \#(F) \leq \aleph_0$.) Then T^+ has been constructed so that

$$\#(T^+(P)) = \#(P) + 1 .$$

We define:

$$T^0(P) = P ,$$

$$T^{m+1}(P) = T^+(T^m(P)) .$$

Thus,

$$\#(T^m(P)) = \#(P) + m .$$

Then, we define

$$T^\infty(P) = (2u+1)\cdot P$$

where u is a variable that does not occur in P, and we note that

$$\#(T^\infty(P)) = \begin{cases} 0 & \text{if } \#(P) = 0 \\ \aleph_0 & \text{if } \#(P) \neq 0 \end{cases}$$

since $2u+1 \neq 0$ for all $u \in R$.

THEOREM. *Let* $C = \{0,1,\ldots,\aleph_0\}$. *Let* $A \subseteq C$ *but* $A \neq \emptyset, C$. *Then there is no algorithm for testing a given Diophantine equation* $P = 0$ *to determine whether* $\#(P) \in A$.

PROOF: It is sufficient to consider the following cases.

Case 1. $A = \{0\}$. This is the analogue for R of Hilbert's tenth problem and our assumption is that there is no algorithm to tell whether an arbitrary Diophantine equation has a solution in R. We will show that the other cases can be reduced to this one.

Case 2. $0,\aleph_0 \in A$. Let $m \notin A$. Then

$$\#(P) = 0 \longleftrightarrow \#(T^m(T^\infty(P))) \notin A.$$

Case 3. $0 \in A$, $\aleph_0 \notin A$. Then

$$\#(P) = 0 \longleftrightarrow \#(T^\infty(P)) \in A.$$

Hence in either Case 2 or Case 3 an algorithm to tell whether $\#(P) \in A$ would also solve Hilbert's tenth problem.

Case 4. $0 \notin A$. Then $C-A$ is either in Case 2 or Case 3, and so has already been treated.

13. DEGREES OF DIOPHANTINE PROBLEMS

In this section we will once again be dealing exclusively with the domain N. We also let $P_0 = 0$, $P_1 = 0$, $P_2 = 0, \ldots$ be an explicit enumeration of all Diophantine equations (e.g., the particular enumeration defined in §5). First, we want to generalize the notion of an algorithm to allow the use of some outside information. We say that A is *computable from* B if there is an algorithm for testing membership in A which however may require answers to (a finite number of) questions of the form: "*Does* $n \in B$?" Notice that this is a broader notion than A is *Diophantine in* B i.e. existentially definable in terms of $+$, $\cdot$, and B. (See §8.) For example, if $A = \{x^2 : x \in B\}$ then A is Diophantine in B as well as computable from B. However if

$$A = \{x: (\exists y)\ [x = \sum_{\substack{b \in B \\ b < y}} b]\}\ ,$$

then A is computable from B but not in general Diophantine in B. Note that A is trivially computable from its complement $\bar{A} = N - A$. We also say that A is *listable from* B provided there is an algorithm for listing members of A which may require answers to a finite number of questions of the form "*Does* $n \in B$?". For example, if

$$A = \{n: (b \in B)(P_n(b, x_1, \ldots) = 0 \text{ has a solution})\}\ ,$$

then A is listable from B. And of course A is computable from B just in case A and $\bar{A}$ are both listable from B. The customary picturesque way of describing the set A being computable from B is to say that A is computable by means of an oracle which can answer correctly questions of the sort "*Does* $n \in B$?".

In discussing these matters, we need to code a finite amount of information about membership (and nonmembership) in a set as a single integer. This may be done in a variety of ways -- all more or less arbitrary. To make matters definite, we call a natural number n the *code* for S where

$$n = \sum_{k \in S} 2^k\ .$$

For any set B, let $\tilde{B}$ be the set of codes for *modified initial segments* of B, i.e. sets of the form

$$\{b: (b \in B \ \&\ b < k) \vee b = k\}\ .$$

Thus,

$$b \in B \leftrightarrow (\exists k, x)[\text{the coefficient of } 2^b \text{ in the binary expansion of } x \text{ is } 1 \ \&\ x \geq 2^{b+1} \ \&\ x \in \tilde{B}]\ . \qquad (1)$$

Clearly, $\tilde{B}$ is computable from B and conversely.

Now instead of asking the original oracle for B a finite number of questions of the form 'Does $n \in B$?' as we proceed in our calculation of A, we can obtain the same information if the oracle for $\tilde{B}$ will disclose a sufficiently large $x \in \tilde{B}$. In order to see how this works, suppose we are given an algorithm for listing A from B. We modify the algorithm to calculate a set R of pairs $\langle a,x \rangle$ so that whenever the original algorithm asks 'Does $n \in B$?', we answer 'yes' if $x \geq 2^{n+1}$ and the coefficient of 2^n in the binary expansion of x is 1; we answer 'no' if $x \geq 2^{n+1}$ and the coefficient of 2^n is 0; if $x < 2^{n+1}$, we go back to the beginning and start over with x+1 in place of x. As the calculation proceeds, whenever the modified algorithm yields that a is in the set being calculated, we put $\langle a,x \rangle$ in R. Thus, R is a listable and hence a Diophantine relation. On the other hand if $\langle a,x \rangle \in R$ and $x \in \tilde{B}$ then $a \in A$. Thus, we have proved:

THEOREM. *A is listable from B if and only if A is Diophantine in $\tilde{B}$ where*

$$\tilde{B} = \{x\colon (\exists n)\ [x = 2^n + \sum_{\substack{k \in B \\ k<n}} 2^k]\}\ . \tag{2}$$

In fact,

$$x \in A \leftrightarrow (\exists x \in \tilde{B})(\exists y_1 \ldots y_n)[P(a,x,y_1,\ldots,y_n) = 0]$$

where $P(a,x,y_1,\ldots,y_n) = 0$ is a Diophantine definition of the relation R above.

The sets A and B are called *Turing equivalent* if each is computable from the other. Now it is easy to see that this is an equivalence relation. The equivalence classes are called *degrees*. For a degree $\underset{\sim}{a}$, we write $\deg(A) = \underset{\sim}{a}$ for $A \in \underset{\sim}{a}$. The degrees are partially ordered (i.e. $\underset{\sim}{a} \leq \underset{\sim}{b}$ if $\underset{\sim}{a} = \deg(A)$ and $\underset{\sim}{b} = \deg(B)$ where A is computable from B); each pair of degrees $\underset{\sim}{a}$, $\underset{\sim}{b}$ has a supremum written $\underset{\sim}{a} \cup \underset{\sim}{b}$, in this ordering; there is a least degree $\underset{\sim}{0}$; and there is a jump operator which maps each degree $\underset{\sim}{a}$ onto the degree $\underset{\sim}{a}'$ which is the maximum degree of a set listable from a set A with $\deg(A) = \underset{\sim}{a}$, so that for each $\underset{\sim}{a}$ we have $\underset{\sim}{a} < \underset{\sim}{a}'$. In particular $\underset{\sim}{0}'$ is the maximum degree of a listable set. (For an introduction to the extensive theory of degrees see Rogers [51] or Shoenfield [56].)

Let

$$S = \{n \mid P_n(x_0,x_1,\ldots) = 0 \text{ has a solution}\}\ ,$$

where we are using the enumeration of Diophantine equations mentioned above. We will show that

$$\deg(S) = \underset{\sim}{0}'. \tag{3}$$

In the first place, S is listable so $\deg(S) \leq \underset{\sim}{0}'$. Let L be any listable set and let $P_n(k,x_1,\ldots,x\ ) = 0$ be a Diophantine definition of L. Then $k \in L$ if and only if the number (in the enumeration) of the equation $P_n(k,x_1,\ldots,x\ ) = 0$ (or any equivalent equation) belongs to S. Since we can devise an algorithm for computing a number of $P_n(k,x_1,\ldots,x_\nu) = 0$ as a function of k (using the numbering of §5, it is enough to calculate the number of $P_n^2 + (x_0-k)^2 = 0$), and can then find out from the oracle for S whether the equation has a solution, we see that

$$\deg(L) \leq \deg(S)$$

for all listable sets L. Hence $\deg(S) = \underset{\sim}{0}'$.

As another example, consider the set F of numbers of equations which have a finite number of solutions. We will show that

$$\deg(F) \leq \underset{\sim}{0}'' \quad (= \deg(S)') . \tag{4}$$

We need to show that F is listable from S. Now all solutions of $P(x_0,\ldots,x_\nu) = 0$ satisfy $x_0 < b, \ldots, x_\nu < b$ if and only if $P(x_0+b,\ldots,x_\nu+b) = 0$ has no solution. Also there is a computable function $h(n,b)$ such that $P_{h(n,b)}(x_0,\ldots,x_\nu) = P_n(x_0+b,\ldots,x_\nu+b)$. Hence $n \in F$ if and only if $h(n,b) \notin S$ for some b. Therefore F is listable from S and (4) holds.

THEOREM. *Let* $C = \{0,1,\ldots,\aleph_0\}$ *and let* $A \subseteq C$, $A \neq \emptyset, C$. *Let* W *be the set of* n *such that* $\#(P_n) \in A$. *Then*

$$\underset{\sim}{0}' \leq \deg(W) \leq \deg(A) \cup \underset{\sim}{0}''. \tag{5}$$

PROOF. The left inequality is a strengthened form of the theorem in the last section, and its proof is easily obtainable by analyzing the proof of the latter. Namely, in each of the cases 2 through 4, we have a computable function g such that either

$$x \in S \leftrightarrow g(x) \in W$$

or

$$x \in S \leftrightarrow g(x) \notin W .$$

In Case 2, for example, $g(n)$ is the number of the equation $T^m(T^\infty(P_n)) = 0$ and the corresponding equivalence can be expressed as

$$n \in S \leftrightarrow g(n) \in W .$$

Hence S is computable from W and hence by (3),

$$\underset{\sim}{0}' \leq \deg(W) .$$

Now for the right inequality in (5), it is sufficient to show that W is computable from A and F. To check whether $n \in W$, we first use F to see whether $P_n = 0$ has a finite or infinite number of solutions. If $n \in F$, then we can find all solutions -- first searching for and finding a bound on the unknowns using S which is of course in $\underset{\sim}{0}''$ and then trying all possible values of the unknowns below this bound -- finally we use A to check whether $\#(P_n) \in A$. If $n \notin F$, then $\#(P_n) \in A$ if and only if $\aleph_0 \in A$. Thus, W is computable from A and F so the theorem is proved.

COROLLARY. *If* A *is a finite set of finite cardinal numbers then* $\deg(W) = \underset{\sim}{0}'$.

PROOF. By the theorem, it is sufficient to show that $\deg(W) \le \underset{\sim}{0}'$. It is enough to consider the case where A consists of a single number, say k, since a uniform upper bound is preserved under a finite union. Now we can easily give an algorithm for listing the set $W_{\ge k}$ of numbers of equations with at least k solutions so

$$\deg(W_{\ge R}) \le \underset{\sim}{0}'.$$

But $W = W_{\ge k} \cap \bar{W}_{\ge k+1}$; so $\deg(W) \le \underset{\sim}{0}'$.

We have presented here all the known information about deg(W) for $A \subseteq C$. In particular, it is not known whether there exists a computable A for which $\deg(W) > \underset{\sim}{0}'$. There is the natural

CONJECTURE. *If* $A = \{\aleph_0\}$, *then* $\deg(W) = \underset{\sim}{0}''$.

Although this conjecture remains open, it is interesting to note that it would follow easily if every Diophantine set has a singlefold or even finitefold Diophantine definition in the sense of §3.

It is well known that there is no algorithm for listing the infinite listable sets. We can weaken the above conjecture slightly and ask

Show there is no algcrithm for listing the Diophantine equations with infinitely many solutions.

14. MODELS OF ARITHMETIC

Here we will be concerned with models of the true sentences of the arithmetic of natural numbers. These models are mathematical structures consisting of a domain together with operations + and $\cdot$, a relation <, and two individuals 0 and 1 defined on the domain so that specified sentences in the language $L(+,\cdot,<,0,1)$ are true. Skolem [59] first showed that there are models whose domains contain *infinite* numbers but for which all sentences true in the arithmetic of natural numbers are true in these models. An element of the domain is an *infinite* number if it is greater than 1, 1+1, 1+1+1, ... If a set S of (true) sentences of arithmetic is sufficiently strong to imply every true sentence (in N) of the form

$$\alpha + \beta = \gamma$$
$$\alpha \cdot \beta = \gamma$$
$$\alpha < \beta$$
$$x \le \alpha \to x = 0 \lor x = 1 \lor \ldots \lor x = \alpha$$
$$x < \alpha \lor x = \alpha \lor x > \alpha$$

where α, β, γ are 0 or sums of 1's, then models of S containing infinite numbers are called *nonstandard* models. The only other models are isomorphic to N itself.

Infinite numbers correspond to sets of natural numbers. In particular, the sentence,

$$(\forall b)(\exists ad)(\forall x)_{\le b}[d(x+1)+1 \mid a \leftrightarrow \phi(x)]$$

where $\phi(x)$ is any formula of number theory with one free variable x, is a true statement of number theory (namely, a form of Gödel's lemma) and hence true in a model of true arithmetic. (This idea was first used by Stanley Tenenbaum.) Let M be a nonstandard model of true arithmetic and b some infinite number in M. Let $\bar{n} = \sum_{i=1}^{n} 1$ for $n \in N$. Then there are a and d in M so that

$$(d(\bar{n}+1)+1)y = a \tag{1}$$

has a solution for y (in M) if and only if $\phi(\bar{n})$ holds in N. Thus, any arithmetically definable set of natural numbers is existentially definable in a nonstandard model of true arithmetic. Scott [55] exploited this remark to sharpen a theorem of Rabin:

The first order theory of countable models of the system of all predicates of natural numbers is categorical.

Note that (1) is a *generalized Diophantine equation over* M, i.e. an equation of the form

$$R_1(x_1,\ldots,x_\nu) = R_2(x_1,\ldots,x_\nu)$$

where R_1 and R_2 are polynomials with coefficients in M (compare with the discussion at the beginning of §11).

We shall consider models of these classes of true sentences of the arithmetic of natural numbers:

(1) models of *true arithmetic*, i.e. all sentences of $L(+,\cdot,<,0,1)$ true in N;

(2) models of Peano's axioms including an axiom identifying $<$ as the ordering induced by addition (called models of *Peano arithmetic*);

(3) models of all $\forall\exists$-sentences true in N, i.e. sentences of the form

$$(\forall x\ldots x_\mu)(\exists y_1\ldots y_\nu)[\phi(x_1,\ldots,x_\mu,y_1,\ldots,y_\nu)]$$

where ϕ is a quantifier-free formula (called models of *$\forall\exists$-arithmetic*).

One aspect of model theory is the study of these structures. The fact that every listable set is Diophantine can be used both to simplify old proofs and to obtain new results. Here we shall give some samples of these results which can be readily understood rather than go into the technicalities of the sharpest theorems.

I. Suppose M and M' are models of true arithmetic and M is a submodel of M'. Then there is a model M^* obtained by *interpolation* from M, i.e. by adjoining elements which lie between elements of M, such that M' is obtained from M^* by *extrapolation*, i.e. by adjoining elements which are greater than every element of M^*. See Adler [1], Čudnovskii [8], and Gaifman [17]. Gaifman generalized the theorem to models of Peano arithmetic.

II. Let M be a nonstandard model of true arithmetic. Then there is a generalized Diophantine equation over M which has no solution in M but which has a solution in an extension of M which is also a model of true arithmetic. (See Rabin [41] and Gaifman [17].)

A model M of $\forall\exists$-arithmetic is called *existentially complete* if every generalized Diophantine equation over M which has no solution in M also has no solution in every extension of M which is also a model of $\forall\exists$-arithmetic.

III. There is a formula $\phi(a)$ in $L(+,\cdot,<,0,1)$ which holds in an existentially complete model of $\forall\exists$-arithmetic if and only if a is infinite. Hence

$$(\forall a)(\sim\phi(a))$$

holds in N but in no other existentially complete model of $\forall\exists$-arithmetic. (See A. Robinson [44].)

REFERENCES

1. Andrew Adler, *Extensions of nonstandard models of number theory*, Zeitschr. f. Math. Logik und Grundlagen d. Math. 15 (1969), 289-290.
2. __________, *Some recursively unsolvable problems in analysis*, Proc. Amer. Math. Soc. 22 (1969), 523-526.
3. __________, *A reduction of homogeneous diophantine problems*, J. Lond. Math. Soc. (2), 3 (1971), 446-448.
4. Jean-Pierre, Azra, *Relations diophantiennes et la solution négative du 10e problème de Hilbert*, Séminaire Bourbaki (1970/71), exposé 383, 11-28.
5. Alan Baker, *Contributions to the theory of Diophantine equations: I. On the representation of integers by binary forms, II. The Diophantine equation* $y^2 = x^3 + k$, Philos. Trans. Roy. Soc. London Ser. A. 263 (1968), 173-208.
6. Paul Cohen, *The independence of the continuum hypothesis*, Proc. Nat. Acad. Sci. USA, 50 (1963), 1143-1148, 51 (1964), 105-110.
7. G. V. Čudnovskii, *Diophantine predicates* (Russian), Uspehi Mat. Nauk, 25 (1970) No. 4, (154), pp. 185-186.
8. __________, *Certain arithmetical problems* (Russian), Akad. Nauk Ukrain. SSR, Preprint IM-71-3.
9. Martin Davis, *Computability and Unsolvability*, McGraw-Hill, New York, 1958.
10. __________, *One equation to rule them all*, Trans. New York Acad. Sci., Series II, 30 (1968), 766-773.
11. __________, *An explicit Diophantine definition of the exponential function*, Comm. Pure Appl. Math. 24 (1971), 137-145.
12. __________, *On the number of solutions of Diophantine equations*, Proc. Amer. Math. Soc., 35 (1972), 552-554.
13. __________, *Hilbert's tenth problem is unsolvable*, Amer. Math. Monthly, 80 (1973), 233-269.
14. __________, Hilary Putnam, and Julia Robinson, *The decision problem for exponential Diophantine equations*, Ann. Math. 74 (1961), 425-436.
15. J. Denef, *Hilbert's tenth problem for quadratic rings*, Proc. Amer. Math. Soc., to appear.
16. Jens Erik Fensted, *Hilbert's 10th problem* (Norwegian), Nordisk Matema-

tisk Tidskift, 19 (1971), 5-14.

17. H. Gaifman, *A note on models and submodels of arithmetic*, Conf. in Math. Logic (W. Hodges, ed.) London 1970, Springer Verlag, 1972, 128-144.

18. Kurt Gödel, *Über formal unentscheidbare Sätze der Principia Mathematica und verwandter Systeme I*, Monatsh. Math. und Physik, 38 (1931) 173-198. English translations: (1) Kurt Gödel, *On Formally Undecidable Propositions of Principia Mathematica and Related Systems*, Basic Books, 1962. (2) Martin Davis (ed.), *The Undecidable*, Raven Press, 1965, pp. 5-38. (3) Jean Van Heijenoort (ed.), *From Frege to Gödel*, Harvard University Press, 1967, pp. 596-616.

19. __________, *Consistency-proof for the generalized continuum hypothesis*, Proc. Nat. Acad. Sci. U.S.A., 25 (1939), 220-224.

20. R. L. Goodstein, and R. D. Lee, *A class of equations in recursive arithmetic*, Zeitschr. f. Math. Logik und Grundlagen d. Math. 12 (1966), 235-239.

21. Ivan Havel, *On Hilbert's tenth problem* (Czech.), Pokr. mat. fyz. a astron. 18 (1973), 185-192.

22. Hans Hermes, *Die Unlösbarkeit des zehnten Hilbertschen Problems*, L'Enseignement Math. II, 18 (1972), 47-56.

23. O. Herrmann, *A nontrivial solution of the Diophantine equation* $9(u^2+7v^2)^2 - 7(r^2+7s^2)^2 = 2$, Computers in Number Theory (Proc. Atlas Symp., 2), Oxford 1969, 207-217.

24. A. E. Ingham, *The Distribution of Prime Numbers*, Cambridge Tract No. 30, Cambridge University Press, 1932.

25. James P. Jones, Daihachiro Sato, Hideo Wada, and Douglas Wiens, *Diophantine representation of the set of prime numbers*, Amer. Math. Monthly, to appear.

26. N. K. Kosovskii, *Diophantine representations of the sequence of solutions of the Pell equation* (Russian), Zap. Naučn. Sem. Leningrad. Otdel. Mat. Inst. Steklova (LOMI), 20 (1971), 49-59. English translation: J. Sov. Math., 1 (1973), 28-35.

27. __________, *On solutions of systems consisting of both word equations and word length inequalities* (Russian), Ibid., 40 (1974), 24-29.

28. Georg Kreisel, *Mathematical significance of consistency proofs*, J. Symbolic Logic 23 (1958), 155-182.

29. Edmund Landau, *Handbuch der Lehre von der Verteilung der Primzahlen-Erste Band*, B. G. Teubner, 1909.

30. Wilhelm Magnus, Fritz Ohettinger and Raj Pal Soni, *Formulas and Theorems for the Special Functions of Mathematical Physics*, Springer Verlag, 1966.

31. A. I. Malcev, *Algorithms and Recursive Functions* (Russian), Moscow 1965. English translation: Walters-Noordhoff, 1970.

32. Yu. I. Manin, *Hilbert's tenth problem* (Russian), Sovremennye Prob. Mat., 1973, I, 5-37.

33. A. A. Markov, *Theory of Algorithms* (Russian), Trud. Mat. Inst. Steklova 42, Moscow-Leningrad 1954. English translation: Ots 60-51085, U. S.

Dept. of Commerce, 1961.

34. Yuri Matijasevič, *Enumerable sets are diophantine* (Russian), Dokl. Akad. Nauk SSSR, 191 (1970), 279-282. Improved English translation: Soviet Math. Doklady, 11 (1970), 354-357.

35. __________, *Diophantine representation of the set of prime numbers* (Russian), Dokl. Akad. Nauk SSSR, 196 (1971), 770-773. Improved English translation with Addendum: Soviet Math. Doklady, 12 (1971), 249-254.

36. __________, *Diophantine sets*, Uspehi Mat. Nauk 27 (1972), # 5, 185-222; English translation: Russ. Math. Surveys 27 (1972), #5, 124-164.

37. __________, *The existence of noneffectizable estimates in the theory of exponential Diophantine equations* (Russian), Zap. Naučn. Sem. Leningrad. Otdel. Mat. Inst. Steklova (LOMI) 40 (1974), 77-93.

38. __________, *On recursive unsolvability of Hilbert's tenth problem*, Proc. Fourth Int. Congress Logic, Method. and Phil. of Sci., Bucharest 1971 (North-Holland 1973), 89-110.

39. __________ and Julia Robinson, *Reduction of an arbitrary Diophantine equation to one in 13 unknowns*, Acta Arithmetica 27 (1974), 521-553.

40. Hilary Putnam, *An unsolvable problem in number theory*, J. Symb. Logic, 25 (1960),220-232.

41. Michael O. Rabin, *Diophantine equations and nonstandard models of arithmetic*, Logic, Methodology, Philosophy of Science (ed. Nagel, Suppes, Tarski), Proc. of 1960 Stanford Congress, Stanford University Press, 1962, 151-158.

42. Daniel Richardson, *Some undecidable problems involving elementary functions of a real variable*, J. Symbolic Logic 33 (1968), 514-520.

43. __________, *Solution of the identity problem for integral exponential functions*, Zeitschr. f. Math. Logik und Grundlagen d. Math. 15 (1969), 333-341.

44. Abraham Robinson, *Nonstandard arithmetic and generic arithmetic*, Proc. Fourth Int. Congress Logic, Method. and Phil. of Sci., Bucharest, 1971 (North-Holland, 1973), 137-154.

45. Julia Robinson, *Definability and decision problems in arithmetic*, J. Symbolic Logic, 14 (1949), 98-114.

46. __________, *Existential definability in arithmetic*, Trans. Amer. Math. Soc., 72 (1952), 437-449.

47. __________, *Solving diophantine equations*, Proc. Fourth Int. Congress Logic, Method. and Phil. of Sci., Bucharest 1971 (North-Holland, 1973), 63-67.

48. __________, *Axioms for number theoretic functions*, Izbrannye voprocy algebry i logiki (volume dedicated to the memory of A. I. Malcev), Novosibirsk 1973, 253-263.

49. R. M. Robinson, *Arithmetical definability of field elements*, J. Symbolic Logic 16 (1951), 125-126.

50. __________, *Some representations of Diophantine sets*, J. Symbolic Logic 37 (1972), 572-578.

51. Hartley Rogers, Jr., *Theory of Recursive Functions and Effective Computability*, McGraw-Hill, 1967.

51a. J. B. Rosser, *Logic for Mathematicians*, McGraw-Hill, 1953.

52. Keijo Ruohnen, *Hilbert's tenth problem* (Finnish), Arkhimedes (Helsinkfors) 1972, 71-100.

53. T. L. Saaty, *Thirteen colorful variations on Guthrie's four-color conjecture*, Amer. Math. Monthly, 79 (1972), 2-43.

54. Edward Schwartz, *Existential definability in terms of some quadratic functions*, Doctoral Dissertation, Yeshiva University, 1974.

55. Dana Scott, *On a theorem of Rabin*, Indag. Math. 22 (1960), 481-484.

56. Joseph Shoenfield, *Degrees of Unsolvability*, North-Holland/American Elsevier, 1971.

57. Carl Ludwig Siegel, *Zur Theorie der quadratischen Formen*, Nachr. Akad. Wiss. Göttingen Math.-Phys. Kl. II (1972), 21-46.

58. Th. Skolem, *Diophantische Gleichungen*, Ergebnisse d. Math. u. Ihrer Grenzgebiete, Bd. 5, Julius Springer, 1938.

59. Th. Skolem, *Über die Nicht-charakterisierbarkeit der Zahlenreihe mittels endlich oder abzählbar unendlich vieler Aussagen mit ausschliesslich Zahlenvariablen*, Fund. Math. 23 (1934), 150-161.

60. C. Smorynski, *Diophantine decision problems; a brief survey* (unpublished).

61. Ernst Straus, Oral communication.

62. Yoshindo Suzuki, *Enumeration of recursive sets*, J. Symbolic Logic 24, (1959), 311.

63. Bryant Tuckerman, *The 24th Mersenne prime*, Notices Amer. Math. Soc. 18 (1971), 60.

64. Alan Turing, *On computable numbers, with an application to the Entscheidungsproblem*, Proc. London Math. Soc. (2), 42 (1936-7), 230-265; 43 (1937), 544-546.

65. Alurin Walther, *Anschauliches zur Riemannschen Zetafunktion*, Acta Math. 48 (1926), 393-400.

OTHER RELATED ARTICLES

66. Gregory Chaitin, *Information-theoretic limitations of formal systems*, J. Assoc. Comput. Mach. 21 (1974), 403-424.

67. Martin Davis and Hilary Putnam, *Diophantine sets over polynomial rings*, Illinois J. Math., 7 (1963), 251-256.

68. Georg Kreisel, *What have we learned from Hilbert's second problem?* This Symposium.

69. Charles F. Miller III, *Some connections between Hilbert's tenth problem and the theory of groups*, Word Problems (Boone, Cannonito, Lyndon ed.) North-Holland, 1973, 483-506.

70. S. Yu. Maslov, *The concept of strict representability in the general theory of calculi*, Trud. Mat. Inst. Steklov 93 (1967), 3-42. English translation: Proc. Steklov Inst. 93 (1967), 1-50.

71. Yuri Matijasevič, *The connection between Hilbert's tenth problem and systems of equations between words and lengths* (Russian), Zap. Naučn. Sem. Leningrad. Otdel. Mat. Inst. Steklova (LOMI), 8 (1968), 132-144. English translation: Seminars in Mathematics, V. A. Steklov Math. Inst., Leningrad, 8 (1970), 61-67.

72. ________, *Two reductions of Hilbert's tenth problem* (Russian), Ibid., 145-158. English translation: Ibid., 68-74.

73. ________, *Arithmetic representation of powers* (Russian), Ibid., 159-165. English translation: Ibid., 159-165.

74. ________, *Diophantine representation of enumerable predicates*, Izv. Akad. Nauk SSSR, Ser. Mat. 35 (1971), 3-30; English translation, Math. USSR Izv. 5 (1971), 1-28.

75. ________ and Julia Robinson, *Two universal three-quantifier representation of recursively enumerable sets* (Russian), Teopiya algorifmov u matematiceskaja logika. Vycislitel'nyi Centr. A. N. SSSR, Moscow 1974 (volume dedicated to A. A. Markov), 112-123.

76. Julia Robinson, *Diophantine decision problems*, MAA Studies in Mathematics 6 (1969) [Studies in Number Theory, ed. W. J. LeVeque, pp. 76-116].

77. ________, *Unsolvable Diophantine problems*, Proc. Amer. Math. Soc. 22 (1969), 534-538.

78. Paul Wang, *The undecidability of the existence of zeros of real elementary functions*, J. Assoc. Comput. Mach. 21 (1974), 586-589.

COURANT INSTITUTE OF MATHEMATICAL SCIENCES OF NEW YORK UNIVERSITY,
STEKLOV MATHEMATICAL INSTITUTE OF ACADEMY OF SCIENCES OF USSR, LENINGRAD,
AND
UNIVERSITY OF CALIFORNIA, BERKELEY

Between logic and arithmetic

Julia Robinson

Fifty years ago, Gödel discovered two theorems which changed our view of mathematics, the Completeness Theorem and the Incompleteness Theorem. Because the subject of these lectures rests so firmly on these two theorems, I will begin with a short review of them.

The Completeness Theorem. Every valid formula of logic is provable using the traditional rules and axioms.

The most familiar statement of the Completeness Theorem is:

Every consistent set of sentences has a model.

Another version is:

Every statement of logic is either provable or there is a countable mathematical structure in which it is false.

Applied to arithmetic, this tells us that:

Given a sentence S and a set A of sentences in the <u>language of arithmetic</u> either S is provable from A or there is a countable <u>model of A</u> in which S is false.

Here the language of arithmetic consists of formulas built up from 0, 1, variables, $+$, $\cdot$, $<$, $=$, $\wedge$ (and), $\vee$ (or), $\neg$ (not), $\rightarrow$ (if ... then...), $\leftrightarrow$ (if and only if), $\exists$ (there exists), and $\forall$ (for all). We will use arithmetic to apply to various systems such as $\mathbb{N}$, $\mathbb{Q}$, $\mathbb{R}$, etc. By a <u>model of A</u>, we mean a structure M denoted by $\langle M,+,\cdot,<,0,1\rangle$ such that every sentence of A is true in M.

Various proofs of the Completeness Theorem yield ways of constructing models satisfying a consistent set of sentences. Rings and integral domains are models of finite sets of axioms. We can think of these as crude approximations to the arithmetic of integers. By means of the Completeness Theorem, we see there are "nonstandard" models of any set whatever of sentences true in $\mathbb{N}$, such as the Peano Axioms or even the set of all true sentences. Indeed, if A is such a set of sentences and we increase the language to include an individual constant, say α, then adjoin to A the sentences

(*) $\alpha>1$, $\alpha>1+1$, $\alpha>1+1+1$, ...,

and call the resulting set B. Then B is consistent since each finite subset of B can be satisfied in $\mathbb{N}$ with α taken as a sufficiently large integer. [Only a finite number of sentences occur in the proof of a contradiction, so B

1

is consistent.] Hence by the Completeness Theorem, there is a model of B, say M. Now look at M in the original language of arithmetic, suppressing those sentences in which α occurs. M is certainly a model of A and it must contain some number greater than all the natural numbers since all the formulas (*) are true for the element represented by α in M as a model of B.

The existence of nonstandard models raises two questions: What are they like? Can they tell us anything about the natural numbers? We shall return to this subject later.

Next we will look at Gödel's Incompleteness Theorem. We will incorporate in the statement an improvement due to Rosser.

The Incompleteness Theorem. A consistent axiomatic theory A which includes enough arithmetic (the Peano Axioms are ample) is incomplete, i.e. there is a sentence ϕ of arithmetic such that neither ϕ nor $\neg\phi$ can be proved from A.

It may seem paradoxical that we require "enough arithmetic" to obtain incompleteness. But to carry out the proof, we need arithmetic to code meta-mathematics into the theory. Of course, leaving out axioms from an incomplete theory will not make it complete. However adding axioms inconsistent with arithmetic may make the theory complete.

In Gödel's proof ϕ is the famous sentence which asserts (in code) its own unprovability and thus is seen to be a true sentence of arithmetic in $\mathbb{N}$.

It is important to understand that an axiomatic theory has a decidable set of axioms - we can check whether a given sentence is an axiom. For example, theories whose axioms consist of a finite number of axioms and axiom schemes are axiomatic.

Clearly, some restrictions on the set of axioms must be made in the Incompleteness Theorem. Otherwise, we could take the set of all sentences true in $\mathbb{N}$ as axioms, which would contradict the theorem.

An immediate corollary is the impossibility of axiomatizing large parts of mathematics. There is more to mathematics than proof in the traditional sense.

How did Gödel manage to code such notions as "unprovable" into formulas of ordinary arithmetic? The crucial step is a method of representing an arbitrary finite sequence of natural numbers by an arithmetical formula.

Gödel's Lemma. Given a sequence $a_1, \ldots, a_n$ of natural numbers, there exist a and d such that

2

$\text{Rem}(a, 1+id) = a_i$ for $i = 1, \ldots, n$.

Here Rem(a, 1+id) denotes the remainder when a is divided by 1+id. The proof follows easily from the Chinese Remainder Theorem and the fact that if n! divides d then 1+d, ..., 1+nd are pairwise relatively prime.

We call a relation ρ arithmetically definable in a structure M if there is a formula ϕ of arithmetic such that $\rho(x_1, \ldots, x_n)$ if and only $\phi(x_1, \ldots, x_n)$ is true in M. For example,

$$\text{Rem}(a, 1+d) = r \leftrightarrow (\exists q)\ [a = q(1+d) + r \wedge 0 \leqq r \leqq d].$$

So the remainder function is arithmetically definable in $\mathbb{N}$.

Gödel used the representation of finite sequences to show that the familiar number theoretical functions and relations are arithmetically definable in $\mathbb{N}$.

He then numbered formulas, sentences, proofs, etc. in a constructive way and used the numbers as codes to represent metamathematical concepts in arithmetic itself. Thus he showed for example that the set of numbers of sentences which are theorems in Peano Arithmetic is arithmetically definable in $\mathbb{N}$. These codes became known as Gödel numbers, and the method of embedding other concepts into arithmetic by working with the corresponding Gödel numbers has been widely applied.

A counting argument shows that not all sets of natural numbers are arithmetically definable since there are only countably many formulas altogether.

Tarski used the self-referential feature of Gödel's proof to obtain:

The set of Gödel numbers of the arithmetical sentences true in $\mathbb{N}$ is not arithmetically definable.

Tarski's theorem is much more general but this version will do here. Thus, even if we start with an arithmetically definable set of sentences true in $\mathbb{N}$ as axioms (waiving the decidability requirement), the resulting set of theorems will be arithmetically defined and hence will fall short of the set of all true sentences and so be incomplete.

What do we mean by decidable or equivalently computable? A set $S \subseteq \mathbb{N}$ is computable if there is a method or program such that for each n, the program specifies a finite calculation which yields either $n \in S$ or $n \notin S$. Here a program is a finite sequence of unambiguous instructions which can be carried out mechanically. We will say more about suitable instructions later.

We wish to compare this notion of computable with that of a listable set.

3

A set S is <u>listable</u> if there is a program which for each n specifies a calculation such that if $n \in S$ the calculation is finite and yields $n \in S$ and if $n \notin S$ either the calculation is finite and yields $n \notin S$ or it is infinite. (We could of course modify the program so the calculation would always be infinite if $n \notin S$.)

We call S listable since we can use the program to make a list of the numbers in S. Namely, by running the program applied to 0,0,1,0,1,2,0,1,2,3,... in turn for longer and longer periods of time, and whenever the calculation stops and yields $n \in S$ putting n on the list.

In a similar way, we define the notion of <u>computable relation</u> and <u>listable relation</u>.

A. A set S is computable iff both S and its complement $\overline{S}$ are listable.

Proof. Clearly if S is computable then both S and $\overline{S}$ are listable for we can easily modify the program which computes S to interchange the answers $n \in S$ and $n \notin S$. Conversely, from the programs P and $\overline{P}$ which terminate when applied to n according as $n \in S$ or $n \notin S$ respectively, we can make up a program which always terminates with the right answer. Namely run P and $\overline{P}$ alternately and stop when one of them stops giving the answer $n \in S$ if P stops or $n \notin S$ if $\overline{P}$ stops yielding $n \in S$.

Assumption. There is a program language L sufficiently strong to contain a program for every listable set.

This assumption is made to prove B. We will defend it later.

B. There is a listable set which is not computable.

Proof. First, make a list of all programs P_0, P_1, ... in L and let L_0, L_1, ... be the corresponding listable sets. This is possible since any sequence of instructions is a program. We can then list the diagonal set:

$D = \{n : n \text{ belongs to } L_n\}$.

D is listable since we can set the nth program computing with input n, and it will halt yielding $n \in L_n$ iff $n \in D$. Now these instructions for listing D are not a program in L but by the assumption there will be some program in L which does list D. If D is computable then $\overline{D}$ will also be listed by a program of L, say P_k. Then

$k \in \overline{D}$ iff $k \in L_k$ by the definition of k,

and

$k \in \overline{D}$ iff $k \notin L_k$ by the definition of D.

4

Hence we have a contradiction, so D is not computable.

After Gödel's paper many logicians including Church, Gödel, Herbrand, Kleene, Post, Turing, etc. investigated the problem of defining computable sets and functions. Each came up with what amounted to a quite different program language. Church was the first to make the bold claim that his λ-calculus was sufficient to obtain all computable sets. This claim is known as Church's Thesis. It turned out that all the languages led to the same class of "computable" sets. However it is the routine method of proving the equivalence of the definitions using Gödel's metamathematical techniques which convinces us of the truth of Church's Thesis and of the Assumption made above.

It should also be noted that while these sets and functions are theoretically computable, space and time limit the number of values which can in fact be computed. The most familiar description of a computable function is simply one whose values can be computed on an electronic computer without space and time limitations.

The <u>decision problem</u> for a set of assertions is the problem of either finding a method of telling whether a particular sentence belongs to the set or showing that no such method exists.

If a method is proposed and it is shown to be correct, everyone will agree that the decision problem has been solved. We expect to recognize a method. However until the notion of a computable set (or function) was satisfactorily defined, it was impossible to show that there was no method for solving certain classes of problems. Now if the set of Gödel numbers of true statements in a model, or of theorems of a particular theory, or of solvable equations of a given family is not computable then clearly no methods exist to tell which statements are true, which are theorems, or which equations are solvable, etc. This is also the case for any class of programs adequate to list all computable functions say of one variable F_0, F_1 ... . For it would then be possible to compute the diagonal function $G(n) = F_n(n) + 1$ which would not be on the list.

Church first showed that there is no decision method for theorems of elementary number theory and then that there is no method of telling whether a formula of logic is valid. Note that by the Completeness Theorem, the standard axiomatization of logic is adequate to prove all valid formulas, so the set is listable but not computable.

Gödel essentially showed in his Incompleteness proof that every listable set is arithmetically definable (although the definition of listable set was not available at the time). In fact he showed more, namely, every listable set or relation ρ can be expressed in the form

$$\rho(x_1 \cdots x_n) \leftrightarrow Q_1 y_1 \cdots Q_m y_m \left[P(x_1, \ldots, x_n, y_1, \ldots, y_m) = 0\right]$$

where P is a polynomial with integer coefficients and the Q's are quantifiers with all the universal quantifiers bounded, e. g. $\forall z < w$.

It is an easy exercise to see that if ρ has such a definition then a systematic search will serve as a program for listing the n-tuples satisfying ρ. Hence there is no decision method for sentences of the sort

$$Q_1 y_1 \cdots Q_m y_m \left[P(y_1, \ldots, y_m) = 0\right]$$

where $P = 0$ is a diophantine equation with the unknowns $y_1, \ldots, y_m$.

Turing made an exhaustive analysis of the possible steps in a computation and reduced them to their simplest components. He then defined a class of computing machines which could carry out these elementary steps and hence could compute any set or function that any machine could. He then showed that there was a single all-purpose or universal machine which could list all listable sets. Thus, the operator wishing to know if n belongs to the kth listable set simply feeds in n and k suitably coded into the universal machine and if n does in fact belong to the kth set the machine will stop; otherwise it would continue computing away endlessly.

We shall be using Turing machines in the second lecture so we must describe in detail a particular formulation of these machines here.

Each machine has a dual purpose head which moves along a tape reading and writing as it goes along. The tape is marked off in squares and extends indefinitely to the right but has only a finite number of squares which are not blank. The machine works in a finite alphabet and has a finite number of states. To start a calculation the machine is put in the initial state and the head is set scanning the leftmost square of the tape.

The machine then scrupulously follows its own particular set of instructions which are of the form:

If the head is scanning a symbol c in state q then it overprints c', moves one square to the right (or left), and goes into state q'.

The machine will stop only if it goes into a special halting state. There is exactly one instruction for each scanned symbol and state, other than the halting state. Note that the direction the head moves and the new state are determined completely by the scanned symbol and the old state.

6

We will make the convention that * is put in the leftmost square of the tape and never removed or printed elsewhere. In order to learn if n belongs to the set defined by the machine, we start the machine in state q_0 placing the head to read the leftmost square of the tape which has * followed by n 1's and is blank elsewhere. If the machine eventually halts, then n belongs to the set listed by the machine.

The second lecture will be on the decision problem for diophantine equations. A particularly simple kind of arithmetical definition is by a diophantine equation:

A relation ρ is <u>diophantine</u> if there is a polynomial P with integer coefficients such that

$$\rho(x_1, \ldots, x_n) \leftrightarrow \exists y_1, \ldots, y_m \left[P(x_1, \ldots, x_n, y_1, \ldots, y_m) = 0\right].$$

Similarly, ρ is <u>exponential diophantine</u> if there are terms E, F built up from natural numbers, $x_1, \ldots, x_n, y_1, \ldots, y_m$ by means of addition, multiplication, and exponentiation such that

$$\rho(x_1, \ldots, x_n) \leftrightarrow \exists y_1, \ldots, y_m \left[E(x_1, \ldots, x_n, y_1, \ldots, y_m) = F(x_1, \ldots, x_n, y_1, \ldots, y_m)\right].$$

We will give a proof due to Matijasevič that a Turing machine like we described above can be coded in an exponential diophantine equation. Thus we obtain:

Every listable set is exponential diophantine. In particular there is no decision method for telling whether an exponential diophantine equation has a solution.

This theorem was first proved by Davis, Putnam, and Robinson by an entirely different method.

As an example of an exponential diophantine definition, we give the following definition of the binomial coefficient. Notice that if $u > 2^n$ then $\binom{n}{i} < u$ for $i = 0, \ldots, n$. Thus the bionomial coefficients in the expansion of $(u+1)^n$ are just the digits when it is written to the base u. Hence $c = \binom{n}{k}$ iff there are natural numbers u, v, w, such that

$$\begin{aligned} u &= 2^n + 1, \\ (u+1)^n &= w \cdot u^{k+1} + c \cdot u^k + v, \\ v &< u^k, \\ c &< u. \end{aligned}$$

7

Since $a < b$ iff there is a natural number d such that $a + 1 + d = b$, it is easy to write out the above conditions as a system of equations. We can then combine the equations in the usual way since $a = 0$ and $b = 0$ iff $a^2 + b^2 = 0$.

Matijasevič has recently improved the theorem by showing that 3 unknowns are sufficient to give an exponential diophantine definiton of an arbitrary listable set or relation.

Matijasevič showed in 1970 that the exponential function a^b has a diophantine definition. However the definition is still too complicated to give here. This yields the remarkable theorem:

Every listable set is diophantine. In particular there is no method of telling whether an arbitrary diophantine equation has a solution in natural numbers.

From the existence of a universal Turing machine, it follows immediately that there is a universal diophantine equation $P = 0$ such that to every listable set S there is a number t such that

$$n \in S \leftrightarrow \exists y_1, \ldots, y_m \left[P(t, n, y_1, \ldots, y_m) = 0\right].$$

Thus there is a fixed number of arithmetical operations sufficient to show that $n \in S$ - the number does not increase with n and the same bound holds for all listable sets. Recently, J. P. Jones using the methods of Matijasevič has shown that 100 arithmetical operations (addition, multiplication, and subtraction) are sufficient to verify that an arbitrary natural number belongs to a given listable set. If n belongs to the set, we consider the search in finding values of the unknowns which will satisfy the diophantine conditions expressed in 100 operations as scratch work and only count the operations needed to check the solution.

The third lecture will be concerned with the following questions for particular rings and fields F.

1. Is there a method of telling whether a sentence is true in F?

2. What sets are arithmetically definable in F?

3. What sets are diophantine in F?

For real closed fields (in particular $\mathbb{R}$), Tarski showed that formulas in the language of arithmetic can be reduced to equivalent formulas without bound variables, so the answer to question 1 is yes. It follows that a set is arithmetically definable in $\mathbb{R}$ iff it is a finite union of intervals with algebraic end points. In $\mathbb{R}$, $x > 0$ iff $x = y^2$ for some y. Hence the sets have diophantine definitions.

Tarski then asked if the integers can be defined in $\mathbb{Q}$. In 1948, I showed that

$$
\begin{aligned}
C \in \mathbb{Z} \leftrightarrow \forall A,B\ \{&\phi(A,B,0) \\
&\forall N[\phi(A,B,N) \leftrightarrow \phi(A,B,N+1)] \\
&\rightarrow \phi(A,B,C)\},
\end{aligned}
$$

where

$$\phi(A,B,C) \text{ is } \exists\, X,Y,Z\left[2 + ABC^2 = X^2 + BY^2 - AZ^2\right].$$

Notice that in these formulas both bound and free variables range over the set of rational numbers. It follows that:

A set S is arithmetically definable in $\mathbb{Q}$ iff the relation given by $\frac{p}{q} \in S \wedge q \neq 0 \wedge (p,q) = 1$

is an arithmetically definable relation in $\mathbb{Z}$.

It is still an open question whether there is a method to tell whether a given diophantine equation has a solution in $\mathbb{Q}$. This decision problem is equivalent to asking for a method of telling whether an arbitrary homogeneous diophantine equation has a nontrivial solution in integers.

Ax and Kochen in a series of beautiful papers in 1965 showed that there is a decision method for p-adic fields and in fact that the elimination of quantifiers can be carried out in a somewhat stronger system. Ershov obtained the same results independently. Based on their work and later methods of Paul Cohen, A. McIntyre has recently determined what sets are arithmetically definable in p-adic fields. We shall also mention recent work of Jan Denef in refining arithmetical definitions to diophantine ones in a number of rings and fields.

In the final lecture, we return to nonstandard models of arithmetic. Let M be a countable nonstandard model and α be an infinite element of M. Then $\ldots, \alpha - 2, \alpha - 1, \alpha, \alpha + 1, \alpha + 2, \ldots$ are all in M. We call this sequence α's string. Suppose α and β lie in different strings and $\alpha < \beta$. Then every element in α's string is less than every element in β's string. So the strings are also linearly ordered. Also either $\alpha + \beta$ or $\alpha + \beta + 1$ is even, say equal to 2γ, and γ's string will lie between α's and β's. Thus, the order type of the strings is η - the order type of the rationals. This is also true for models of weak fragments of arithmetic, even for arithmetic without multiplication.

9

The usual correspondence between natural numbers and finite sets of natural numbers yields a correspondence between infinite integers and some infinite sets. If n represents the set $S_n = \{x : 2^x$ occurs in the binary expansion of $n\}$, then

$$x \in S_n \leftrightarrow (\exists yz)\left[t = y \cdot 2^{x+1} + 2^x + z \wedge z < 2^x\right].$$

Since 2^x can be defined, this correspondence can be extended to infinite integers:

$$\alpha \in S_\tau \leftrightarrow (\exists \beta,\gamma)\left[\tau = \beta \cdot 2^{\alpha+1} + 2^\alpha + \gamma \wedge \gamma < 2^\alpha\right].$$

Looking at the intersection of S_τ with $\mathbb{N}$, we see that τ may code an infinite set of natural numbers. Also Gödel's lemma on finite sequences shows that a pair of infinite integers codes a number-theoretic function F. Thus,

$$F_{\alpha,\delta}(x) = y \leftrightarrow y = \mathrm{Rem}(\alpha, 1 + (1+x)\delta).$$

There is also a theorem due to Gaifman on extensions of models of Peano arithmetic which is reminiscent of Steinitz's theorem on field extensions.

Let M_1 and M_2 be models of Peano arithmetic such that M_2 is an extension of M_1. Then there is a model M_0 of Peano arithmetic such that M_0 is a cofinal extension of M_1 and M_2 is an end extension of M_0.

A model B is a <u>cofinal extension</u> of A if B is an extension of A and if every element $b \in B$ is less than some element in A.

A model B is an <u>end extension</u> of A if B is an extension of A and every element in A is less than every element in B which is not in A.

Now that we have mentioned a few facts about nonstandard models, we take up the question: what are they good for?

The most natural use is to prove independence results. In the last few years, much progress in this direction has been made by Paris and Harrington, followed by others.

Namely, there exist quite natural variants of Ramsey's theorem which are independent of the Peano Axioms. Because these are statements very close to interests of mathematicians, we see that the Peano Axioms are less satisfactory than we had thought even after Gödel's Incompleteness Theorem.

Now a provably computable function F is a function such that for some diophantine definition of F, say

$$F(x) = y \leftrightarrow \exists\, u_1, \ldots, u_k\ [P(x,y,u_1, \ldots, u_k) = 0],$$

the formula

$$\forall x\, \exists\, y, u_1, \ldots, u_k\ [P(x,y,u_1, \ldots, u_k) = 0]$$

10

is provable in Peano Arithmetic. In other words F is provably computable if it has a diophantine definition which can be proved to define a function. For example, all primitive recursive functions are provably computable.

As a corollary to the Paris-Harrington theorem, we have:

There is a particular algorithm for computing a function G which eventually majorizes all provably computable functions.

Acknowledgments

The American Mathematical Society gratefully acknowledges the kindness of these institutions and individuals in granting the following permissions:

J. W. Addison

The decision problem for fields, The Theory of Models: Proceedings of the 1963 International Symposium at Berkeley (J. W. Addison et al., eds.), North-Holland, Amsterdam, 1965, pp. 299–311.

Annals of Mathematics

An iterative method of solving a game, Ann. of Math. **54** (1951), 296–301.

The decision problem for exponential diophantine equations (with Martin Davis and Hilary Putnam), Ann. of Math. **74** (1961), 425–436.

Association for Symbolic Logic

Definability and decision problems in arithmetic, J. Symbolic Logic **14** (1949), 98–114.

An introduction to hyperarithmetical functions, J. Symbolic Logic **32** (1967), 325–342.

Computing Center for Russian Academy of Sciences

Axioms for number-theoretic functions, Selected Questions of Algebra and Logic (A. I. Shirshov et al., eds.), Siberian Academy of Sciences, Novosibirsk, 1973, pp. 253 263.

Izdatet'stvo Nauka

Two universal three-quantifier representations of recursive enumerable sets (with Yuri Matijasevich), Theory of Algorithms and Mathematical Logic (B. A. Kushner and N. M. Nagornyi, eds.), Vychisl. Centr. Akad. Nauk SSSR (1974), 112–123, 216.

Mathematical Association of America

Diophantine decision problems, Studies in Number Theory (W. J. LeVeque, ed.), MAA Stud. Math., vol. 6, Math. Assoc. America, Washington, DC, 1969, pp. 76–116.

Mittag-Leffler Institute

Reduction of an arbitrary diophantine equation to one in 13 unknowns (with Yuri Matijasevich), Acta Arith. **27** (1975), 521–553.

National Academy Press

Julia Bowman Robinson 1919–1985: A Biographical Memoir, Solomon Feferman, Biographical Memoirs, Vol. 63, 1994, the National Academy Press, Washington, DC.

Stanford University Press

The undecidability of exponential Diophantine equations, Logic, Methodology, and Philosophy of Science: Proceedings of the 1960 International Congress (E. Nagel et al., eds.), Stanford Univ. Press, Stanford, 1962, pp. 12–13.

On the decision problem for algebraic rings, Studies in Mathematical Analysis and Related Topics, (D. Gilbarg et al., eds.), Stanford Univ. Press, Stanford, 1962, pp. 297–304.

Patrick Suppes

Solving diophantine equations, Proceedings of the Fourth International Congress for Logic, Methodology and Philosophy of Science (P. C. Suppes et al., eds.), North-Holland, Amsterdam, 1973, pp. 63–67.

University of California Press

A note on exact sequential analysis, Univ. California Publ. Math. (N.S.) **1** (1948), 241–246.

ISBN 0-8218-0575-4

9 780821 805756